LEONARDO DA VINCI'S CODEX LEICESTER | *A New Edition*

LEONARDO DA VINCI'S CODEX LEICESTER

LEONARDO DA VINCI'S CODEX LEICESTER *A New Edition*

DOMENICO LAURENZA & MARTIN KEMP

INTERPRETATIVE ESSAYS AND THE HISTORY OF THE CODEX LEICESTER

VOLUME II

OXFORD
UNIVERSITY PRESS

Acknowledgements

THE PRIME ACKNOWLEDGEMENT must go to Bill Gates, the owner of the Codex, who has fostered the research and publication of the Codex with great devotion to Leonardo and stoic patience in the face of a long-winded process. Curtis Wong, who was responsible for the edition of the Codex on a CD-ROM in 1996, and who invented the Codescope as a digital tool for deciphering Leonardo's difficult text, has been a source of unwavering inspiration throughout. Fred Schroeder has assumed responsibility for the many stages of work during which the edition has been generated and published; his resolution that it should come to completion and be brought into the public domain according to the highest standards has been crucial throughout. Robert Faber of Oxford University Press was the parent of the first scheme for a digital and print edition. Cathryn Steele has seen the project to completion in a resolute and skilful manner. Ela Kotkowska has devoted extraordinary care to the production of the final texts, not least the intricate implementation of formats for the transcription and translation that would accurately signal the nature of Leonardo's strange folios. Without her remarkable efforts, the quality and utility of our edition would have been greatly diminished. Emma Slaughter provided vital input and assistance at the stages of proofing. The complex design of the volumes has been accomplished with panache by Anna Scully.

Thanks to librarians/libraries, archivists/archives who facilitated research. In particular: Kimerly Rorschach, Nicholas Dorman, and Lauren Mellon of the Seattle Art Museum; Christine Hiskey, Suzanne Reynolds, Lucy Purvis of Holkham Hall Archive and Collections; Ursula Verena Fischer Pace, Herman Mildenberger, Ingrid Arnhold, and Margarete Oppel for the search for the Weimar copy of the Codex Leicester in the Herzogin Anna Amalia Bibliothek in Weimar; Roberto Marcuccio of the Biblioteca Panizzi in Reggio Emilia, Margherita Breccia of the Biblioteca Nazionale Centrale, Rome; Maria Rascaglia, of the Biblioteca Nazionale, Naples; Françoise Bérard, Françoise Viatte, and Pierre Rosenberg for their help in the research in the Bibliothèque de l'Institut de France and other libraries and archives in Paris; Marchis Niccolini and Rita Romanelli for the research in the Niccolini Archive, Florence; Sonja Brink of Graphische Sammlung, Department of Prints and Drawings, Düsseldorf; Luigia Grillo and the Istituto Italiano per gli Studi Filosofici for the research in the Serra di Cassano family archive; the staff of the British Library in London, of the Biblioteca Nazionale in Florence, and of the Capitoline Archive, Biblioteca Corsiniana-Accademia dei Lincei, Biblioteca Angelica, Vatican

Library in Rome. Edward Worth Library, Dublin, Trinity College Library and Trinity Long Room Hub, Dublin.

We are grateful for the encouragement and input of many scholars to whom we have spoken at various events at which we have aired some of our findings. Others have contacted us as we proceeded and made valuable suggestions. We owe a historic debt to our predecessors as editors, Girolamo Calvi and Carlo Pedretti, who have provided foundations on which we have built.

Martin Kemp's agent, Caroline Dawnay, and his personal assistant, Judd Flogdell, have as always played indispensable roles at key stages in the process. Tania Correia played a key role in proofing the commentary and paraphrase. The authors thank each other for the complementary skills they have brought to the edition. No document in the history of science is more difficult than the Codex Leicester, and our harmonious collaboration as friends has accomplished more than we could have hoped to do on our own.

Contents

Symbols and Abbreviations

A/5, A/25	Account books, Holkham Hall Archive
Arundel	*Il Codice Arundel 263 nel Museo Britannico*, ed. C. Pedretti and C. Vecce, Florence, 1998
CA	*Il Codice Atlantico di Leonardo da Vinci nella Biblioteca Ambrosiana di Milano*, ed. A. Marinoni, 12 vols of plates and 12 vols of text, Florence, 1973–5 and 1975–80
Forster I to III	Codices Forster I–III, Victoria & Albert Museum, London
Cod. Urb. Lat.	Codex Urbinas Latinus
FT/C4	Account book of Thomas Coke's grand tour, Holkham Hall Archive
FT/C5	Account book of Thomas Coke's grand tour, Holkham Hall Archive
FT/C6	Account book of Thomas Coke's grand tour, Holkham Hall Archive
Huygens	Codex Huygens, The Morgan Library, New York
Leicester	Codex Leicester, this Edition
Madrid I & II	Codices Madrid I (8937) and II (8936), Biblioteca Nacional, Madrid
MS A to MS M	Manuscripts A–M, Institut de France, Paris
RS	Royal Society, London
Windsor	Windsor Castle, Royal Library

Other bibliographical references can be found in the Bibliography in Volume IV of this Edition.

List of Figures

List of Plates

THE CODEX LEICESTER ‘IN ITSELF’

PART I

Introduction 1

MARTIN KEMP AND DOMENICO LAURENZA

THE CODEX LEICESTER is the greatest of all Leonardo's compilations of his writings on science, and exercised a considerable if unrecognized impact, particularly in the eighteenth century. It deals primarily with three major topics: the nature of the moon in terms of the optics of light it receives directly or indirectly from the sun; the properties, motion, and management of water; and the long history of radical changes in the 'body of the earth'. The last of these was truly revolutionary, as recognized by those who had access to it or to its copies over subsequent centuries.

This new edition includes high quality reproductions of the Codex, a new transcription and translation, accompanied by a paraphrase in modern language, and a page-by-page commentary. We have also provided interpretative essays, including a substantial study of the *fortuna* of the Codex and the surprising number of copies that have been discovered to date. The four volumes constitute the most comprehensive edition of any of Leonardo's manuscripts.

The name of the Codex reflects its ownership from the second decade of the eighteenth century until 1980 by successive Earls of Leicester, housed for much of that time in the library of Holkham Hall, their very grand Palladian mansion in Norfolk. It was exported from Italy by Thomas Coke, for whom the title of the Earl of Leicester had been revived. He almost certainly acquired it in Rome from the former owner Giuseppe Ghezzi, who in turn had found the Codex among the belongings of the heirs of the sixteenth-century sculptor and engineer, Guglielmo della Porta. It therefore has a history separate from the main bulk of Leonardo's manuscripts, which had passed into the hands of Francesco Melzi, Leonardo's aristocratic pupil.

In the eighteenth century, the Codex was handsomely bound in a cover of red Moroccan leather stamped in gold, and in the nineteenth century, it was in turn enclosed in a clamshell box stamped in gold with Thomas William Coke's (1754–1842) family crest, and, on the spine, the abridged title in Italian: 'L. DA VINCI DELLA NATURA &C. DELLE ACQUE MS. ORIGINALE DELL'AUTORE ALLA MANCINA. 699'.[1] Next came the first of the flyleaves with a note added in the nineteenth

1 699 was the inventory number of the Codex in the Holkham Hall library.

century by William Roscoe, Liverpool collector and historian.[2] Roscoe observes that the Codex 'has never been printed' but that 'a very legible transcript of this original MS. is in the Collection at Holkham'. This first flyleaf also carries what seems to be the signature of Thomas William Coke, who died in 1842.[3] These inscriptions date from the reordering of the library between 1813 and 1830.[4] The watermark (a flower with three petals) is the same as the following two flyleaves. After a blank flyleaf, comes the title page from the beginning of the eighteenth century, recording information generated by Padre Sebastiano Resta and Giuseppe Ghezzi, both of whom were important in transmitting knowledge of the Codex.[5] On the reverse of the title-page, there are notes regarding Leonardo, dating back to the same period.[6]

After the flyleaves comes a second and more ancient cardboard cover with notes on the verso of the left half,[7] on the recto of the right half,[8] and on the verso of the right half.[9] The poor condition of the first sheet of the Codex (the outer folios, 1r–36v), which has been worn and stained by repeated handling, suggests that it did not have a cover for some time—during, and probably after, Leonardo's lifetime—and therefore this is not the original cover of the Codex.[10]

The Codex Leicester was not directly cited in Giambattista Venturi's pioneering *Essai* on aspects of Leonardo's science, printed in Paris, 1797. However, there were earlier, printed and written, references to the content of the Codex, based mainly on the legible manuscript copies, such as that recorded by Roscoe. The first printed anthology of excerpts from the Codex was included in Jean-Paul Richter's

2 Calvi 1909, pp. xxvii–xxxviii mentions seven flyleaves (*fogli di custodia*) including the sixteenth century cover, four of them as blank; Christie's 1994, p. 40, apparently mentions six flyleaves, some of them as blank. In any case, what is here described as flyleaves bears a numbering from 1 to 3.

3 'Tho.s Will.m Coke' (upper left corner); '696' (in pencil or black chalk). Roscoe's note follows:

> This Treatise on the nature weight and motion of water, written by Leonardo da Vinci alla mancina or from right to left had never been printed. On the reverse of the modern Title may be found an extract from the Life of Lion. da Vinci, by du Fresne, in which this volume is particularly mentioned. It appears from the Title page (altho the name of the possessor has been obliterated) that it has belonged to Giuseppe Ghezzi, an eminent painter at Rome; who, according to the inscrip.n 'had obtained it, by a vast sum of money, to ~~perfect~~ compleat the laborious collection of his library'. A very legible transcript of this original Ms. is in the Collection at Holkham. W. Roscoe.

On the recto of second half of this flyleaf are visible notes, apparently from the same Roscoe's period, not included in Calvi: 'A 3 B [or β?] Rupio Casi [or Rupic Casi?]'.

4 Reynolds 2015, pp. 19ff.

5 Libro Originale Della Natura, peso e moto delle Acque, Composto, scritto e figurato di proprio Carattere alla mancina Dall'Insigne Pittore e Geometra Leonardo da Vinci In tempo di Ludovico il Moro, nel condur che fece le Acque del Naviglio della Martesana dall'Adda a Milano. Si autentica cone la precisa mentione che ne fa Raffaelle du fresne nella Vita di detto Leonardo, descritta nel suo Libro stampato in Parigi da Giacomo Longlois l'Anno 1651. intitolato Trattato della Pittura di Leonardo da Vinci. Acquistato con la gran forza dell'oro, per sublimare le fatigose raccolte del suo Studio da ~~Giuseppe Ghezzi Pittore in Roma~~.

6 'Soleva il Vinci scrivere Alla mancina, secondo luso degli Ebrei, nella qual maniera erano scritti quei tredici Volumi de quali de già abiamo fatto menzione, et esendo il carattere buono, si legeva assai facilmente mediante un spechio grande; è probabile ch'egli facessi questo accio tutti non legessero cosi facilmente i suoi scritti. L'impresa del naviglio di Mortesana gli diede ocasione di scrivere un libro della Natura, peso e Moto delle Acque, pieno di gran Numero di disegni di Varie rote, Machine per Molini, e regolar il Corso dell'acque, e levarle in Alto.'

7 '4. Libro scritto da leonardo Vincio che tratta del Sole della Luna del corso dell'acqua, de i punti, et de i moti' and 'carte n.° trentasei'.

8 '4' 'carte n° trentasei' and, bottom left, by a later hand (Francesco Ducci?), '37'.

9 '40', '49'.

10 Pedretti 1987, pp. 165–6, Christie's 1994, p. 4c, and apparently Bambach 2007, p. 15 ('le copertine di cartoncino (databili al primo Cinquecento)') consider it as the original cover. However, Calvi 1909, p. xxviii, apparently does not.

remarkable *The Literary Works of Leonardo da Vinci*, London, 1883, accompanied by the reproduction of some of the drawings. Drawing upon Richter's work, Mario Baratta was able to include some passages from the Codex Leicester in his pioneering monograph on Leonardo's geology in 1903. However, Richter published only the first part of the text from sheet 36r in which Leonardo explains his fundamental theory of an active uplift of the continents from the sea.[11] Baratta therefore overlooked the most advanced and fascinating part of Leonardo's theory, and was led to present Leonardo as a 'Neptunist', that is to say akin to those later geologists who supported a passive process for the uncovering lands by the progressive lowering of the sea.[12]

In the same year, 1903, Gerolamo Calvi began his study for the first complete edition of the Codex Leicester, and the Codex was transferred from Holkham Hall to the British Museum, in order that Calvi could examine and transcribe it.[13] Calvi's edition was published in 1909 as *Il Codice di Leonardo da Vinci della Biblioteca di Lord Leicester in Holkham Hall*. It contains a comprehensive introduction, critical and diplomatic transcriptions, and the complete reproduction of the manuscript, providing a fundamental model for later editions of Leonardo's codices. The second edition of the Codex, after its sale to Armand Hammer, was Carlo Pedretti's *The Codex Hammer of Leonardo da Vinci* in 1987, which includes a diplomatic transcription, an English translation, and a large study of the philological and chronological relationship of the Codex with other writings and drawings by Leonardo. The work of Calvi and Pedretti has indelibly marked the development of scholarship on the Codex and more generally on Leonardo's legacy. In 1996, a digital edition of the Codex was the prime focus of a CD-ROM, *Leonardo da Vinci*, issued by Corbis, for which Martin Kemp served as the general consultant and author, with Claire Farago contributing crucially to the digital edition and interpretative material.

In the 1970s, the Trustees of the Holkham Estate needed to raise money. Rather than selling items from their distinguished art collection, they chose the Codex, which was held in the library. On 12 December 1980, it was auctioned by Christie's and sold for US$5.12 million. The purchaser was Armand Hammer, President and Chairman of Occidental Petroleum. Hammer dealt and collected on substantial scale, concentrating on Impressionist and Post-Impressionist paintings, Russian art and old masters, together with a substantial collection of Daumier's satires.

Leonardo's own assembly of the double sheets, laying one inside the other before folding, did not obey conventional logic, and he certainly did not systematically begin on the first folio (1r) of sheet one and work his way through to the last page—or vice versa. Reversing Leonardo's process of assembly, Hammer arranged for its unbinding. In this unbound form, it could be displayed as separately mounted, double-sided sheets, each of four pages. It was renamed the Codex Hammer, and Pedretti published his new edition with the folios renumbered. Leonardo's own preferred order has subsequently to be reconstructed.

After Hammer's death in 1990, it became apparent that the museum he founded in Los Angeles needed to raise a substantial sum of money. In 1994, amidst a flurry of legal actions, the University of California came to the rescue and agreed that the Hammer collection would in effect become the university's art

11 Richter 1883, § 938.

12 Baratta 1903, pp. 206ff.

13 Calvi 1909, p. xxxii.

museum. To raise an adequate endowment they decided to sell the Codex at auction, and it was duly was sold again at Christie's on 11 November, this time for US$30.8 million. The purchaser was Bill Gates, who reinstated its historic name.

Our new endeavour is able to take advantage of the high scholarly achievement of the two previous editions. It benefits from a large volume of recent research into Leonardo's legacy, and utilizes high-resolution digital images. It also introduces important new research into the interpretation of the texts and images, on the setting of Leonardo's ideas in the context of ancient and mediaeval theories (which have only been sketchily researched) and, above all, into the notable fortunes of the Codex within the sciences of astronomy, water, and the history of the earth.[14]

14 The important book of essays edited by Paolo Galluzzi (2018a), published to accompany the exhibition of the Codex Leicester at the Uffizi in 2018–19 became available just as the present edition was entering production. For a synthesis of the main themes of the Codex, see also Laurenza 2018a.

Compilation and Chaos | 2

MARTIN KEMP

NO HISTORICAL FIGURE has left a more difficult, puzzling, and apparently disorderly legacy of works on paper than Leonardo. No one had ever used paper in the way Leonardo did, and there is no obvious later parallel. The difficulties do not just involve his handwriting, his vocabulary, his punctuation, the eccentric progress of his thought, and the inherently demanding nature of the content. The problems extend to the *types* of material that have survived. We talk of Leonardo's manuscripts and his drawings, but these general terms cover an extraordinarily wide variety of different kinds of works on paper. This variety reflects both the different ways that he generated and assembled his works and the accidents of the ways that they have survived.

It is possible to determine at least six types of work on paper that have come down to us.

1| There are separate sheets, especially drawings for works of art, that have never been part of a bound volume. They are of varying size and colour. A number are no longer their original size, having been trimmed or more radically cut from parent sheets. Collectors in the past performed acts of what we would regard as vandalism to 'improve' the look of the drawings.

2| Some originally separate sheets were assembled or inserted by later owners into bound volumes, and either remain in these volumes or have been removed and mounted separately (like the drawings for works of art in Windsor). Many of these have also been radically altered in size and format.

3| Some of his bound notebooks or volumes, consisting of sheets gathered by Leonardo and stitched within covers (either by Leonardo himself or soon after his death) have survived. A very few remain in more or less in their original form. Some have been bound together but retain the original sequence of their folios.

4| Some of the notebooks or volumes gathered and/or bound by Leonardo have subsequently been disbound and survive incompletely as scattered, separate sheets.

5| Sheets from notebooks or volumes gathered and/or bound by Leonardo have been dismembered in relatively recent times and mounted separately (like the Codex Leicester).

6| Sheets from notebooks or volumes gathered and/or bound by Leonardo have been subsequently rebound together, sometimes including separate pages or sheets, in assemblages of miscellaneous folios (like CA and Arundel, both of which have now been disbound). The rebinding may have retained some of the original order of the parent manuscripts, or the pages may have been re-distributed in a way that makes the original order difficult or impossible to reconstruct.

Recognizing the nature of what we have in each case and its relationship to its original function is important if we wish to construct a framework within which we can interpret those of his works on paper that have survived the ravages of time.

The bound volumes in Leonardo's workshop range widely in type. At one extreme are assemblages of large sheets that were folded before being interleaved. Each sheet, folded vertically in the middle, comprises two folios and four pages or sides. He generally seems to have written and drawn on the four pages before the sheets were interleaved. At the other extreme are very compact and fat 'pocket books' with small pages, which he carried around to record observations and thoughts as they came to him. Leonardo's 'default mode' for filling his notebooks is to begin at what we would regard as the back, writing on the last with his left-hand mirror script and working his way towards the front of the book. Sometimes he also begins at the other end, having turned the notebook upside down so that what we would originally have seen as the first page becomes another last page. This highly unusual method of back-to-front filling of pages is shared with some of the larger folded sheets. He would (in theory at least) begin at page four, passing to three, two, and then one. Sets of such folded sheets could then be laid on top of each other and folded over to create a quire, or a gathering of sheets in the manner that was traditional for manuscripts and printed books. This kind of gathering of sheets is what we see in the Codex Leicester.

However, as always with Leonardo, as soon as we try to lay down a 'default mode'—a typical way of proceeding—we find that no such wholly reliable norm exists. The filling of the pages in the notebooks or the folded sheets, as is apparent in the Codex Leicester, was never a straightforward, linear process. A single page may well contain material added at quite different times, sometimes on the same subject but not necessarily so. A page filled, or more or less filled, at one time may deal with topics of a widely different kind—or, at least, they look very different to us. A compilation of sheets folded together could be rearranged and new sheets or pages inserted at any stage. (Detailed observations about sequences within and across pages are contained in the commentary.) The process was ceaselessly dynamic, and it is reasonable to say that not one of his surviving manuscripts ever reached a wholly definitive state. This provisional nature does not just involve a lack of finish but more profoundly his congenital inability to arrive at a definitively fixed and classified order for the subjects he wishes to treat.

Leonardo himself was well aware of the problems posed by lack of order. On a number of occasions in his manuscripts from after 1506 or so (when he was fifty-four years old), he scribbled notes acknowledging the need to bring order into his writings. On the first page of the Codex Arundel in the British Museum he admitted that,

> This will be a collection without order drawn from many pages that I have copied here, hoping to put them in order in their places, according to the subjects with which they will deal, and I believe that before I am at the end of this I will have to repeat myself many times.

On f. 2v in the Codex Leicester he steps aside from the main discussion to tell us,

> I shall omit here the proofs that will be presented afterwards in the ordered work, and I shall pay attention only to finding cases and inventions, and I shall note them successively as they present themselves,

> then I shall give an order by putting together those of the same kind; therefore, for now, you will not wonder nor will you laugh at me, reader, if here such big leaps are made from one matter to another matter.

Similarly, on the next folio, having discussed how water might penetrate the earth, he tells us that:

> Here I shall follow and make a little discourse about finding waters, although it appears rather out of order, and then I shall put them into order in their places as I compose the work.

And on f. 25v, having become tangled up in a convoluted discussion of intersecting rivers, he reminds himself:

> Here you must proceed in an orderly manner, that is distinguishing your propositions, member by member of each part, and so you will be well understood without confusion.

His promised order never wholly emerges.

These statements about order are in one sense addressed to himself, but they also seem to be speaking to a potential reader of his manuscript—as a kind of insurance in case a definitive version was never to emerge, which he must have suspected to be the case. They serve as a kind of apology to those who were destined to grapple with his texts after his death.

In what has come down to us in his manuscripts as a whole, his variousness is more apparent than any immediately systematic order. The underlying pattern is that of his fluid thinking across categories rather than being restrained by standard classifications. The 'repetitions' and 'jumps' that he lamented in the Codex Arundel are a prominent feature of the Codex Leicester.

In the list of types of surviving work, the Codex belongs in the third category, that is to say it consists of sheets (eighteen in number, thirty-six folios, seventy-two pages) that were gathered together by Leonardo and folded over to form a unit. Whether it was stitched in Leonardo's lifetime is uncertain, but we can be reasonably confident that we do have a unit that he would have recognized. Its pages are amongst the largest in his manuscript, varying slightly in size from 29.9 x 44.7 cm to 29.8 x 44.6 cm. It is his least illustrated manuscript, comprising mainly large sections of text with marginal illustrations. Exceptionally for Leonardo, there are eleven pages devoid of illustrations. The number of words in our transcription is 57,572, length enough to be properly called a 'book'.

There are more than enough internal signs in the Codex to show how complex were its writing and assembly. We will look first at how the form and content of the pages reflects the Codex's assembly and editing, including indications of how some of the material was transcribed or adapted from other of Leonardo's manuscripts. We will then look at various pieces of physical evidence about its compilation. These two bodies of evidence will lead us to suggest a framework for its assembly in terms of what we are calling an 'inner set' and 'outer set'. Finally, we will use Leonardo's own cross-references to his 'books' to determine he was developing a systematic programme within which to fit the material in the Codex.

Assembly and Editing

Looking at the last page, f. 36v, which in Leonardo's 'default mode' would correspond to where he began, we can see that it does not really have the character of a beginning. At first sight it looks reasonably promising. It contains two blocks of text, one devoted to the question of whether there are seas on the moon, arguing against an unknown 'adversary', and the other dealing with waves in water. Water, its behaviour and its role in the 'body of the earth', is the main topic in the Codex; and the related question of how the waters of the moon work with the light of the sun and the reflected light of the earth is prominent in the outer sheets. However, neither of the two sections function adequately as conclusions, introductions, or summaries of what is contained in the Codex. The section of the moon can best be seen as an incomplete continuation of discussions on ff. 1r and 1v on the same folded outermost sheet.

The other side of this folio in the outer sheet, f. 36r, deals for the most part with a different topic: the vast changes that the earth has undergone over the ages, looking at the immense caverns into which the crust of the earth collapses, and how new portions of the earth come to protrude through the sphere of the water. This discussion continues in the lower part of the main text on the facing page, f. 35v. He also discusses one of his recurrent themes, namely why the Biblical Deluge could not have been responsible for the presence of shells high on mountains. There is also a short note on why the air appears blue in the sky, which is discussed more fully elsewhere in the Codex.

Such recurrences feature throughout the Codex. For example, his most sustained account of the blueness of the air occurs on f. 4r, which is unusual in its focus on a single topic, and the same subject is discussed more briefly as a 'case' on f. 20r. Other issues, like the presence of seas on the moon, the reason why shells are found high on the mountains, and the 'veins' of water in the body of the earth, are discussed repeatedly, sometimes in a very similar manner and in other places with analyses that have been refined. He clearly considered it was better to repeat something than to assume it was present in the Codex and risk it being left out. The other form of repetition is of basic principles. For example, when discussing a detailed question about the movement of water, he frequently steps aside to remind us that water moves of its own volition only when it is out of its sphere, that is to say further from the centre of the earth than it should be. The centres of the sphere of water, the centre of gravity of the earth, and of the whole system of the elements is another basic theme that is much repeated.

The issue of the massive transformations in the body of the earth that have resulted from changes in its centres of gravity as discussed on f. 36r is closely related to the texts and diagrams on f. 35v, which faced it in the Codex as it was finally assembled if not stitched. However, lest we should think that we are encountering Leonardo in his default mode, working from back to front with each of the interleaved sheets, the other side of folio 35 deals primarily with tides and the ebb and flow of waters (with no illustrations), while the next page in this reverse order, f. 34v, contains a substantial section on siphons, which relates not only to f. 34r but also f. 3v (both pages of the same double sheet, no. 3). Whatever this may suggest about the order in which he filled the sheet, it certainly tells us that the 'default mode' of back-to-front filling does not adequately explain the sequence of contents on each page, folio, or sheet.

If we look at one of the double folded sheets further into the Codex, sheet 12 for example, we find a comparably complex picture, which is hard to describe. Beginning with the last or 'first' page, f. 25v, we can see that he opens with a horribly complex description of the meeting of two or even three rivers, with a labelled diagram that is difficult to follow in relation to the text, before reminding himself that he 'should proceed in an orderly manner'—which he then does to some extent before breaking off in mid-sentence. He then turns to some nice analyses of waves and obstacles to the flow of water. The remaining marginal illustrations relate only generically to the texts, and the one that is labelled with letters is not discussed directly. The three lower diagrams are so sketchy as to serve no obvious use in relation to the texts. None of the topics on f. 25v obviously follow on from f. 26r. Nor do the subjects on the other side of folio 25, the recto, which is dominated by a long and unusually continuous discussion of bubbles and the 'cohesiveness' of water, represent a logical continuation of the verso. On the other hand, f. 12v, which is the other page of the inner side of this same double sheet, does contain one section of the 'architecture' of bubbles. However, the other topics on f. 12v do not relate particularly closely to those on f. 25r. The final (or first) page of sheet 12, is one of the most complex and diverse of all the pages, and cannot be seen to be continuous with other pages on the double sheet or with f. 11v, which is the abutting page in the interleaved sequence. However we may reconstruct the order in which the four sides of sheet 12 were filled, we cannot adduce any kind of logic based on a neatly linear process of sequential thought.

Folio 12r will also serve to show how complex the formatting of an individual page can be. To quote from the commentary:

> The opening section of seven lines written with a thicker pen deals in a complex way with how weights behave in relation to the gravitational pull of the centre of the world. The discussion and related diagrams are then continued later, down the right margin and in two lines squeezed below the main text. He presumably returned to the issue of the weights after he had filled most of the page with texts on water. The small diagram three-quarters of the way down the margin seems to have been sketched before the text above it was completed, since the writing flows around it.

We can also see that one of the diagrams is crossed out and labelled '*non valido*'—not valid.

The other topics on the page involve variously the speed of water though channels and sluices, the fall of water through air, storms at sea, winds and currents at deltas, water in pipes, the generation of waves, the diverting of a river, and a further series of questions about falling water.

This way to fill a page is not typical. The only typical feature is that he obeys no standard procedure for the compilation of the seventy-two pages. The notes we have provided for each successive page reveal the extraordinarily varied sequences in which Leonardo laid down his written and drawn material. On some occasions, especially in the 'inner set' (e.g. ff. 22r, 23r, and 28v), he seems to insert the first part or parts of the main text with illustrations in the margin, accompanied by short notes, before adding the later sections of the main text.

Even the marginal illustrations and notes, which we might expect to have a clear function in relation to the main text, are extraordinarily various. Occasionally, the material in the margins is functionally related to main text, but more often they assume a life and character of their own. For example, the marginal illustration and note about an experiment on f. 10r bears no clear relationship to the main text. A topic in the margin that engages him may surge downwards with its own impetus, and the marginal illustrations often begin to explore ideas on their own account. This is certainly the case on f. 12r. Sometimes, again for example on f. 10r, the marginal illustrations are labelled with letters that do not feature either in the main text or marginal notes.

At the top of f. 12r he records that it contains twenty-one 'cases' or topics. Each 'case' is marked by an exaggeratedly large initial letter at its start. As he undertakes the counting he re-emphasizes some of the initial letters at the start of what he wants to denote as a 'case'. Just below the middle of the page the 'C' of *Come* has been enlarged. We can also see how a capital 'E' of 'and' near the start of the last line of the main block of text has been drawn heavily over a lower case 'e'. Throughout the Codex we can readily observe where he either writes a conspicuous capital letter at the start of each 'case' or re-emphasizes a capital or capitalizes a lower-case initial. In the top left-hand corner of f. 12r, he then tells us that 'in these seven sheets [there] are 657 cases about water and its beds'. His acts of counting numbers of 'cases' in sets of sheets testifies that he intervened editorially when many if not all of the sheets had been interleaved and folded inside each other.

The term he uses most often for the topics that he retrospectively counts is *casi* ('cases'). However, on five pages he uses the term *propositioni* ('propositions'). Three of these are pages from the same double sheet, number 16, but the fourth page, 16v, reverts to 'cases'. The other two are the recto and verso of folio 15. When he begins to total the numbers across a series of sheets, he resorts on three occasions (ff. 9r, 26v, and 31r) to the term *conclusioni* ('conclusions'), having previously designated the topics on f. 26v as 'cases'. Two other terms appear more rarely. *Invenzioni* ('inventions') is apparently used as a synonym for 'cases' on f. 2v. *Conceptioni* ('conceptions') appears on ff. 21v and 22r.

It is tempting to grant a special significance to each of these terms. We might imagine that a 'case' is an example or instance or case study, that a 'proposition' involves the formulation or application of a rule, that a 'conclusion' is indeed a conclusion, that an 'invention' is a novel discovery, and that a 'conception' involves a fundamental theoretical concept. However, detailed comparison of how he uses these terms provides little encouragement for their systematic differentiation. What seems more likely is that the terms succeeded each other in more or less chronological order, with 'cases' first, followed by 'propositions', and then by 'conclusions'. The only term that does seem to have a very specific meaning is 'conceptions', which is indeed used for foundational principles.

The largest number of 'cases' or 'propositions' or 'conclusions' on a single page is 39 on f. 22v, and numbers in the 20s and 30s are common. The smallest number is 7 on f. 9r. It is generally possible, if not simple, to reconcile the number inscribed at the top of the page with the emphasized capitals in the main text on each page.

His decision to count his 'cases' and 'conclusions' gives a good idea of the character of many of the 'inner set' of sheets. The 'outer' set of seven sheets, the evidence for which will be assessed later, do not generally divide so readily into 'cases', and are for the most part immune from his counting. Folio 5v very obviously contains fourteen clearly demarcated statements, but they are not totalled at the top of the page. The 'cases' in the 'inner set' often assume something of the nature of lists, sometimes on closely related topics but frequently switching subject quite abruptly more than once on a single page, even when the text looks continuous. The lists are typically of questions to be answered and topics for his planned treatise. Some have the character of subject headings of topics he had already discussed, rather than serving as desiderata in a research programme. They are frequently prefaced by a series of stock introductory words, each granted its emphatic capital. 'How' (***Come***), 'On' or 'Of' (***De***), 'Why' or 'Because' (***Perché***), 'What' or 'Which' (***Quale***), 'When' (***Quando***), 'Where' (***Dove***), 'If' (***Se***), and 'And' (***E***) regularly punctuate the pages of the 'inner set'.

The listing of questions as 'cases'—asking 'why?', 'how?', 'when?', 'where?', 'what happens if?', and so on—is even more prominent in the Codex than in his other manuscripts. Many of his questions find no ready answers, given the sheer complexity of the phenomena he was striving to master.

Some of the 'cases' are short statements or questions, a line or two long, and belong in compact series. However, even when he is generating a relatively telegraphic list, his attention typically becomes captured by one of them, diverting his text into a more detailed and sometimes speculative exposition of that particular topic, which itself may be subdivided into a set of issues. The original listing may then be resumed, but not always. On other occasions, the majority of the cases on a page are more discursive. For example, only eight are counted on f. 28r, which contains more than 1000 words in the transcription, and one of the 'cases' occupies almost one third of the page. Once again, it is difficult to generalize about what is typical. His system, if it can be called such, seems endlessly flexible and responsive to his continually shifting patterns of thought.

There are four occasions in which he counts the totals across a number of sheets. His bouts of counting always include the innermost sheet, no. 19, i.e. ff. 18v to 29r, and extend progressively outwards, but never include the outer five sheets. On f. 12r he writes that 'on these seven sheets are 657 cases about water and its beds'. If we add the numbers at the top of each page of these inner seven sheets the total is 655. It is easy to make a mistake when adding the numbers from twenty-eight pages. On f. 26v he writes that 'on these eight sheets are 730 conclusions about water'. The total seems to be 728. On f. 9r he writes 'ten sheets and 853 conclusions', whereas the total is 855.

Finally, on f. 31r he records '905 conclusions'. In this case, he includes sheets 8 to 18 plus f. 6v itself, but neither sheet 7, on which no 'cases' are counted, nor the other three pages on sheet 6 that have no numbers. None of the sheets 1 to 5 are brought within the counting system. As we will see, the exclusion of sheet 7 from this final act of counting has significance for what seems to be a late change of order in the interleaved sheets.

A number of the compilations of 'cases' feel as if they have been assembled from other pages in his notebooks. On at least three occasions there are clear signs of his having broken off the text he is writing (ff. 24v, 25v, and 26v), as if he has garnered all his wants from the particular source he is using. More telling in this respect are his explicit references to sources in his own papers—leaving aside for the moment his references to the titles of 'books' and sections in his planned treatises. The best documented of his source manuscripts is the one he designates as 'Libro A', which has been lost but has been ingeniously reconstructed by Carlo Pedretti in his *Leonardo da Vinci on Painting. A Lost Book (Libro A)* in 1965. There are no less than eight pages that specifically reference Libro A (ff. 13r, 14r, 16r, 19r, 21r, 24v, 28r, and 28v), drawing upon its text and sometimes on its illustrations. Each reference is cited with a number. The lowest number is 54 on f. 28v and the highest, on the same page, is 95. Most of the references are clustered in the 80s. These could be page numbers, but it seems more likely that they are sections that he has numbered for his own reference during the process of transfer. On f. 26r he makes a cluster of references to a series of at least ten *quaderni*, that is to say folded gatherings or quires of sheets rather than formal notebooks with bindings.

Frustratingly, none of the texts designated by cross-references in the Codex can be identified as close transcriptions of passages in surviving manuscripts. However, it is very likely that Leonardo never transcribed his earlier statements in a mechanical manner. It may be that some of the passages in other manuscripts such as the Codex Atlanticus and Codex Madrid II that are quite similar to those in this Codex are the sources on which he founded some of his 'cases'. As noted in the commentary on f. 26v, one text (which breaks off) has clear affinities with a passage on f. 42v in Madrid II, which has been marked with a cross to designate it as intended for transcription or as having been transcribed. Given Leonardo's ways of proceeding, the term 'transcribe' may be inappropriate. It may be that 'transfer' or 'adapt' would more accurately convey how he used his own sources. In this matter, as in others, Leonardo subverts any attempt to impose retrospective restrictions on how he operates.

The Physical Evidence

As is becoming apparent, the eighteen double sheets were worked on in an ad hoc manner as separate sheets, sometimes folded over in one direction and occasionally in the opposite manner, as well as being worked on when laid out flat. They were then assembled, one folded inside the other, and almost certainly edited and rearranged more than once.

There are a number of detailed pieces of physical evidence that throw some light on the compilation of the treatise from its double sheets, as Carlo Pedretti cleverly noted. This evidence falls into three categories; the offsetting or transfer of fresh ink, blots, smudges, and water or ink stains from one page onto another; compass holes from geometrically constructed diagrams; and the character of Leonardo's script with respect to the size of letters, thickness of the nib, and type of ink. Let us first look at the 'blotting', which seems to have arisen in various ways: from accidental blobs of ink; from contact between sheets before the ink had dried, most notably if the ink was heavily applied in a particular place; and from a damp patch that moistened the ink and spread it.

The evidence of blots indicates clearly that the sheets were worked on separately when they were folded and before they were interleaved with each other in their final arrangement. There are a number of blots or smudges that were offset when the sheets were individually folded. A blot in the second line of the upper marginal note on f. 31r has left a mark in the fourth line of the main text on f. 6v. The lower margin on f. 27r carries signs of the still wet ink on the lowest line of f. 10v. On f. 25r a blot on the fifth line from the bottom has marked the second line from the bottom on f. 12v. On the same pair of pages we see that the left end of line 22 (the word *sotto*) on f. 12v has left a mark below line ten in the second marginal note on f. 25r. This means that the offsetting occurs in both directions on the inside of this sheet, and it may be that one of the facing pages was not wholly finished before the other. The penultimate line of text on f. 23r has left a series of small smudges below the final marginal drawing on f. 14v. The stylus indentations on straight lines of the geometrical diagrams on f. 23r have pressed through to f. 14r. The indentations on f. 14r appear to have preceded the ink drawings, suggesting that the filling of the sheet was at least in part left to right. It was common for Leonardo to use a stylus to incise lines for his precise geometrical diagrams before adding any ink. The most conspicuous set of incised lines is on 1r, not all of which are followed by the final inked lines.

There is at least one instance of offsetting when a sheet was folded over in reverse. Nineteen lines from bottom on f. 32v the blotted words *che ssi* have left marks in the space above the last main block of text on f. 5r.

One of set of compass holes confirms that the sheets could be folded in reverse while he was working on them. There are traces of compass holes and lines on folio 14 from diagrams at the lower left on f. 23r. On the other hand, folio 36 has compass holes from the lowest diagram on 1r, and was therefore folded in the 'normal' way.

More surprisingly, it seems that when the sheets were interleaved they were not always in their present order. A little cluster of marks above the line *q b* and below the word *luna* or f. 7r corresponds to a smudge on ff. 2r–v. This indicates that the second sheet was once folded around the seventh, which has the advantage that the astronomical drawings and related texts on ff. 7r, 7v, and 30r follow analyses of sunlight on the earth and moon on folios 1 and 2. We can therefore suggest that the sequence of sheets at one point ran 1–2–7–3–4, and that sheets 5 and 6 were later inserted, separating the astronomy on sheet 7 from that on sheets 1 and 2. Such a complex process of assembly and insertion can be witnessed in the Turin Codex on the Flight of Birds.

We can also trace a large, if pale, stain that migrates across ff. 36v to 33r, that is, on sheets 1 to 4. Its position on each sheet corresponds to lines 3 to 9 on f. 33r, centred about one inch and a half from the inner edge of each page. This persistent and penetrative stain (perhaps caused by water on the outermost sheet) is not apparent on ff. 32v and 31v, suggesting that sheets 5 and 6 were not folded inside the four others when the stain was generated. There is a trace of the stain in the upper middle of f. 30v and even on f. 30r, though the latter is too elusive to be fully confirmed. This would support our hypothesis that sheet 7 formed an original set with 1–4, before the addition of 5–6. The presence of the stain on sheets 1 to 4 supports the idea that sheets 5 and 6 were inserted quite late in the day.

In summary, we have seen that one page was not necessarily finished before its facing page or reverse. It is also apparent that a number of the marginal notes and drawings, including the book list on f. 15v, which was added late to the page, are so close to the inner edge that they must have been written before the sheets were interleaved. It is evident that some of the sheets, most notably the 'outer set' of seven, were not always interleaved in the same order. Late in the day—how much later we cannot tell—after the sheets had been interleaved, a few final additions were made to the assembled pages. The clearest sign of this happening occurs on f. 30r (the seventh sheet), where the lower left note refers to a marginal drawing on f. 29v (the eighth sheet), which would only face it when the sheets were arranged as they were in the Codex before it was disbound. We have also seen that the bouts of counting across sets of sheets occurred after most of them had been laid one on top of another or folded together.

An Inner and Outer Set

Looking at the sheets, passing from the outer to the inner, it seems visually apparent that sheets 8 to 18 exhibit a more consistent layout than the outer sheets from 1 to 7. The script is generally small and arranged in continuous lines that are relatively regularly spaced with straight margins. There are few line breaks, and the 'cases' are invariably marked with emphatic initial letters. Although the inner sides of sheet 2 (ff. 2v and 35r) and the outer faces of sheet 5 (ff. 5r to 32v) and folio 3r share something of the character of the inner sheets, the outer sheets as a whole are more varied in script, looser in layout, and restrict the illustrations less exclusively to the margins. Later pages in the outer sheets, most notably ff. 32r, 33r, 33v, 34r, and 36v are incompletely filled, whereas the inner sheets are filled from top to bottom in a way that is unusual for Leonardo.

We have already noted from the evidence of blots that the sequence of outer sheets may at one point have been 1–2–7–3–4, perhaps with 5 and 6 inserted later, separating the astronomy on sheet 7 from that on sheets 1 and 2. This order would help explain why sheet 7 did not participate in the counting of 'cases'. It may be that the location of sheet 7 was changed when the lower section on water engineering on f. 7r was added. Indeed, if sheet 7 was for the most part compiled rather separately from the other outer sheets, or produced in conjunction with sheets 1 and 2, we can understand why three of its four pages contain discussions of astronomy. It does not seem well placed in its present position. There is a very faint red chalk '3' between the pulley system and the lowest of the optical studies, which could conceivably be Leonardo's reminder that this was previously the third sheet.

Although there are broader theoretical discussions in the inner sheets, and he is always prone to adduce basic principles at key points, they are predominantly devoted to detailed observations of water in motion, with a recurrent emphasis upon how the engineer can understand what water 'wishes' to do, and how to work in practice to exploit its behaviour. Alongside these practical concerns, he looks on a much wider and longer scale at the history of the earth as determined by water. The prevailing intellectual tone of the 'outer set' is that of theoretical issues, such as the nature of the body of the earth and the passage of sunlight directly and by reflection on to the earth and moon. As always with Leonardo, it is parlous to generalize too confidently, but there does seem to be discernible differences in tone between the 'outer set' and the 'inner set'.

On balance there is a strong case for thinking that there is an 'inner set' of eleven sheets from 8 to 18 that were generated in a relatively uniform manner in a sustained campaign. The outer sheets may have arrived later and in a more piecemeal manner, with at least one late adjustment of order and perhaps late additions. There are, however, signs that the outer sheets were subject to at least one of the editorial procedures applied to the inner sheets. On ff. 3r, 6v, 30r, 31r, 31v, and 33r we can see that some of the initial capitals of 'cases' have been re-emphasized, even when no cases have been recorded at the top of the pages. Only f. 31r has a 'case' count, together with his recording of 905 'conclusions'.

It is therefore possible that the 'inner set' and 'outer set' were produced relatively independently at different times, with the 'outer set' compiled in a more piecemeal manner, before they were brought together and edited by Leonardo as a seventy-two-page *quaderno*.

Putting Things in Order

If the Codex itself does not manifest anything like a definitive order, is it possible to determine how it contents might have been reassembled in such an order. Leonardo's own cross references to numbered sections in his own 'books' encourages us to think that such an order was in the process of being formulated.

There are twenty-five occasions in the Codex on which Leonardo records cross-references to his own 'books', generally citing numbered sections or propositions in 'books' that have also been assigned numbers. The scheme of 'books' within a 'book' or volume follows an established convention, and is similar to way that we now compile a book from a series of numbered chapters. A typical reference is to 'the fourth of the second', and the citations often but not invariably allude to statements of basic rules about the behaviour of the elements in the world.

The contents of his projected 'book on water' outlined on f. 15v contains fifteen books, which begin with 'waters in themselves' (the basic properties and behaviour of water) before moving on to a series of topics about water in the world, and finishing with a series of practical issues about the management of water. This pattern of progress from basics to practicalities is not seemingly reflected in the cross-references in the Codex.

On two occasions in the Codex, on ff. 19r and 27v, he refers to the 'fifth of the first'. The citation on f. 19r might just refer to a book on 'waters in themselves', since it concerns the way that currents excavate the centre of their beds more deeply than the sides. The citation on f. 27v, however, refers to how a lock that is dammed up high will fill its bed with mud, and it would best belong in book 12 in his list, which deals with canals. The 'fourth of the second' is cited on ff. 10v, 27v, and 32r, but the citations do not refer to the same subject. On f. 10v, Leonardo is dealing with the diminution of the power of a blow if the object it strikes is bent, diverting the impetus; on f. 17v, he is citing a rule about the flowing of rebounding water from river banks; while on f. 32r he refers to the basic principle that an object endowed with impetus will continue to move in its original direction. The 'fifth of the second' features twice, on ff. 25r and 30v. On the former occasion, it refers to the principle that an element that is out of its natural level will try to move to its proper position along the most direct route, which stands reasonably alongside the 'fourth of the second' on f. 32r. However, on f. 30v he is dealing with how the widening of a support by four times does not mean that it can sustain four times the original weight or pressure.

It may be that some of the references are to treatises on basic dynamic phenomena and not to a volume specifically devoted to water. However, when he notes the title of 'Book 9, On the Encounters of the Rivers and their Flux and Reflux' on f. 5r, he is certainly referring to a treatise on water, but is does not correspond to book 9 in the list on f. 15v, which is devoted to things moving in or on water.

The lack of any discernible consistency discourages any quest to reconstruct a firmly established list of books and topics to which he regularly made reference. It would be nice to think that he had a single master scheme at his elbow, but this seems not to be the case. Indeed, the references in the Codex best support the idea that he was not convinced that a single, consolidated treatise on water was a practical proposition, given its vast number and diversity of topics. He may well have intended to compose a series of treatises on and around the element of water. We might imagine that each of the fifteen 'books' on f. 15v could potentially comprise a treatise in its own right, each divided into numbered 'books' and sections. If we assume that he was not making up the numbering system as he went along, we can think that he had various scattered lists amongst a chaos of papers to which he referred in a hit-and-miss manner. He must have complied some handy reference lists of 'books' to which he could refer as he went along. Very few have survived.

The longest list of 'books' on water is on CA, ff. 201r–v. The double sheet (originally folded to make four pages) contains eight columns of lists of topics on water that comprise well over two hundred items, many garnered from a lost 'Libro' that he designated as 'M'. Since he also draws, as we have seen, on Libro A, we may suppose that there were at least thirteen notebooks each accorded an identifying letter. The topics on the double sheet are for the most part of a detailed nature, belonging later in the scheme on f. 15v, and some similar items are clustered together. However, there is not a prevailing sense of a pre-determined order in the whole or in the parts. For example, the third column in CA, f. 201r, begins with a topic on the percussion of water, followed by a section on how to use canals to raise boats up mountains, which is in turn followed by items about waves in the sea and about the courses of rivers. Even the keenest aficionado of Leonardo's genius cannot but be disheartened by his apparent inability to devise and sustain any systematic order of compilation, either in his planned treatises as a whole or within a single manuscript, like the Codex Leicester. It was an enduring problem of which he was well aware himself, as his intermittent apologies attest.

A Heroic Agenda

The sources of frustration in tackling the content of the Codex Leicester are therefore clear. The frustration is integral to a process that Leonardo never completed. However, as we will see when we set the

Codex in the context of his thought, the inventive fertility of his mind is exhilarating as he radically probes received wisdom, inundates real or invented 'adversaries' with torrents of argument, journeys from the minutest scale of 'atoms' in smoke to the whole universe traversed by light, cross-fertilizes topics that only he sees as relevant to each other, and develops extraordinary visions of nature in extreme action over a vast number of years. Reading each page, we gain a sense of the impulsive urgency of its compilation, even a certain desperation at being able to envision so much and yet having so little time in the later phase of his life. The awesome visual and theoretical complexities of the topics in the Codex, centring on the motion of water, were deeply compatible with Leonardo's mind and seemed to promise to unlock the secrets of the heart of life in the body of the earth. Water set him a heroic agenda in scale, visual intricacy, and organization. But it was ultimately an impossible agenda. As he wrote on a number of occasions (in variant forms), 'tell me if anything was ever done?'

Mirror of his Mind

The Codex in Leonardo's Thought

3

MARTIN KEMP

WATER, THE MAIN subject of the Codex, occupied a special place in Leonardo's thought.[1] Indeed, of all the subjects with which he was involved, its dynamic complexity resonated most deeply with the shape of his imagination. Water was a matter of infinite visual fascination, scientific wonder, practical magnitude, artistic potential, and emotional potency.

In the Codex Arundel in the British Library, he composed an extraordinary prose poem in praise of water as the dominant element in the extreme dynamics of life and death.

> Water . . . is sometimes sharp and sometimes strong, sometimes acid and sometimes bitter, sometimes sweet and sometimes thick or thin, sometimes it is seen to bring pain and pestilence. Sometimes health-giving, sometimes poisonous. So one would say that is suffers change into as many natures as are the different places though which it passes. As the mirror changes with the colour of its object, so water changes with the nature of the place through which it passes: health-giving, noisome, laxative, astringent, sulphurous, salt, incarnadined, mournful, raging, angry, red, yellow, green, black, blue, greasy, fat, thin. Sometimes it starts a conflagration, sometimes it extinguishes one. It is warm and cold, transports away or deposits, hollows out or raises up, tears down or consolidates, fills up or empties, raises itself up or tunnels down, speeds along or is still. It is the cause at different times of life or death, or increase or privation, nourishes at times and at others does the contrary. At times it has the tang of salt at others it is without savour. At times it submerges the wide valleys with great floods. With time everything changes.[2]

In one of his earlier drafts for the opening of his projected book 'on water', he begins to define his vocabulary in a sober manner—'open sea . . . whirlpool . . . river . . . stream . . . canal . . .'—but when he turns to the description of motion his orderly classification is rapidly inundated by an expressive

1 For additional discussion of the topics in this essay see the important studies published in Galluzzi 2018a.

2 Arundel, f. 57r.

cascade of terms: 'rebound, circulation, revolution, rotation, turning, repercussing, submerging, surging, declination, elevation, depression, consummation, percussion, destruction . . .', and so it rushes onwards until over sixty categories of motion have poured forth.[3] Elsewhere he asks, 'with what language and with what vocabulary will I be able to express or tell of the vile ruination, the incredible desolation, the inexorable rapacity of rapacious rivers in flood, against which no human resource can avail'.[4] This expressive tone is characteristic of his accounts of water in action, both in itself and its impact on other elements.

Even in our age of apparent technological command over so many aspects of nature, the raging waters of floods, hurricanes, tornadoes, and tsunamis serve to remind us of our vulnerability in the face of the overwhelming power of the elements. What we have lost, at least in part, is a powerful sense of the crucial *utility* of water in the age of Leonardo. Water was a major mode of inland transportation, providing the best means to move heavy goods around and between urban centres. Rivers were not always navigable, especially at times of flood or drought, and a number of cities acted as hubs for networks of man-made canals, now largely filled in or covered over. Successive rulers of Milan had invested in an extensive network of canals, not least the Naviglio Martesana, Naviglio Grande, and Naviglio Sforzesco, which were to engage Leonardo's attention in the Codex. Teams of specialist water engineers were employed by the Sforzas and their French successors, and by the late fifteenth century, the Martesana to the east had been linked to the canals within the city, which in turn lead to the River Ticino in the west.

An investor in canals could expect a healthy return from tolls. As will be seen, Leonardo was fully alert to such financial issues. Canals were also major sources of water for irrigation, and farmers could be charged for drawing off a quantity of water (measured in 'ounces') for their crops. Leonardo was granted a water concession in Milan by the French king, Louis XII. He did not consume the water itself, but it would be sold on to those who needed to use it. Needless to say, he busied himself with improving the system of measurement.

Water for drinking and other domestic purposes, something that developed countries now take for granted, was not widely available at the turn of a tap. The provision of unpolluted water from local outlets, such as wells and fountains, was a significant priority for rulers and governments of cities and states. Vivacious fountains also served the needs of courtly and popular delight. Engineers who specialized in the transmission of water, such as Leonardo's colleague, Francesco di Giorgio of Siena, were famous and well rewarded. Sanitation and drainage were also major concerns. Although the mechanisms of disease were not adequately understood, the need to carry away 'fetid substances' in underground channels was well recognized, and was fully reflected in Leonardo's designs for ideal cities, for which there are a series of studies in MS B, where the new town is criss-crossed with a network of channels.[5] The project to drain the Pontine Marshes south of Rome, with which Leonardo was involved in 1515, was driven not least by the need to combat the scourge of malaria.

Alongside the construction, regulation, and economics of man-made canals, there was no less a need to manage natural watercourses. The greatest priority was to mitigate the damage caused by floods and

3 Madrid I, ff. 72r–71v.

4 CA, f. 302r.

5 e.g. MS B, ff. 16r, 17r, 37v, 38r.

lesser inundations. Dams, weirs, sluices, locks, bridges, banks, and beds were all under perpetual threat from the mobile and weighty violence of water on the move. A great deal of space in the Codex is devoted to these issues, analysing what is happening and why, and suggesting how the engineer can work collaboratively with water to reduce the problems and to achieve human ends. What distinguished Leonardo's efforts from those of his fellow engineers was that he aspired to bring the laws of motion into play when manipulating water. In times of war, as Leonardo emphasizes, watercourses could act as lines of defence, and the violence of water could be harnessed to inundate invading enemies. The engineering of water was therefore of enormous importance in Renaissance society. To be regarded as a *maestro d'acqua*—'master of water'—counted as a major achievement, and such 'masters' were very attractive to potential employers.

The destructive power of water was the subject of some of Leonardo's most memorable passages of writing. In his era, and for him in particular, it was water in violent motion rather than the eruption of volcanoes that loomed largest in the transformation of the body of the earth. On a large sheet in the Codex Atlanticus he wrote:

> Amongst all the causes of the destruction of human property, it seems to me that rivers, on account of their overwhelming and impetuous inundations, occupy the first rank. And if anyone wishes to adduce fire versus the fury of impetuous rivers, it seems to me that such a person would be devoid of good judgement, for fire remains spent and dead when it lacks nourishment, but against the irreparable flooding of proud and swollen rivers no protection devised by human contrivance can suffice. With a combination of turbulence and tumult, water gnaws and ruins the high banks, growing turbid with earth from ploughed land, destroying houses and uprooting tall trees. It carries these things as prey to the sea, its lair, conveying within itself men, trees, beasts, houses and properties, ruining every bank and any other barrier, bearing with it light objects, and ruining and laying waste to those that are heavy, rendering small fissures as great ruptures, filling up the low valleys with floods of noxious and harsh waters. How many must flee from the flooded locality? O how many cities, how many lands, how many castles, villas and houses has it consumed? How many labours of farm hands have been rendered vain and without product? How many families have been dismembered and overwhelmed? What shall I say of the herds of cattle that have been drowned and lost?[6]

The context for this impetuous and accelerating passage, added late on the page, is very telling for our understanding of Leonardo's engagement with water (Figure 1). The recto of the sheet (Figure 2) is dominated by three large designs running horizontally across the length of the page. They show a canal that steps upwards via a series of locks to pass over a river that flows through two arches in the aqueduct. We know that he was thinking of such a scheme for his projected canal in Tuscany to bypass the non-navigable sections of the Arno.[7] He notes that a boat passing over the bridge will not increase the net weight bearing down on the bridge. In the lower corner of the page, he depicts vortices and erosion caused by the construction of a channel, probably based on something he has seen. Then in a small diagram squeezed between two canal studies he represents, in a schematic manner, the Mediterranean Sea and its constricted outlet into the ocean via the Strait of Gibraltar. The diagram shows that only a narrow central current can pass through the outlet, while all the other outflowing waters become jammed up.[8]

6 CA, f. 1007v.

7 CA, f. 126v. Compare the scheme on CA, f. 46va.

8 See also MS A, f. 57r.

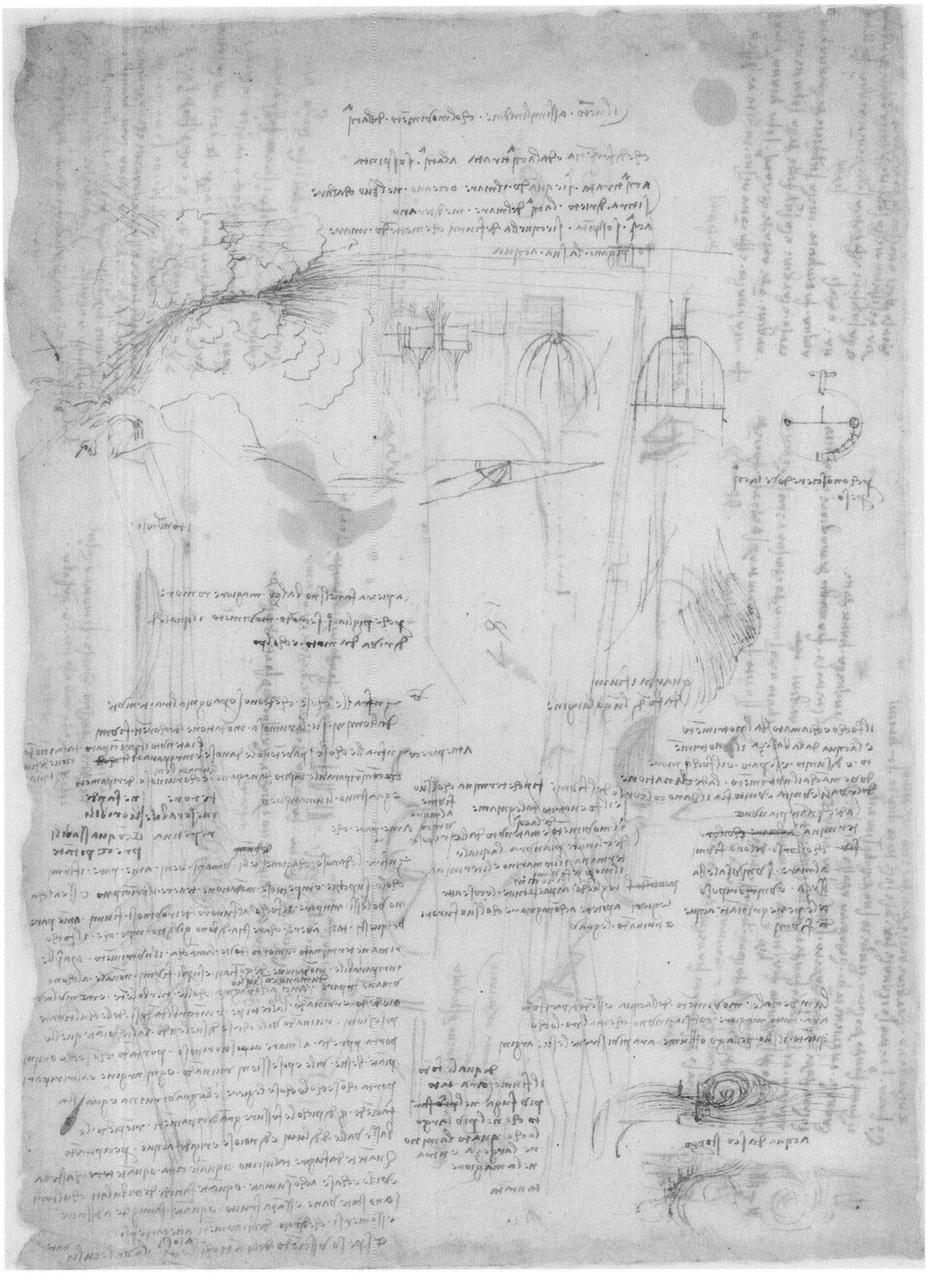

FIG. 1
Notes on the power of water, the passage of water through channels and clouds. Biblioteca Ambrosiana, Milan, Codex Atlanticus, f. 1007v.

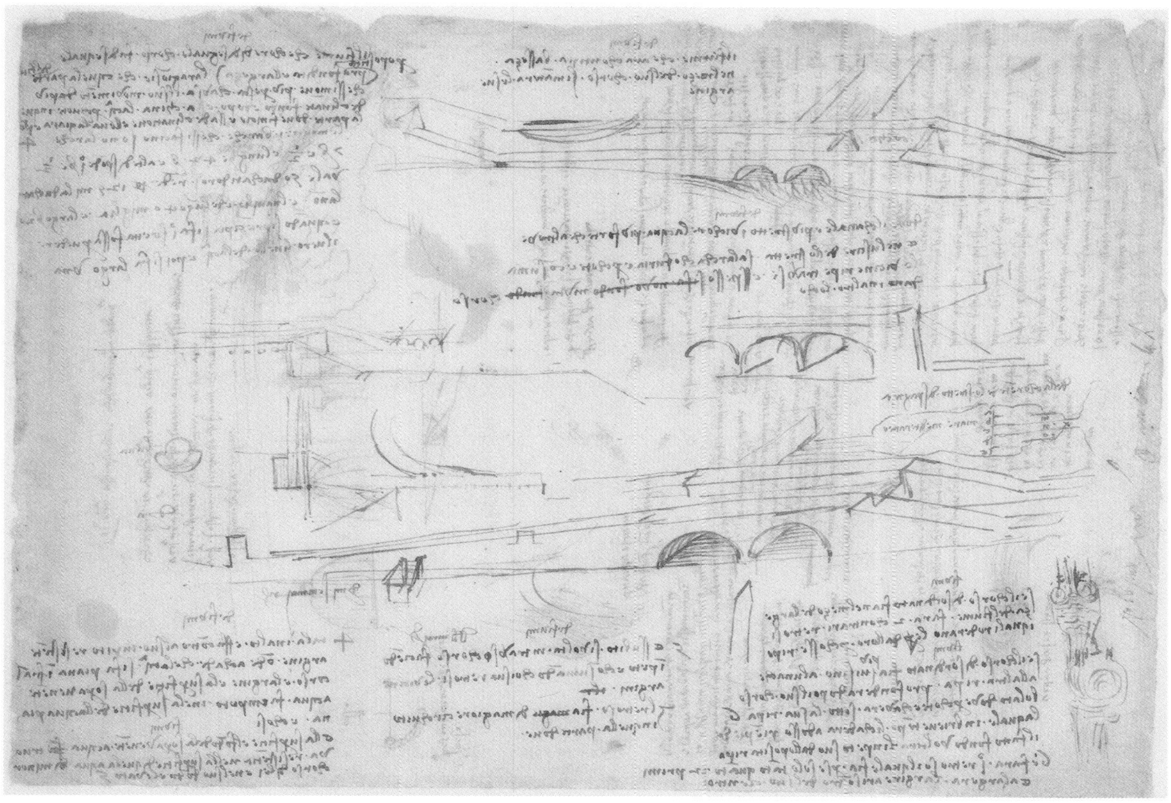

FIG. 2
Designs for a canal passing over a waterway. Biblioteca Ambrosiana, Milan, Codex Atlanticus, f. 1007r.

His exploration of the passage of a large body of water through such narrow straits is continued on the verso of the sheet, which is now orientated vertically. Leonardo adds a small sketch of a circular instrument with a balance arm to measure the 'weight' of water disposed around portions of its circumference. To the left of the diagram of the straits he draws a quivering cloud being rent asunder by a current of powerful wind, penetrating through a narrow gap just like water surging through the Strait of Gibraltar. He notes that 'wind in its motion resembles that of water'. On the lower right, he draws a damaging vortex occasioned by water squeezing through a narrow channel that he records as having seen at 'Santo Strorzo', that is near the church of S. Eustorgio in Milan.

It is the implicit violence of such pent-up water that seems to have triggered his powerful account of the havoc caused by floods. His narrative is prefaced by two columns of words that he seems to be testing for their potency:

> terror / vile
> inexorable / the ghastly and the shattered
> sudden / precipitate[9]

Finally, returning to the recto, there is a tiny, faint drawing of a sphere half surrounded by a vase-like form. It is labelled 'moon' and is related in some way to some highly theoretical diagrams of the relationships of the spheres of earth and water on the penultimate folio of the Codex Leicester (f. 35v). For good measure, he inscribes an abbreviated version of his recurrent refrain in small neat script below the

9 *terrore / nefande*
insesorabile / li orribili e conquassabili
repentina / precipitante.

third of the drawings of the elevated canal: '*Dimmi se mai [fu fatto alcuna cosa]*' ('tell me if ever [anything were done]').

It has been worth looking in some detail at this sheet because it signals almost all the main themes in the Codex Leicester and demonstrates the essential continuity of Leonardo's thought within each sheet and throughout the whole Codex, as his analyses flow across what we regard as essentially different topics. Not least this sheet from the Codex Atlanticus blends science and sensibility in a way that is specially Leonardo's own.

On one side of the coin, water in motion is a vision of death and destruction. On the other side is life and levitation. Water brings life to the animate body of the earth: 'water is that which serves the vital humour of this arid earth'.[10] The framework for this vitalist vision of water is the theory of the microcosm and macrocosm, which underlies the central themes in the Codex.

He laid out the basic premise of the microcosm / macrocosm around 1490 in MS A, acknowledging that it was an ancient concept:

> By the ancients man was termed a lesser world and certainly the use of this name is well bestowed, because, in that man is composed of water, earth, air and fire, his body is a similitude for the world: just as man has in himself bones, the support and framework of his flesh, the world has the rocks that support the earth; just as man has in himself the lake of the blood, in which the lungs increase and decrease during breathing, so the body of the earth has its Oceanic sea which likewise increases and decreases every six hours with the breathing of the world; just as in that lake of blood the veins originate, which make ramifications throughout the human body, similarly the Oceanic sea fills the body of the earth with infinite veins of water. The nerves are lacking in the body of the earth. The nerves are not to be seen there because the nerves are made for the purpose of movement, and the world being perpetually stable, movement does not occur, and movement not happening, the nerves are not necessary there. But in all the other things they are very similar.[11]

Leonardo later marked this statement as the 'beginning of the treatise on water'. It is immediately followed by an account of the 'the veins of water on the summits of mountains'. He explains that 'if the body of the earth were not similar to that of man, it would be impossible for the water of the sea, which is so much lower than the mountains, to be able to rise naturally to the summits of the mountains'. Water circulates in the world, within 'veins' that originate from the sea beds and rise into the mountains where they give rise to high springs: 'the water passes from the rivers to the sea, and from the sea to the rivers, ever making the same round'.

Over the course of the next three folios in MS A, Leonardo points to analogous processes that provide evidence for this vascular system. He notes the way that blood spurts from a severed vessel in a man's head, and how sap spouts from the top of a pruned vine, just as water emerges on the upper slopes of mountains. Heat is the driver of this ascendant motion of 'humours' in the earth and the human body. We see that heat can drive sooty smoke up a chimney. In the body of the earth, 'the same heat that holds up so great a weight of water as is seen to rain from the clouds takes it up high from the bases of the mountains and draws it up and confines it within the summits of mountains, and there the water finds crevices, and continuing to exit it creates rivers'.[12]

10 Arundel, f. 210r.

11 MS A, f. 55v.

12 MS A, f. 54v.

He typically adduces an experiment to show how fire can attract water upwards:

> Heat a vase and place it in a bowl [of water] with its mouth downwards, and put a burning coal onto it. And you will see that the humidity goes directly towards the heat, rising to replace the air in the vase, and the vase will fill with water, and the air that was enclosed in the vase will escape though its mouth.[13]

When he came to compose the Codex Leicester, more than a dozen years later, he drew a very different conclusion from the same experiment.

The folios in MS A that deal with the circulation in the lesser and greater worlds, accompanied by spirited drawings of observations, demonstrate both the continuities and the changes in Leonardo's thinking. The Codex Leicester, for all its focus and elaboration, is one step, however mighty, in what is a recurrent process of profound thinking about the element of water, geography, and hydraulic engineering.

Before 1482: Archimedean Screws and Perpetual Motion

In his letter of self-recommendation to Ludovico Sforza, written in the early 1480s as he was moving to Milan at the age of thirty or so, Leonardo lists nine categories of military engineering in which he claimed to proffer innovatory designs. Amongst the abilities itemized in the tenth category—useful 'in time of peace'— he promises to 'give complete satisfaction . . . in conducting water from one place to another'.[14] Although the Florentine workshop of Andrea Verrocchio, in which he had been trained, was extremely versatile and was involved in architectural engineering, there is no direct evidence that the young Leonardo had practical experience of canal design or other aspects of the management of water. Indeed, Milan was well equipped with engineers who specialized in hydraulic engineering.

However, Leonardo had begun to study the kind of treatises and sets of illustrations that were the stock-in-trade of Renaissance engineers. In Florence, Filippo Brunelleschi had assumed legendary status for his inventions, which included boats for transporting marble, in addition to the famed machines he designed to erect the dome of Florence Cathedral. In Leonardo's time, Buonacorso Ghiberti (grandson of the great Lorenzo) compiled a 'scrapbook' of notable machines, including Brunelleschi's lifting devices. Neighbouring Siena produced a succession of major engineers, most notably Brunelleschi's colleague, Mariano Taccola, and Francesco di Giorgio, with whom Leonardo worked in Milan and one of whose manuscripts he owned.[15]

A large sheet in the Codex Atlanticus bears testimony to the range of ambitions Leonardo was developing.[16] They show various mechanisms for raising water, presented in drawings that adopt much of the visual repertoire of the earlier and contemporary treatises. Leonardo concentrates on means for raising water to higher levels, which would serve a variety of needs when a good head of water is required. He illustrates a variety of pumps in sophisticated drawings that convey a notably good idea of the spatial structure of the mechanisms, together with machines to raise water from wells. The most striking of the devices are the 'Archimedean screws', named after the famed Greek mathematician and inventor. A spiral pipe or a screw, often within a cylindrical casing, is set at such an angle that water can be wound 'uphill' from a reservoir. On the recto (Figure 3), Leonardo shows a water wheel driven by

13 MS A, f. 56r.

14 CA, f. 1082r.

15 Galluzzi 1991, 1996.

16 CA, f. 1069.

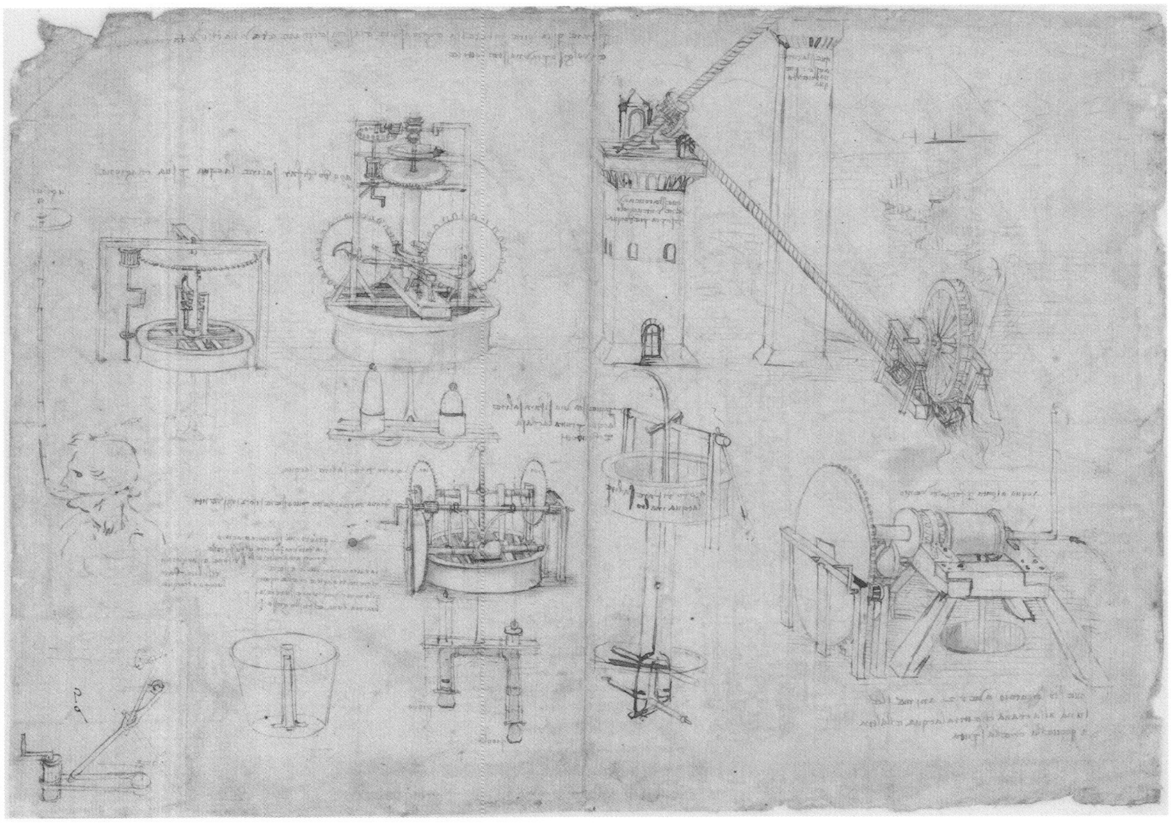

FIG. 3
Screws for raising water. Pumps and breathing device. Biblioteca Ambrosiana, Milan, Codex Atlanticus, f. 1069r.

a surging stream that emanates from distant mountains. The wheel rotates a long screw that rises to a reservoir at the top of a tower. The rotation of this first screw in turn drives a lantern gear that turns a second spiral tube, which feeds a lanky tower that is 'completely full of water'. There is no indication of how this ramshackle arrangement of exceedingly long and heavy screws could be supported.

On the left of the verso of this sheet (Figure 4), Leonardo demonstrates a single screw driven by a waterwheel and at the top of a squat tower. There is no indication of how the screw is rotated other than by the waterwheel, which is itself fed by the water pouring from the top of the screw. As drawn, it appears to be a perpetual motion device. The quest for perpetual motion had been a long-standing quest from antiquity onwards. In the set of four screws mounted in parallel on the lower right of the sheet, he also seems to be speculating about a self-driven system. The screws that feed into the top of the tower are driven by four wheels, the topmost of which is fed by a pipe that emerges from the internal reservoir in the tower. Water then successively flows from stepped tanks to turn the lower three wheels. The four sketchily indicated screws in the drawing at the upper right feed a tower that supports a thick axle rotated by a bucket with a counterweight. This in turn seems to turn a wheel, over which a chain of buckets passes, tipping water into an elevated tank. It is difficult to see how the whole system works and whether the screws are coupled with the mechanism of buckets, but it looks as if he is again speculating about a self-driven system.

As if to underline that he is speculating visually with mechanisms certified in the manuscripts of the engineers, he shows on the recto a device for breathing underwater—a jointed tube that passes through a circular float—which is of a kind familiar in earlier treatises. The two sides of the sheet share a quotient of implausibility with the stock illustrations. Some designs in the treatises are of practical utility, while others might best be described as 'visual boasting', designed to impress a patron with the engineer's

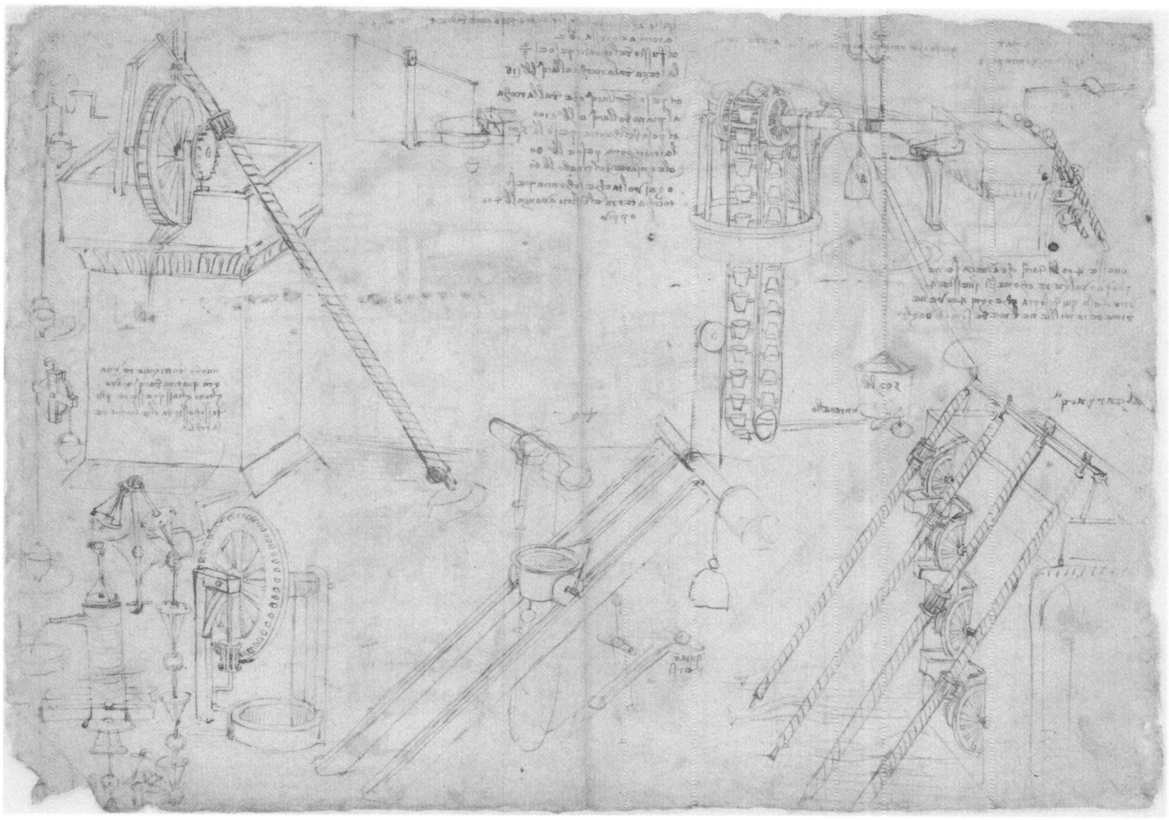

FIG. 4
Screws and devices for raising water. Biblioteca Ambrosiana, Milan, Codex Atlanticus, f. 1007v.

inventive potential as a technological visionary. If this inventiveness could be pitched in relation to the famed technicians of antiquity, as with the screws attributed to Archimedes, so much the better. The screw designs belong in spirit with his early drawings of ancient-style weapons, such as scythed chariots. Pedigree tends to outweigh practicality.

These early essays in hydraulic engineering do not seem for the most part to have been followed up substantially in his later work. The screw makes a rather inglorious later appearance in the early 1490s. In the last fourteen folios in Codex Forster I in the Victoria and Albert Museum, Leonardo speculates about various perpetual motion mechanisms constructed from spiral tubes and rotational systems.[17] The least implausible involve double Archimedean screws in which a wider outer spiral, down which water causes a strong enough rotational motion to allow an inner narrower spiral to lift water upwards to fill the wider spiral—and so on.[18] It looks plausible enough in a drawing, but would not work.

Hugely attractive though the prospect of perpetual motion may have seemed to someone as inventive as Leonardo, by 1500 he had convinced himself that the quest to create perpetual motion using water was futile:

> I remember having seen men from various countries brought by their infantile credulity to Venice with great hope of gain by making mills in dead water. Being unable, after much expenditure, to make such a machine move, they were compelled by great fury to escape from this debacle.[19]

17 For further discussion of such devices, and his proofs of their failure, see Martin Kemp, 'Leonardo's Book of Negative Demonstrations', in Juliana Barone, ed., *A Mind in Motion*, London, 2019, pp 216–23.

18 Forster II, f. 43v.

19 Madrid I, flyleaf.

By the time he returned to the problem in the second of the Forster Codices, perhaps a little after 1500, he was making a sustained effort to establish a scientific base for hydraulic engineering. Accordingly, he dismisses all such attempts, telling the 'speculators on perpetual motion' to 'go and join up with the seekers after gold'—that is to say with the alchemists, whom Leonardo despised.[20]

1482–1500: Water in Sforza Milan

The first substantial indications of Leonardo's engagement with the science of water, with its role in the body of the earth and how water behaves in a series of actual situations, are found in MS A from the early 1490s, which we have already quoted for his first definition of the microcosm / macrocosm analogy and his discussion of springs in mountains. A substantial proportion of the manuscript is devoted to his theory of art, including perspective and bodily proportions, and he is clearly assembling material for a treatise on painting. The sections on water, though smaller in extent, show that a treatise on water is also on his mind. Working from 55v, where the analogy appears, through some ten folios (twenty pages), we find an agenda that prefigures that of the Codex Leicester in a surprising number of respects.

For four folios from 55v, Leonardo continues to deal specifically with water in the body of the earth, especially the vexed issue of high spring, always resorting to the circulatory explanation of the water-filled 'veins' and how heat draws water upwards. In addition to the experiment of the inverted vase, he proposes one that involves two small copper balls suspended from a balance arm. If one of the balls is heated until it is red hot, the balance will rise on that side since it will have 'been made light by the heat', proving that heat makes things rise.[21] Presumably, this results from the action of the heated air around the ball not from any change in its density.

We can identify a series of topics that reappear in the Codex Leicester. On the same folio as the experiment with the heated ball, Leonardo draws a small, schematic map of the Mediterranean devoted to the question of how sea squeezes through the Strait of Gibraltar, an issue that occupies a good deal of space in the later Codex. He also rejects the idea that the moon causes the tides, as he was to continue to do. On the verso of folio 57, he notes that when surface waves move in a certain direction, as towards the seashore, there will be a contrary motion of water at the bottom. He subsequently looks at water falling from conduits and the way it consumes banks and beds, affecting stones of various sizes in a graded manner. Then on 58v he analyses the distinct centres of the sphere of water and of the irregular earth, taking up theoretical themes discussed in the Middle Ages. Looking at the portion of the earth that rises above the surface of the sphere of water, he surmises that the plains crossed by major European rivers must be much higher than the tops of mountains by the sea, but he firmly refutes the traditional idea that the surface of the middle of the ocean can rise higher than the loftiest mountains as the explanation for high springs (an idea he castigates as 'vain and puerile').

Typically, he sees universal rules at work behind all the phenomena, however complex. These mathematical rules govern all forms of motion. They are visibly shared by water and air. When he

20 Forster II, f. 89v.

21 MS A, f. 57r.

describes the 'tremors' or 'shaking' in water when concentric ripples are generated, he is alert to the similar propagation of light in the space around an illuminated object.[22]

The fundamental rule is that of impetus, that is to say the potential for motion infused in a body by the force exerted by its original mover:

> All things universally desire to maintain themselves in their natural state. Thus a moving current of water strives to maintain its course according to its force, and if it encounters any opposition it completes the length of its initiated movement with circular and eddying motions.[23]

Impetus was the potency for continued motion that had been adduced by Jean Buridan and other mediaeval philosophers to explain why a body continued to move when it was no longer actively pushed by the agent that had set it in motion. The body endowed with impetus must actively complete its assigned motion. Thus as he explained on f. 24r, the path of bouncing ball will inexorably equal that when it is thrown freely through the air. He invites us to confirm this rule through an experiment in which a glass marble in bounced on a hard surface. In the case of water, compared to a thrown marble or moving air, the impetus is more destructive because water is a 'continuous quantity', that is to say it is a unified substance in space and is incompressible and cohesive. The performance of impetuous water is the most unyielding of all. Our fear of water on the move has a scientific basis.

Contemplating the experiment with the thrown marble, he is moved to exclaim, 'how admirable is your justice Prime mover. You have willed that no power lacks the order and qualities of its necessary effects.' The Prime Mover—the ultimate motive power established by God in the outer boundary of the universe—is the principle that governs all motion in the finite universe. As the 'final cause' it subjugates every natural cause and effect in such a way that they are obedient to the dictate of mathematical law.

On the last sheet in MS A, Leonardo looks at the passage of light from the sun to the earth and the moon, which is the prime subject of the outer folios of the Codex Leicester.[24] Already he is arguing in favour of the moon being covered by sea and land, like the earth, though at this point he is more concerned with the apparent size of the celestial bodies to the observer than the nature of the luminosity of the moon. The issue of apparent size is of obvious relevance to the optical issues in his art theory elsewhere in the manuscript.

Given the close parallels between the subjects in these folios in MS A and the Codex, it may well be that he looked over them when he was beginning to compile his later treatise. None of the topics are pursued as extensively as in the Codex Leicester or with such intensity of focus. The greatest difference is that MS A contains no direct discussion of historical geography, that is to say the vast transformations in the body of the earth over long eras of time. After Leonardo's return to Florence, the redirection of his interests towards the development of the body of the earth involved a series of interlocking factors, including his work on the Arno canal, his dissection of the 'centenarian' and of a child, and the revelations of Strabo's *Geography*.[25]

22 MS A, f. 60r.

23 MS A, f. 60r.

24 MS A, f. 64r.

25 See Chapter 4 in this volume.

There is little discussion in MS A of the practicalities of hydraulic engineering, but this lack is remedied by MS H, which was compiled in and around 1493–4. We can readily reconstruct why Leonardo turned to more the practical matters of the canal system around Milan. As he records, in the early months of 1494 he was in Vigevano, little more than twenty miles south-west of Milan, where Ludovico was notably active, constructing a grand castle and an arcaded Renaissance piazza. In the surrounding countryside Ludovico promoted advanced schemes of canalization and irrigation, establishing two ideal villa-farms, one of which, La Sforzesca, he presented to his wife Beatrice. On the first folio of MS H, Leonardo observed 'a locking gate for a sluice at Vigevano', and on 70v he made drawings of what seems to have been a portable 'pavilion at Vigevano', constructed from jointed wood. On 2 February 1494 he 'drew a set of 25 steps, sixteen inches deep and fifteen feet wide'.[26] The 'steps of Vigevano', which were designed to conduct water to a lower level without destructive effect, were to be the subject of more than one analysis in the Codex Leicester. On 20 March 1494, he was still in Vigevano, where he made observations in the vineyards.[27]

A series of steps with concave faces is shown on folio 80r (Figure 5).[28] Above the steps is an extraordinary design for a raised canal with two lateral passages, which passes through a vaulted tunnel with apertures in the roof to admit air. It seems that Leonardo is considering how to design a lock with a massive change of level. The steps conduct water downwards from the two lateral channels until the water level is such as to allow the boat to float across the top of the lock. This clearly tackled an extreme problem, as he was later to face in his designs for a canal, massive dam, and lock at Tre Corni on the Adda east of Milan. It is possible that the tunnel and steps represent his first ideas for the Tre Corni bypass.

It was in the rural paradise around Vigevano that he turned his attention to mills and other hydraulic engineering, again prefiguring a number of the topics in the Codex Leicester. His observations range from the speeds and angles of water that strike millwheels to the way that obstacles in flowing water cause the erosion and deposition of sediment. Some of his observations are distilled into lists of 'propositions' like those in the later Codex: 'where the motion of water is slower, the surface of its bottom

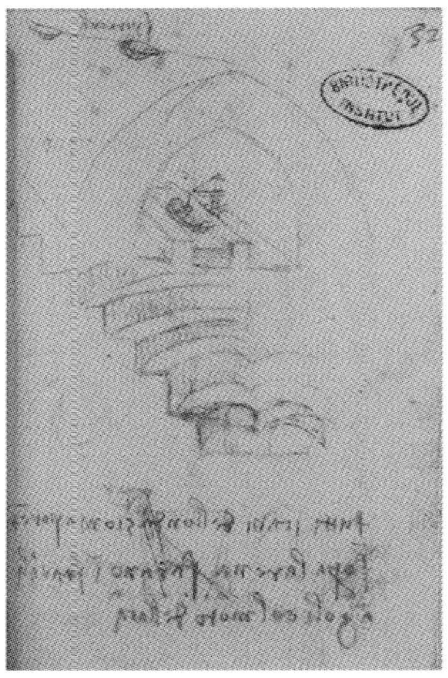

FIG. 5
Canal passing through a tunnel and a series of concave steps. Institut de France, Paris, MS H, f. 80r.

26 MS H, f. 65v.

27 MS H, f. 38r.

28 The note on MS H, f. 80r does not directly engage with the drawing.

will have finer sand'.[29] A typical rule is that 'the percussion of water on to a wheel will be at the highest degree of power when it strikes at equal angles [perpendicularly]'.[30]

At this time, he was also looking more widely at the grand canal schemes in and around Milan, which were so vital to the city's economy. On 65r, he notes that the Martesana Canal has a drop of two inches for about every thousand feet of length. The Martesana, one of the most sophisticated canals of its era, runs alongside the Adda before bearing east into the city. It also was to be discussed in the Codex. He also made detailed drawing of the 'Lock of S. Marco' within the city, which played a key role in the linking of waterways entering from the east, which previously terminated in the basin of S. Marco, with those entering from the west. His drawing in the Codex Atlanticus (Figure 6) shows one of the angled or mitre gates of a pound lock, with a low sluice unlocked by a system of rods.[31] He makes a detailed drawing to show how the door of the sluice pivots on a hinge two-thirds of the way across the aperture. He also indicates the ingenious system of paving with stakes across the floor of the lock.[32]

Leonardo himself has been taken as the designer of the lock, which is documented as being constructed in 1496–7, but the tone of the notes that accompany his drawings suggests that he was recording a scheme devised by one of the ducal engineers, probably Bartolomeo della Valle, successor to famed Bertola da Novate, who had been responsible for the Martesana and other Milanese canals. It is in keeping with this spirit of study that he reminded himself in the Codex Atlanticus (f. 611ar) to find a *maestro d'aqua* to tell him about 'repair and cost—the repair of a lock, canal and a mill in the Lombard manner'.[33] In the same memorandum about topics to be investigated, people to pester with questions, and books to read, he notes that 'Gian Angelo the painter has a book on water that was owned by his father'. Giovanni Angelo Mirofoli da Saregno with his brother had painted the organ shutters in the Cathedral. He also mentions a copy of Aristotle's *Meteorology* translated into Italian, though whether this is wishful thinking is unclear. Leonardo was trying to build up a stock of knowledge from any sources that came his way. He also notes enigmatically that 'master Giovanni the Frenchman' (probably Jean Perréal the painter) had promised to show him 'the measurement of the sun'.[34]

Leonardo's official role as a 'ducal engineer' in Sforza Milan has been overstated. A document from the 1490s lists ten 'ducal engineers' of whom the first four are also 'bombardiers' (military experts) and ten 'communal engineers'. On the other side of the sheet are listed 'Bramante engineer and painter, Doclebuono engineer and sculptor, Giovanni Batagio . . . engineer and builder, Leonardo of Florence engineer and painter'. The four on the reverse of the sheet fell into different categories than the twenty official engineers. Leonardo was clearly recognized as having some expertise in engineering but was not one of the cadre of professional engineers at the court.[35]

By the time that Ludovico's regime fell, in the face of pressure from the invading French armies in 1499, Leonardo had studied some of the practicalities of hydraulic engineering, and had characteristically

29 MS H, f. 30r.

30 MS H, f. 63r.

31 CA, f. 656ar. Compare CA, f. 408v.

32 Compare CA, f. 935v for the paving of the floor of a lock.

33 CA, f. 611ar.

34 CA, f. 611r: 'la misura del sole promessarmi da maestro Giovanni Francese'.

35 The document is reproduced in *La Vera Immagine*, and discussed by Maffioli 2016b, p. 212.

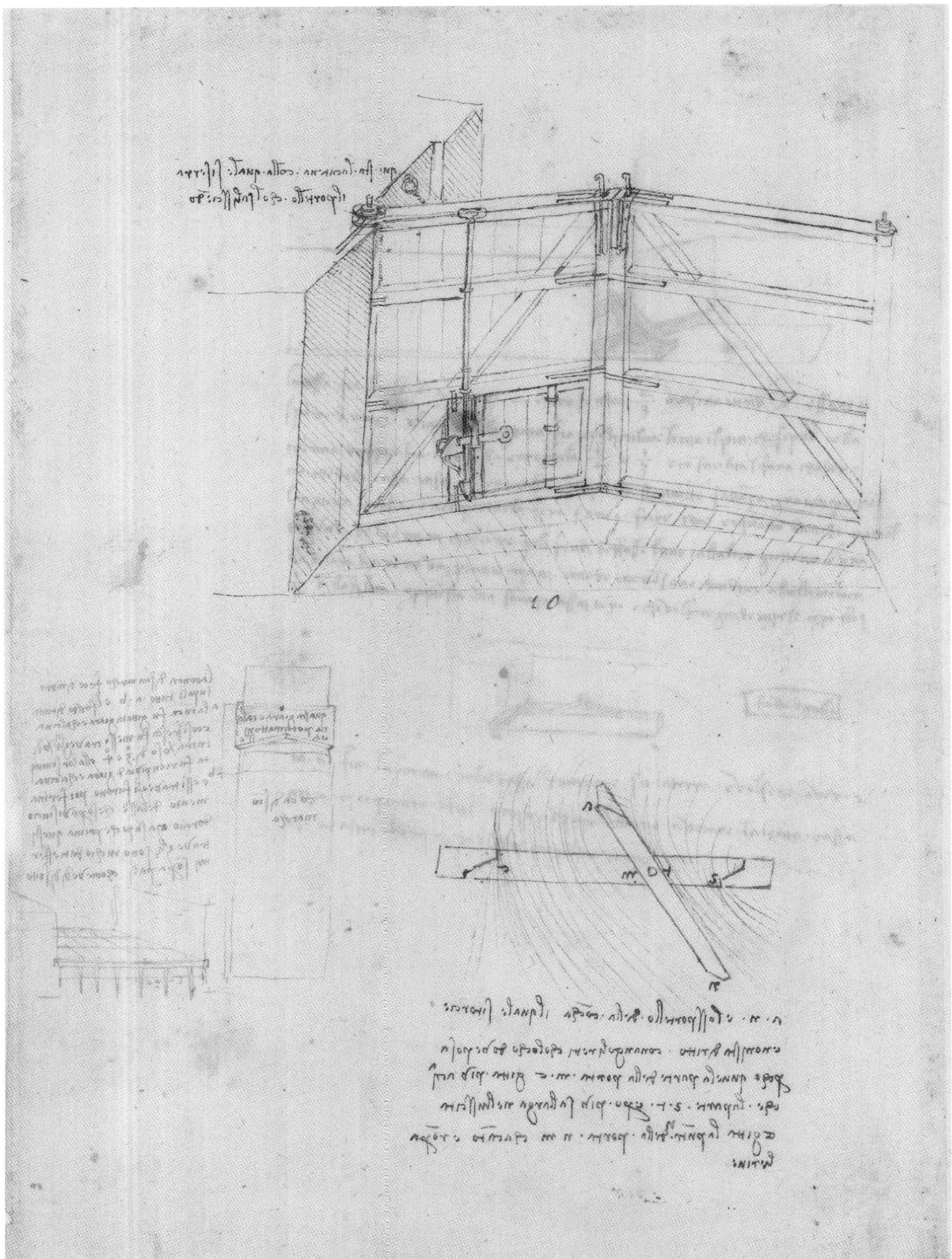

FIG. 6
Studies of the pound lock at S. Marco in Milan. Biblioteca Ambrosiana, Milan, Codex Atlanticus, f. 656ar.

blended such practical concerns with the theory of water in motion in a way that was different from standard engineering practice. He had also begun to make inroads into the natural philosophy of water in the body of the earth. A treatise on water was already in his mind.

1500–8: The Arno Canal

The next major landmark in Leonardo's engagement with water came when he became involved in Florentine schemes for canalization in the early years of the next century. On the fall of Ludovico's regime, he did not return directly to Florence in 1500. He spent some time in Venice—we have already noted his witnessing of a context to make mills revolve in 'dead water'—and he was hosted in Mantua by Isabella d'Este, sister of Ludovico's late wife. His cartoon for a portrait of Isabella is in the Louvre, but the Marchioness of Mantua was never to receive from Leonardo either a finished portrait or any other painting that she craved.

We have a later note by Leonardo himself testifying that he had served the Venetian republic as an expert on the management of water. In the Codex Arundel, when he was planning some canalization for Francis I (i.e. after 1513), he reminded himself, 'let the sluice be mobile like the one I arranged in Friuli, where the water excavated the bed when the channel was opened'.[36] A mobile dam and sluice that could move progressively along a canal is discussed and illustrated in action in the Codex Leicester on folio 22v. The Venetian scheme involved the River Isonzo (the Soča in Slovenia), which provided a natural line of defence against the aggressive Turks, as it did for Italy in World War I in the face of the invading Austro-Hungarian armies. Leonardo's mobile sluice, which excavated successive sections of the river, was designed to provide greater control of its flow. Rough, fragmentary drafts of a letter to the Venetian authorities survive in the Codex Atlanticus, in which he characteristically underlines his advice about practical matters with statements about the theory of water in motion.[37] Leonardo also noted more generally in the Codex Arundel that he intended to write books on 'how to drive back armies by the fury of floods caused by the release of waters', and 'how to inundate armies by the closing of the outlets of valleys'.[38]

On his return to Florence, Leonardo's first involvement with an Arno canal was also military in nature. On 24 July 1503, a Florentine government official in the camp outside Pisa reported that Leonardo and a member of the 'Ten of the Balia' (the ten-man committee of war) had presented plans for the diversion of the Arno around the city of Pisa, which had acceded from Florentine rule.[39] Pisa, as a port, was of vital strategic interest to Florence and its recapture was a high priority. 'After many debates and doubts', the authorities decided to pursue the plan to deprive Pisa of its river. On a map in Codex Madrid II (Plate A; Figure 7), Leonardo notes where he intended to 'break the Arno' out of its natural bed southwards from the first meander east of Pisa to link with an existing canal that feeds into a large lake south

36 Arundel, f. 270r.

37 CA, f. 638dv; Villata 1999, pp. 131–2.

38 Arundel, f. 35r. See also MS B, f. 64r.

39 Vezzosi 1984, p. 52. See also A. Vezzosi, in collaboration with A. Sabato, 'The Florence and Pistoia Canal: deviating the Arno through Prato and the Val di Nievole', *Scientific Itineraries in Tuscany*, Museo di Galileo, http://brunelleschi.imss.fi.it/itineraries/itinerary/TheFlorenceAndPistoiaCanalDeviatingTheArnoThroughPratoAndTheValNievole.html.

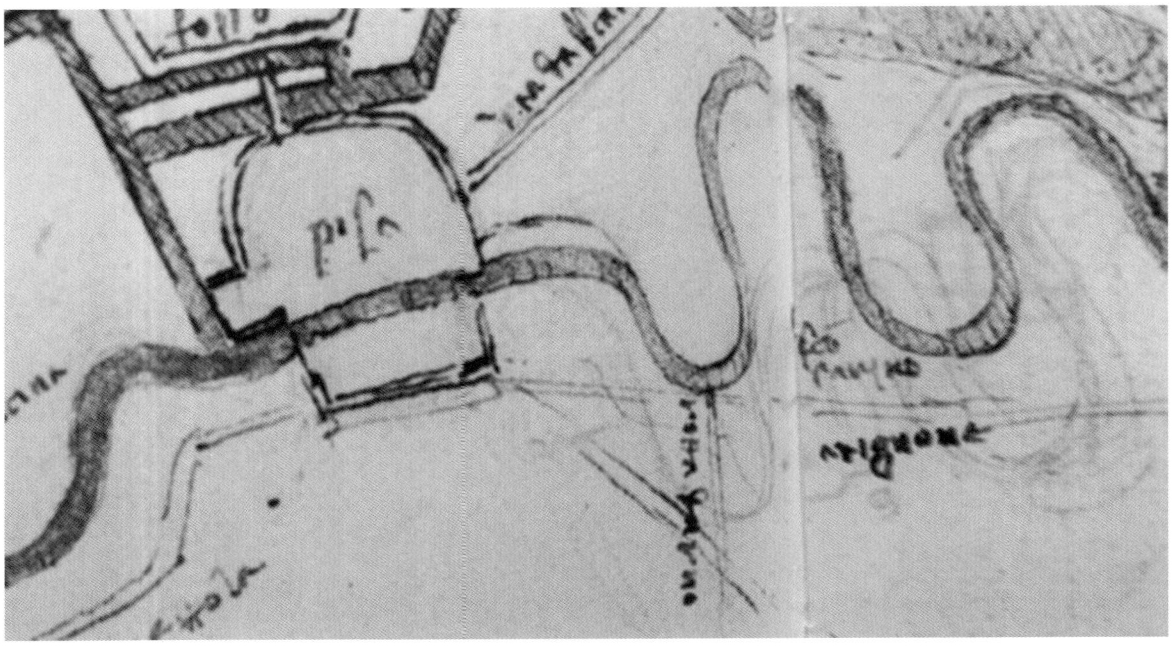

FIG. 7
Map of the Arno showing Pisa and the proposed 'breaking' of the Arno. Codex Madrid II, ff. 52v–53r (detail).

of the besieged city.[40] The map is labelled in left-to-right script, so that the names were more readily legible. Niccolò Machiavelli, who was Secretary to the Florentine government at this time, was a keen supporter of the project. Two days later, expenses were granted for a six-horse carriage and personal subsistence to enable Leonardo to pursue his goal. A year later, work was actually underway, on the basis of a compromise scheme agreed by the various factions. It involved the deployment of one thousand labourers. The architect Antonio da Sangallo was involved as a collaborator with Leonardo. However, in October, it became evident to the Florentines that the river would not be pushed around and the project was ignominiously abandoned.

Alongside this short-term fiasco, which cannot have done Leonardo's reputation any good, there are signs that he did productively serve the republic on less grandiose schemes.[41] On a beautiful coloured drawing (Figure 8), again labelled left to right so that it could be read by the authorities, he records how two weirs were resulting in serious damage to the bank.[42] Water surging through an aperture in each of the weirs (whether designed as orifice's or accidental ruptures) results in a turbulent current scouring out the bank under and behind its immediate wall, accompanied by a raging anti-clockwise vortex. He notes that the rupture in the bank is over 110 feet long. The uncontrolled current then rebounds in an arc to excavate a second cavity which is considerably longer than the first. It vividly demonstrates the continuity of impetus in a stream of moving water. A further map at Windsor allows us to locate the weirs on the Arno east of Florence and downstream from the entry of the tributary of the Mensola.[43] No doubt he recommended how this damaging situation could be rectified, probably using more enduring weirs placed with greater strategic awareness—as advocated in the Codex.

40 Madrid II, ff. 52v–53r.

41 For Leonardo's studies of the Arno in and around Florence, see Di Teodoro 1992.

42 Windsor 12680.

43 Windsor 12679.

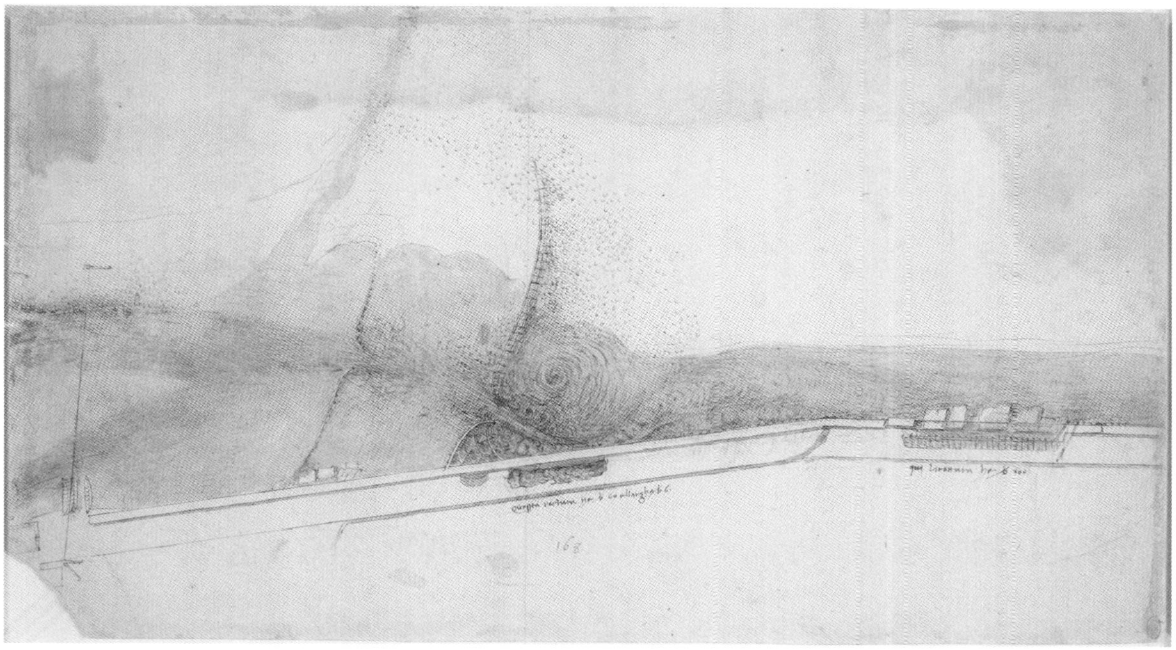

FIG. 8
Weirs on the Arno and damage to the bank, West of Florence. Royal Library, Windsor 12679.

Alongside the Pisa canal and other more limited projects for the republic, Leonardo was continuously thinking about a major peacetime project for a canal to bypass the non-navigable sections of the Arno west of Florence. Here the idea was not to bypass Pisa, but to provide a great arching canal north of the present river to re-join it east of the port city. Such a canal was a long-standing aim of the Florentines, but the nature of the Florentine regime—a republic before 1512 even when dominated by the Medici—made the undertaking of such massive and costly schemes more problematic than in ducal Milan. Leonardo's concept is demonstrated in a series of brilliant maps (Plate B).[44] Although scaled for precision, they are far from inert charts of geographic features but always breathe the spirit of the living body of earth, as expressed so vividly in the Codex Leicester. Indeed, the most experimental of the maps exude a dynamic sense and even a passion not normally associated with the mapmaker's art. His route, essentially that of the first half of the present motorway (A-11) to the west of the city, arcs past Prato and Pistoia, before curving southwards into the valley south of Altopascio, then occupied by Lake Sesto. He does not seem to have intended to follow the whole of the southerly axis of the lake, but planned to pursue a more south-westerly path across relatively low-lying ground. The apparently extravagant curve of his canal is his response to the problems of the intractable terrain along more direct routes, and he claimed that it is actually shorter than the meandering course of the river. In Codex Atlanticus, f. 127r, which includes a small sketch-map of the canal, he notes that it will be almost forty feet wide at its bed and sixty wide at the surface.

Even before his return to Florence, Leonardo had been thinking about the best way to finance such an enterprise: 'the Arte della Lana [the wealthy Wool Guild] makes the canal and takes the income, the canal passing through Prato Pistoia, emptying into the lake. It will be without locks and more durable and give more income from the places through which it passes.'[45] He even gives some figures for the expected revenue:

44 Windsor 12279 and 12685r. See also Madrid II, ff. 22v–32r and 52v–53r, and the small sketch-map on CA, f. 127r.

45 CA, f. 117r.

> [T]his will fertilise the country, and Prato, Pistoia and Pisa together with Florence will have an annual revenue of more than two hundred thousand ducats, and they will supply labour and costs for this work, and the Lucchesi similarly. Because the lake of Sesto will be navigable, let them make the route from Prato and Pistoia and cutting through Serravalle and exit into the lake, because it will need no locks or dams, which are not enduring but on the contrary always call for work on their operation and maintenance.[46]

On CA, f. 126v, Leonardo also begins to estimate costs:

> [T]his canal cannot be excavated for less than four *denari* per *braccio*, paying each labourer at the rate of four *soldi* per day. And this canal is to be made between the middle of March and the middle of June because the peasants, being free from their ordinary work, will be able to be hired, and the days are long and the heat does not prove exhausting.

On this same page, he designs an impressive scheme for a large three-arch aqueduct with a lock that allows his projected canal to pass securely over the river. He labels his design 'canal for Florence'. He also explains that 'if this river ordinarily occupies the width of one arch, make the bridge with three of them, and this is because of floods. . . . So that the arches of the bridge will be as high as possible because of the flooding of the river that passes under the bridge.'

There is no sign that this grand project progressed beyond Leonardo's drawing board, and the Florentine authorities would not have been encouraged by the failure of the Pisa canal. However, from our point of view, his intensive studies of the Arno and the topography around Florence yielded an extraordinarily positive result. Looking at the valleys, the mountains, the cuttings of the rivers, the strata of rocks and clay, and the deposits of shells in high places, his sense of the dynamic body of the earth—always potently apparent throughout his career—crystallized into a vision of historical geography. He came to envisage a magnificent picture of the huge transformations that had occurred over vast tracts of time, as lands once underwater became fertile valleys and even mighty mountains, while previously dry regions of the earth's crust sank below the waters. The earth itself bore vivid witness to its ancient and violent history.

Looking specifically at the Arno valley west and south-east of Florence, he concluded that the present valleys were once under water. The first of Leonardo's prehistoric lakes occupied the rather low-lying ground that runs in a north-western arc from Florence through Prato, Pistoia, and Serravalle. This is the route for his projected canal from Florence to bypass the non-navigable reaches of the Arno as mapped out in Windsor 12279, 12685, and the related maps in the CA, f. 127r, and Codex Madrid II, ff. 22v–32r and 52v–53r. The second lake ran from S. Jacopo al Girone just to the east of Florence in a south-easterly direction to Arezzo, extending for 40 miles. The Pratomagno range of mountains now marks the watershed between the upper and lower Arno valleys. The upper lake was once joined to Lake Trasimeno, west of Perugia and to the south of Arezzo.

This historical reconstruction of the ancient lake south of Arezzo, encouraged him to devise a visionary project for the Val di Chiana. On a map of the valley at Windsor (Plate C) he recorded in light blue the wide spread of inhospitable bogs on either side of the course of the ancient River Clanis, which had over

46 CA, f. 127r.

time separated as two separate rivers.[47] These are delineated in darker blue in the middle of the bogs. Leonardo advocates the re-instatement of the original watercourse in the context of a substantial re-balancing of the main rivers, canals, and lakes in Tuscany and Umbria. On CA, f. 127r, he proposes that 'sluices are to be made in the Chiane of Arezzo so that, since there is no water in the Arno in summer, the channel would not dry out'. His grand scheme utilized tunnels that connect Lake Trasimeno to the Arno (via a channel regulated by a gate), and link Lake Trasimeno to the Tiber as it passed north of the Lake and Perugia.[48] The conjoined aim of these mighty projects was to coordinate the motion of waters all the way from the Tiber to the sea at Pisa. He realized that the earlier, piece-meal approaches to the draining of the valley would not achieve the desired ends.

Leonardo's illustration of these massive schemes involved sophisticated mapping. He was entering a cartographic territory in which Florence was at the cutting edge of Renaissance map-making. In the fifteenth century, the humanist revival of Ptolemy's *Geography* found expression in illustrated editions of an expensive and sophisticated kind. One of those involved, Piero del Massaio (born 1420 and deceased before 1480), seemingly of modest achievements as a painter and miniaturist, became something of a specialist geographical illustrator, drafting regional maps and making more detailed studies of cities and their environs for military purposes.[49] These two types of map belonged to the Ptolemaic categories of *geographica* (a general survey map from a high or abstracted viewpoint) and *corographica* (a more pictorial portrayal of a particular location). Leonardo certainly knew of Piero's work; three of his maps of the Val di Chiana are demonstrably based on Piero's map of Etruria or 'New Tuscany', albeit with a vitality of touch that was beyond his predecessor's command (Figure 9).[50]

Leonardo's emphasis in the majority of his maps is clearly on watercourses, which has a practical dimension, but also speaks to his sustained fascination with the *vene d'acqua* in the living body of the earth. One unusual feature of a number of his maps of watercourses is the tracing of the main streams of water using a sinuous dark blue line within a wider zone of flooded land, described in light blue. The wider zone is apparent in Massaio's map, indicating that this records the spread of the waters at that time. Leonardo is tracking what had happened or might happen if the wider marshes were to be drained by a free-flowing river. On Leicester f. 21r he speaks of the central current of a river and the spreading of its water across valleys. He is thus embedding an element of time into his description in an entirely novel manner.

Piero del Massaio's successor was Francesco Rosselli, miniaturist, printmaker, and cartographer whose production culminated in the great world map published in 1506 with the Venetian Giovanni Matteo Contarini, which was the first to incorporate the discoveries of Columbus.[51] In 1508 Rosselli produced a striking world map using an oval projection. It is difficult to think that Leonardo was unaware of Rosselli's activities in Florence and his innovatory mapping of the new and old worlds.

47 Windsor 12682. See Zammattio 1981, p. 23; and the convincing account by Camerota 2018, pp. 102–5.

48 Windsor 12278r. The tunnel link between the Tiber and the lake is in the upper right corner of the map.

49 Kent and Elam 2015.

50 Windsor 12278; Kent and Elam 2015, figs 5 and 6. See also Woodward 2007, vol. 2, Part 1, pp. 910–1 and 916. Windsor 12277 is similar. It may be that the more linear map on CA, f. 910 is directly taken from Massaio's map and the other two developed from it. Windsor 12682 shows the Val di Chiana from a more perspectival viewpoint.

51 Levenson, Oberhuber, and Sheehan 1973.

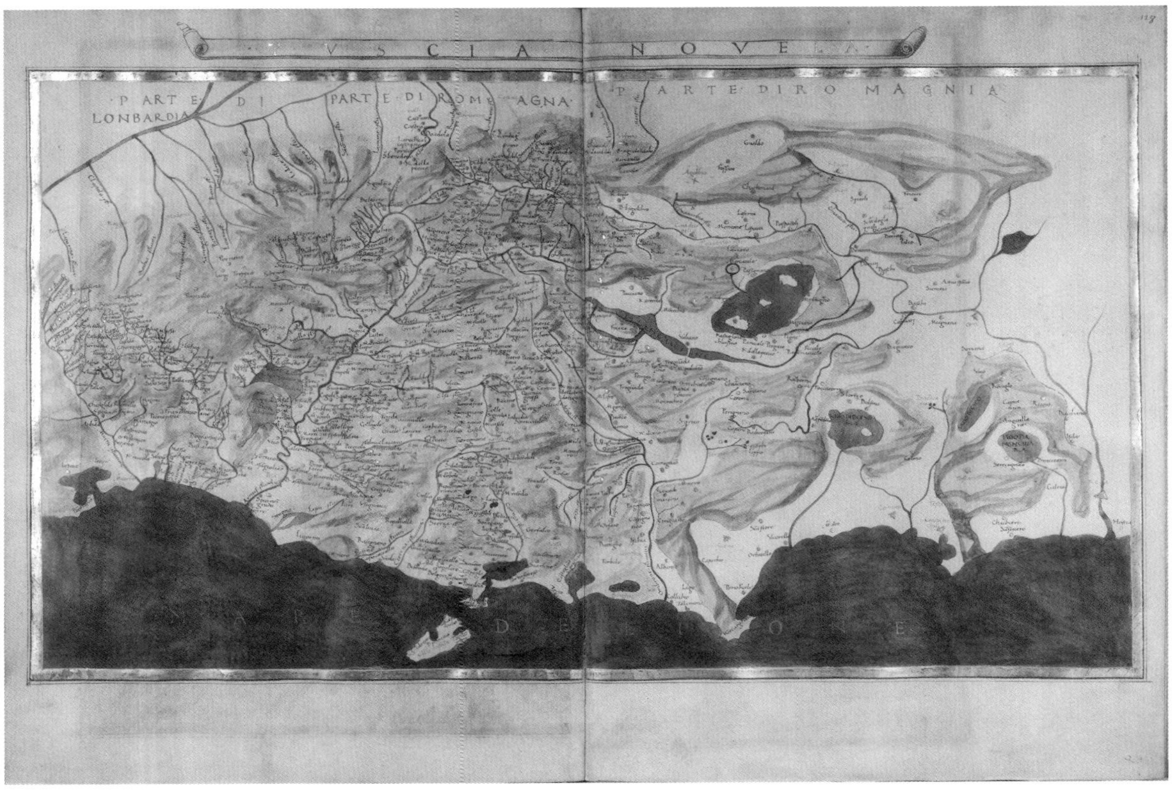

FIG. 9
Piero del Massaio, with lettering by Ugo Comminelli, *New Map of Etruria (Tvscia novela)*, 1475–8. From Ptolemy, *Cosmographia*, Bibliothèque Nationale, Paris, MS Lat. 4802, ff. 127v–128r.

The studies that Leonardo undertook of the topography of Tuscany, and more widely of central Italy, in 1503–7 may have exploited techniques that were known to his fellow mapmakers, but the feel of his images is very distinctive. None of his contemporaries convey a vision of life and change in the body of the earth. This vision was to be extrapolated onto the largest possible scale in the Codex, not only to embrace the great valleys of the major rivers in Europe but also to encompass the extraordinary idea that the Mediterranean would eventually become the biggest valley of all, drained by a massively extended Nile and series of tributaries from Italy and other Mediterranean countries.

It is to the Codex and its vision that we now turn.

When Was the Codex Made?

The chronological position of the Codex in the aftermath of Leonardo's work on the civil canal for the Arno is reasonably secure. It contains no notes that are specifically dated by Leonardo himself. The rich series of references to phenomena or structures that he has observed first-hand in both Tuscany and Lombardy might suggest that the central period for the compilation of the Codex was 1505–8, years during which he resided off and on in both Florence and Milan, moving back and forth between the two cities in response to demands from the French rulers of Milan and pressure from the Florentine authorities to complete his wall painting of the *Battle of Anghiari* for the Council Hall. However, we need to bear in mind that some of the items in the Codex were transcribed from other notebooks and that the same may be true of the eyewitness records. We have seen that the compilation of the Codex was a complex business, and it probably took some years for it to assume its final form. The 'inner set' of sheets is reasonably homogenous, and seems to correspond to other attempts around this time to 'to arrange things in order according to the subject they treat'—as he reminds himself on the first folio of the Codex Arundel, dated 22 March 1508.

In their distinguished editions of the Codex, both Girolamo Calvi and Carlo Pedretti have plotted the links with other manuscripts and notebooks with a view to defining its date of compilation, with the former arriving at a somewhat earlier date than the latter. The key issue in dating is the relationship with MS F in the Institut de France, the contents of which are to be discussed later. The sections on water in MS F, which was begun on 12 September 1508 and is dated at the back in October of the same year, overlap substantially with the Codex, and some of its analyses can best be read as refinements and revisions of statements in the thirty-six-folio compilation. However, we cannot assume that all the pages in MS F were entirely filled over the course of those two months, and it cannot therefore be used to provide definitive evidence for an earlier date for the Codex. We also have to bear in mind that Leonardo's frequent recapitulations of earlier material, changes of mind, and transcription of notes from elsewhere work against our ability to calculate a clear chronological sequence in notebooks from approximately the same period. It is therefore best not to be dogmatic about when he first wrote on one of the sheets that went to make up the Codex and when he made his very last addition to it. There is no obvious sign of his experiences in Rome, the city to which he moved in September 1513. Later, the Roman area is included among the regions once covered by the sea in a small drawing of the Mediterranean in the Codex Atlanticus (f. 901r), following his study of the remains of ancient sea creatures on Monte Mario in Rome, probably in 1515.[52] On folio 253v, he tells himself to 'make drawings to show where the shells are on Monte Mari[o]'. It is probably safest to say that some of the material, or source material, may be as early as 1505, and that he was still making additions after 1509, but not after 1512.

In the three succeeding sections, in which we present an overview of some of the key themes of the Codex Leicester, we will be doing so in a broad way, since detailed analyses are provided in the commentaries and in the subsequent essay on Leonardo's relationship with his sources.

The Codex 1: The Ancient World

All the topics in the Codex fall under the embrace of 'water', providing we define the study of water in keeping with the expansive nature of Leonardo's thought. The great underlying theme of the sections that concern the nature and position of water in the body of the earth, which occupies most space in the Codex, is the pulsing dynamism of the whole system that comprises the four elements in the world, in particular the fluctuating relationship between earth and water. The scale of this relationship ranges from minute observations of waves transporting sediments in rivers in particular situations (and sometimes in named places) to the vastest transformations in the earth's crust, the Oceans, and huge water-filled cavities within the earth. The fundament of the system is the standard arrangement of the concentric elements according to their relative weights: earth surrounded by water; water surrounded by air; and air surrounded by fire. If each of the successive elements were to adopt a spherical configuration, as Leonardo noted, the whole ensemble would be stable. However, earth is irregularly shaped, sometimes protruding above the waters in 'our hemisphere' (as he terms it) and sometimes sinking below the waves of the Oceans and supporting the higher waters of lakes. This situation is inherently unstable since portions of heavier elements were outside their designated sphere and there-

52 Codex Atlanticus, f. 901r. See Chapter 4 in this volume.

fore outside their natural level. This is particularly important in the case of water, which becomes mobile and heavy once it is at a level that raises it into the sphere of the air. Leonardo recurrently stresses, even when looking at its detailed behaviour, the basic rule that water out of its own sphere strives or 'desires' to return to its natural level. Thus water trapped in a high lake or springing forth as the source of a river in the mountains or surging upwards in waves is inexorably drawn to find its true level. Both on the large and small scales it can do so with evident violence. As it moves with irresistible impetus it alters the earth through which it moves, eroding the subterranean and surface channels though which it courses, transporting and depositing sand and stones, rolling over rocks, cutting through hills and mountains, and destabilizing even the grandest features of the earth's crust.

The vision of water in motion is framed by the parallel between the macrocosm and the microcosm. Like the human body as the 'the lesser world', the larger body of the earth is 'vivified' by 'veins' or vessels that transport those substances that endow things with the qualities of life. Leonardo repeatedly refers to the *vene d'acqua*, the underground and superficial 'veins of water' that share so much in common with the vascular systems that bring life to plants and animals. He again notes, as analogies, the way that sap oozes from the top of a cut vines and blood spouts from a severed blood vessel in the temples of a man's head. The surging of springs high in the mountains is explicable via the analogy of the macro- and microcosms, in which analogy is not just a convenient way of thinking but is in itself an explanation.

In some ways Leonardo goes further in his study of the 'veins of water' than he does in his account of the role of blood vessels in the human body. In the human vascular system he does not envisage a complete circulation or 'round' as he calls it. Unlike William Harvey in the next century, Leonardo does not see blood as passing from the arteries to the veins though invisible vessels but adheres to the traditional idea that blood could pass though pores in the septum that separates the two ventricles. In the earth, he envisions a complete circulation: veins of water originate from the beds of the seas, where the water is fresh rather than salty; some of the veins pass through salt mines within the caverns of the earth, dissolving some of the salt and returning their heavier waters to the seas, rendering the upper levels of the oceans salty; other veins containing lighter water rise into the bodies of the mountains in a ramifying system, bursting out where they find fissures and rushing downwards in the form of rivers that eventually flow back into the sea. There are aspects of this earthly 'circulation' that continue to bother him. What is the role of the water that disperses into clouds, later falling as rain? How does water actually rise against its nature within the mountains? Is the microcosmic explanation adequate in itself or should it be possible to adduce a specific physical process, such as the distillation of water in high caverns, an idea to which he gives serious consideration in the Codex? It is the nature of Leonardo's arguments, assertive as they might seem as he is stating them, that they are never rigidly fixed, as we will see when we look at his views after the Codex was compiled. It is by no means easy to arrive at a conclusive statement of what he is arguing in the Codex, since he is re-shaping his thinking as he goes along. This is analogous to the restless changes made during the execution of his paintings, even when he seems to have arrived as a definitive composition.

Whatever the actual mechanisms of 'circulation', the consequences of the inherent instability of the earth and the bodies of water in the world were huge and undeniable—as Leonardo stressed in his arguments with 'adversaries'. Only on one occasion is his 'adversary' named in the Codex. On folio 1v, he cites the philosopher and physician Andrea Cattaneo, with whom he clearly argued in person about water on the moon. Sometimes he may have had a specific authority in mind, but more often the 'adversary' seems to be a device to review both sides of a question—providing that Leonardo emerges victorious at the end. The pattern of argument and counter-argument shares much in common with mediaeval dialectic, in which an issue and proposition are defined, objections are formulated, counter-arguments and proofs of a conclusive nature are established, and a resounding conclusion is drawn. However, Leonardo is not predisposed to settle on definitive conclusions. His predominant use of the dialectic mode of argument in the Codex may well reflect that it is unusually concerned with traditional debates of great theological moment. Indeed, nowhere is Leonardo more verbal and philosophical in his means, with eleven of the pages devoid of illustrations. Most of the drawings he does supply on other pages are literally marginal.

The most contentious issue, to which he repeatedly returns, is whether the present appearance of geological features in the earth is consistent with the Biblical Deluge. At the centre of this question is the presence of shells of marine animals in high places. A series of sharp observations and analyses made where rivers have cut through mountains, such as those of successive strata containing deposited and petrified shells between layers of sedimentary material, testify to a succession of long-term immersions of much of our present 'hemisphere' by ancient seas. One of his most insistent arguments is that the rising waters of the Deluge could not have carried all the heavy shells to the present levels. He also noted evidence that the animals inhabiting the shells once thrived in those high locations, forming colonies. Nowhere are Leonardo's insistent powers of observation and remorseless reasoning better exemplified.

The fact that wide and elevated tracts of land had previously been under the waters for long periods means that the relative distribution of land and water must have been mightily transformed over the ages. The mechanism relies upon the relative weights of water and earth, in particular as they affect the centres of gravity of each of them.[53] If heavy earth is eroded by waters, descending to a lower level and even being cast under the sea bed, the centre of gravity of the earth would shift downwards. Water, the thickness of which is continuously variable, remorselessly retains its spherical surface where it lies above the level of the land, and its mobile nature determines the shared centre of gravity of the conjoined system of earth and water. The process of erosion effectively lightens that portion of the earth from which the soil has been removed, paradoxically causing a general move of the earth upwards. In the latter folios of the Codex, Leonardo amplifies the transformations of mass caused by erosion with truly radical accounts of the earth permeated by immense caverns of water, or even by a huge internal reservoir, into which portions of the earth's crust periodically collapse. It is unclear how this watery interior relates to the inner fires that he describes as responsible for volcanoes.

In terms of the actual history of the earth as Leonardo witnesses it in his own time, this vision translates into astonishing narratives of how the valleys of great rivers like the Danube, and in all probability the Rhine and the Rhone, were once covered by seas. Erosion of the protruding 'islands' of those mountains that were high enough to rise above the seas, gradually and inexorably resulted in the rivers depositing material and extending their courses so that they would drain the plains until they assumed their present configurations. Projecting this process into the future, Leonardo envisages the severe constriction of the Mediterranean at the Strait of Gibraltar being opened in such a way that its bed will become

53 In addition to the discussions in the Codex, see MS A, f. 58v; Forster III, f. 7r; Madrid I, f. 16r; MS L, ff. 17r and 76r; and CA, f. 412v.

a plain drained by the Nile as the dominant river, with the other major rivers that drain into the sea serving as its tributaries. Along the way, he argues that the relatively low tidal fluctuations of the Mediterranean compared to those of the 'Ocean' outside it are attributable to the damming up of water in the Strait, which to his mind indicates that the moon is not the chief cause of the 'ebb and flow' of the tides. His sources and the arguments of his 'adversaries' in these debates about water in the body of the earth will be examined later.

Leonardo's vision is never seen on a more magnificent scale than in his account of the body of the earth in the Codex Leicester. Beneath the apparently chaotic surface of the notes is one of the grandest accounts ever formulated of how the great elementary systems of the world have functioned in the remote past, how they do function now, and how they will function in the distant future.

The Codex 2: Heaven and Earth

It is strange that Leonardo, who was so committed to the geometrical orders of creation, paid such limited attention to astronomy. We could argue that most of his writings on astronomy have been lost, but the way he compiled his notebooks and the manner in which they have come down to us indicates that his surviving pages would have contained at least substantial glimpses of any major discussions of the celestial system that he might have undertaken. Perhaps the planets and stars were too remote for his hands-on system of analysis. His discussion of moonlight, which is the major theme of the first folios in the Codex, involves turning the moon into something like the earth, and the earth was something he could of course study directly. It can be said that these pages are less focused on astronomy as such and more concerned with the optics and perception of observed light as its impinges on the earth from the sun and moon. His studies of light on spherical and irregular bodies provided a good interpretative model. As he wrote, 'There is no part of astronomy that is not a function of visual lines and perspective'.[54] In practice, his actual diagrams of the earth, moon, and sun treat them as if they were normal-sized objects placed at relatively close distances from each other. He was aware that his diagrams were 'demonstrations' of the optical principles involved, rather than scaled illustrations. In this they resemble the diagrams in standard books of astronomy. His mode of demonstration cannot accommodate the fact that rays of light from the very distant sun are in effect parallel.

There was also the vexed question of predictive or judicial astrology. In the Renaissance, the terms *astronomia* and *astrologia* were more or less interchangeable. Ludovico's Milanese court was something of a hotbed of astrological practice in the medical and political realms.[55] Ambitious astrologers jostled with other court members for the Duke's attention. Leonardo was dismissive of the metaphysical claims of astrology, as of other predictive systems. On the other hand, two of the six books on astronomical themes in his library centre upon astrology, namely Latin versions of Arabic texts by Alcadbitius (Al-Qabisi) and the highly regarded Albumasar (Abu Ma'shar).[56] It may well be that he respected the

54 Cod. Urb. Lat., f. 7v.

55 Azzolini 2013.

56 Azzolini 2013, pp. 22–3; Kemp and Pagiavla 2014; and Nanni 2005.
The list that follows proposes identifications for Leonardo's laconic records in Madrid II. It may be that he purchased key texts in Venice in 1500.

'Spera Mundi': Johannes de Sacrobosco, *Sphera Mundi*, Venice, 1499.
'Calendario': Regiomontanus, *Calendarium*, Venice, 1476.

precision with which the astrologers characterized the motions of the celestial bodies while rejecting their metaphysics. The illustrated books of astrology also provided a good crib for the iconography of the planetary deities. In any event, he seems not to pursue detailed studies of the planets and fixed stars on his own behalf, astronomically or astrologically.

The folios dedicated to lunar light are all in the 'outer set', and are characterized by a looser page organization than is apparent in the sheets concerned with the body of the earth. Indeed, two of the pages, 1r and 7r, are driven by the geometrically drawn diagrams, which were the first items to be laid down. The drawings on 2r are vital to the content, and the incredibly subtle drawing of the glimmers of light in the shaded portion of the new moon makes the point through pictorial means in a way that is not equalled elsewhere in the Codex.

The running theme of the folios is what we can adduce about the moon from optical effects visible on earth, extrapolating from what can be demonstrated directly in the terrestrial realm. Leonardo had expended considerable effort in studying the passage of light by direct illumination and by rebounds from light sources and on to bodies of various shapes, not least in MS C of *c.*1490. He posited that the intensity of illumination of a surface was proportional to its angle of impact, and he had gone to some trouble to distinguish between illumination on a matt surface and the gleam of a highlight on shiny body, which moved across the body in response to the position of the observer's eye. He also looked carefully at how shaded portions of a body could receive light that rebounded in a secondary way from adjacent surfaces. In this he was adopting a threefold division in mediaeval optics, which was well known to Dante. Everything begins with the actual source of light—*lux* or *luce*, in Latin and Italian. On passing through the air and striking a surface this becomes illumination—*lumen* or *lume*. If the body is shiny, a bright highlight is created—*splendor* or *lustro*. The impacts and rebounds of light are analogous to the behaviour of a bouncing ball and are governed by the same rules of 'impetus' that were manifested in the motion of water. Light, as he knew from his studies of mediaeval optical texts like John Pecham's *Perspective communis*, had come to be seen as the radiant manifestation of God's spiritual and mathematical rule over nature.

When Leonardo is drawing diagrams to determine the divine geometry of the passage of light to the earth from the sun, either directly or involving the moon (his main concern), he is determined to define the nature of impacts and rebounds, and above all how these impinge on the eye of the observer. It is on the basis of what we can directly see on earth that he argues against his 'adversaries', including the named Andrea Cattaneo. His four main arguments run as follows. Firstly, the moon has no inherent light of its own but relies on reflecting rays from the sun, as can be implicitly verified from the way that the light and shade on the moon are distributed as their relative 'aspects' or positions change. Secondly, the moon cannot be a uniformly shiny body, since such a body would reflect a lustrous and brilliant highlight from a limited area of its surface. Thirdly, the nature of the relatively uniform illumination across the lit surface of the moon indicates that its surface is wrinkled or corrugated, reflecting light in a more diffuse manner. Fourthly, this wrinkling is primarily the result of waves in the

'Alcabitio Vulgare del Serigatto' [owned or once owned by Francesco Serigatti of Florence]: Italian (manuscript?) version of Alchabitius, *Introductorium Alchabitii Arabici ad Scientiam Judicialem Astronomiae*, various editions.
'Albumasar': Albumusar, probably *Flores Astrologie* (Venice, 1500) or perhaps his *Introductiorum maius*, Augsburg, 1489.
'Quadrante': a manuscript of the treatise on the quadrant by Carlo Marmocchi?
'Libro d'Amadio': Amadeus Derthonensis, *Ratio dierum, calendarum, nonarum, iduum, mensiumque*, nd.

seas of the moon, with light reflected from innumerable points of the crests of waves, with the shadier portions of the troughs dimming the overall radiance from certain angles.

His definition of light on the moon's waves depended on his researches into how sunlight reflected into the eye off the earth's wavy seas to observers in a variety of positions and for different angles of sun. He looked at this phenomenon in the Codex Arundel and Codex Atlanticus.[57] Folio 28r of this codex contains analyses of sunlight on a wavy sea and of light on the moon that are very much part of the same debate he is conducting at this time in the Codex Leicester. On Codex Arundel, f. 104r, he justifies his argument that the heavy moon stands in the centre of its own system of the elements by reference to the yolk of an egg remaining in the centre of its albumen regardless of which way it is turned.

On the basis of the four 'propositions' defined above, Leonardo is able to explain the phenomenon of the 'ashen light', the *lumen cinereum*, which is visible on the shady side of the moon when the directly illuminated crescent is at its thinnest. This glimmer results from yet another rebound, this time of light from the earth's seas that is reflected as far as the moon. The moon receives earthlight! The wonderfully refined illustration at the bottom right on folio 2r shows the reflected glimmer, not only in terms of the direct optics of the reflected light but also in terms of the subjective phenomena of perception. Leonardo explains that the glimmer of light looks least bright to the eye where it borders on the luminous crescent, while it appears brightest where it borders on the black darkness of the night sky. The physical explanation is notable in its own right, and the attention he pays to how it is perceived as an *apparent* effect is very remarkable. He had discussed this effect of edge contrast from the standpoint of an artist in draft notes for his treatise on painting. This attention to the proclivities of the eye in interpreting optical phenomena 'out there' is very unusual, only rivalled by the emphasis of the Arabian scholar, Ibn Al-Haytham (Alhazen) on the *deceptiones visus* (visual deceptions or illusions) that are inherent in our visual system.[58]

A consequence of Leonardo's optical arguments that the moon possesses its own wavy seas is that he has to defend the radical idea that the moon has a system of elements essentially the same as that on our terrestrial globe. One of his 'adversaries' claims that any water in the moon would cascade downwards on to the earth. Leonardo counters by saying that the solid body of the moon would also fall downwards were it not a stable, self-contained system that was sustained in space in the same way that happens with our world. Needless to say, these are radical claims, not least since they effectively destroy the traditional distinction of the earth from the celestial bodies. His insistent repetitions of the key arguments in favour of the moon's earth-like nature indicate his overt awareness of the originality of his position with respect to both ancient and modern authorities.

The Codex 3: Working with Water

Although Leonardo did not arrive at a definite plan for the ever-increasing detail and expanding scope of his book 'On Water', we can be confident that its dominant direction was from theory to practice. The Codex, for all its lack of resolved order (or even because of its lack of order), probably contains examples of all the main *types* of topic he intended to include.

57 Arundel, ff. 94v and 25r; and CA, f. 331v

58 Kemp 1992.

What Leonardo was attempting to accomplish was very novel in its seamless fusion of basic principles—the very nature of water and the geometry of its complex motions—with highly specific examples of water engineering. Earlier engineers had demonstrated only limited interest in applying mathematical theories of dynamics to their practice. This was beginning to change in the later fifteenth century, and had both scientific and social dimensions. It served to promote the status of engineers as men of intellect, much as art theory (especially the optics of perspective) was helping to lift leading artists out of the ranks of artisans—a cause very dear to Leonardo's heart. Francesco di Giorgio, Leonardo's one-time companion and one of whose treatises he possessed and annotated, was a key pioneer. He laced the opening of his texts with natural philosophy, although the more abstract concerns do not feed directly into the detailed accounts of his splendidly drawn machines. Leonardo aspires throughout the Codex, as in all his work, to make theory the foundation of practice and to use the experience of practice to devise and certify theory in a quite new way. There are almost no highly conceptual concerns in the Codex that do not feed at one or more steps of remove into engineering practice, and every detailed practical action takes place in the context of the basic rules of water and mechanical components in action. This does not mean that each topic, theoretical or practical, is not engrossing in its own right. Rather, it means that every fascinating facet of his thinking is related to a central unity in a way that denies the discreteness of what we regard as different disciplines.

Leonardo requires us to wade through some very tortuous narratives of turbulent water in motion. The dominant impetus of each current is overlaid by competing complexities: contrary currents are generated at the bed; waves break violently, either forwards or collapsing backwards on successive waves; water raised into the air by waves, plunges into the depths, sometimes rebounding from the bed; vortices mutually percuss contrary vortices and the original current; sediment of various grades of coarseness is snatched up and then deposited; and various resistant obstacles interfere turbulently with the flow. At points we may wonder whether the effort is worth it. However, the level of understanding of such complex phenomena that Leonardo is striving to accomplish through observation and deduction refracts in equal measure into theory and practice. Each observed incident of impetus and percussion is woven into his theory of force and motion, expressing it and casting light upon it. The patterns of vortices and bed transport inform his inspection of what is happening in rivers and canals, particularly at the junction of colliding currents or at a weir. When he looks at formations of water in the Arno at bridges, mills, weirs, and junctions he does so with an eye deeply informed by his more general narratives of water in action.[59] When he advised the Florentine authorities about problems with weirs and erosion he brought theory to bear on practical experience, even it may not have been evident to the people with whom he was dealing that theory was relevant to the problem in hand. After the debacle of the canal around Pisa they may have been wary of being 'blinded by science'.

Leonardo was aiming to formulate basic rules governing the flow of water in rivers and canals, emphasising that the current is fastest at the centre in a straight river, corresponding to the deepest part of the bed, while meanders cause erosion to the banks and uneven deposition of sediment, while the deepest part of the bed wanders from side to side. At the time, as he is involved with the Codex, he notes, in connection with his dissection of the 'old man', that the meandering blood vessels of the elderly interfere with efficient flow and cause silting up in an entirely comparable manner.[60] He plans to show how

59 Detailed analyses of the Arno in and near Florence are in Arundel, ff. 271r–278v. See Di Teodoro 1992.

60 Laurenza 2009, p. 105ff.

the relationship of the speed of flow to the cross-sectional area of a water channel and to its gradient are matters of mathematics, alongside rule-of-thumb experience. The banks of free-standing canals could be constructed with full awareness of how water pressure increases with depth. The same understanding of pressure guides us as to the height at which a sluice of a given aperture should be placed to draw water off canals in the required quantity—ideally without depleting the source excessively. This would also help in judging how to feed appropriate amounts of water from rivers into canals, as with the Adda and the Martesana canals. The behaviour of flow also finds expression in his schemes to drain marshes by diverting incoming waters into straight or gradually curved channels that run beside the undesirable bodies of turbid water and mud.

The basic patterns of impetuous flow and percussion are manifested in the erosion of banks, bridges, and weirs. The hollows and sandbanks generated by obstructions to the free flow of canals and rivers are to be understood on the basis of observations in the field and theories of turbulent motion. The basics of geometrical obstacles are to be classified in a novel way with respect to their causes and effects, individually and in various compound arrangements. If Leonardo can tabulate all the variables in any of the various phenomena that govern water flow, he has a tool with which to dissect the anatomy of each of the actual set-ups that he encounters as an engineer. He can also forecast what effect a newly placed obstacle will have on the flow. As always, his understanding is designed to inform us how to work with rather than against water. His mobile weirs (probably of the kind he deployed for Venice in the River Isonzo) progressively excavate the riverbeds as they are moved downstream. The graded steps that he recommends for controlling the descent of water without violent falls, like those he observed in the countryside near Vigevano, deposit sediment at their foot in such a way that they are able to create 'meadows out of a swamp of great depth'.[61] Leonardo also shows, in a more general demonstration, how judiciously placed weirs can deposit sediment beneath a building that is being undermined by a curving current.

In designing specific structures to manage water, Leonardo brings his knowledge of architecture, his earlier studies of water engineers and engineering, and his extraordinary inventiveness to bear fruitfully on practice. The stone 'chains' (long stones that link or bind others together), the lead joints that secure one stone to another, and the elaborate swallow-tail interlocking of adjacent stones that he used in his weirs are adapted from the architectural techniques he had first acquired when working with Verrocchio on the great ball atop Brunelleschi's Florentine dome. It is likely that Leonardo's pile-drivers were based on the sophisticated devices developed over the years by water engineers, not least in Venice, a city founded on piles. But his mathematical calculation of the extra energy needed to drive piles diagonally signals his personal approach. It is also likely that his orchestrated ritual of men climbing ladders and leaping off in stirrups to act as counter-weights in order to lift a mighty pile-driver is his own eccentric conception.

Alongside the heavily engineered constructions he also proposes simpler structures composed from wooden poles, like the portable, prefabricated, self-supporting bridges that needed no binding or nails.[62] He recommends using tree trunks with branches, placed horizontally in rivers and tied to piles in a pyramidal formation, to create a simple kind of weir. He argues that the branches facing the current will gather silt from the flowing water and would therefore consolidate a barrier that otherwise might

61 Leicester, f. 32r.

62 CA, ff. 69ar and 71v.

seem too flimsy and impermanent. Here as elsewhere it is impossible to determine if this particular invention is Leonardo's own or if it is part of known practice, but the idea of exploiting the capacity a water to carry sediment in order to serve the engineer's ends is very much his style. His use of siphons for practical purposes in moving water is based on standard practice, but his determined formulation of the rules governing siphons with different kinds and numbers of inlets and outlets is entirely characteristic of his personal way of thinking.

Of the recurrent themes in the Codex, the configuration and dynamics of concentric surface ripples is perhaps the least directly applicable to practical matters. Leonardo clearly derives much visual and intellectual delight from the way that sets of ripples appear to intersect without interference and retaining their geometrical integrity. He is particularly pleased to see that a triangular object dropped into water generates ripples that rapidly tend to adopt a concentric formation. For the most part he regards the ripples as tremors or vibrations in the surface, like the waves in a rope oscillated at one end and secured at the other; but on one occasion he develops an explanation based upon serial internal reflections that are exceedingly intricate to visualize. Even if the geometrical dynamics of ripples are peripheral to the concerns of the water engineer, they have very wide implications, since sound and light comparably radiate outwards from their sources in graded manners carrying their 'images' into the surrounding space. Multiple sources of sound and light can be seen independently as they transmit their impulses to our ears and eyes. Just as the vortices of water performed like those of air, so ripples in the 'skin' of water spoke of the commonality of cause and effect in nature and of the principle of 'necessity' that shapes every phenomenon in conformity with the mathematics of God's design.

The Codex 4: Experiment

Leonardo repeatedly stressed his reliance upon 'experience' rather than 'ideas that begin and end in the mind'. His revered 'experience'—that is to say observed reality—is present in one form or other on every page of the Codex, even when the arguments are translated into mathematical abstractions. The term he uses for 'experience', *sperienzia* (with variant spellings), is also used to mean 'experiment' (sometimes as *sperimento*). Although he was not an empiricist in the seventeenth-century manner of Francis Bacon, he was a dedicated believer that verifiable knowledge came from sensory experience that is then subject to proper analysis and testing. Leonardo did not develop a systematic theory of experimental science in the Baconian manner, but he repeatedly and very inventively contrived artificial 'experiences' that acted as 'proofs', not least by setting up experimental models of physical set-ups of phenomena he has observed in nature. His methods and insights in this respect, as in others, are very novel.

Actual physical experiments are proposed on no less than fifteen of the pages of the Codex. Some are of a kind that had been or could be staged relatively readily, while others appear more as if they were conducted in his mind as visualized surrogates for actual set-ups. There are also a series of more abstract 'thought experiments', in which he in effect says, 'what if we imagine this or that situation . . .?' This happens when he speculates what would happen if there were no earth and only the spheres of water, air, and fire on 25r, or when he envisages a well drilled through the centre of the earth to emerge on the other side on 21v.

The most notable 'thought experiments' are those on 35v and 36r, in which he not only uses highly schematic demonstrations to illustrate his theories of the relationship of water and earth in the world, but also to construct highly theoretical alternative scenarios. In one striking instance on the lower right border of 35v, he envisages the earth as a kind of hemispherical 'casing' that lies entirely outside and around the sphere of water.

With respect to more physical experiments, we have already noted that Leonardo adduced an experiment with an inverted vessel in a basin of water in MS A, f. 56r, to demonstrate how heat attracts water upwards, which is used to explain how water can rise inside mountains. In the Codex Leicester, f. 3v, the same experiment is cited and illustrated, with a piece of hot coal being placed on the vessel to heat the air inside it, expelling air from its inverted rim, after which the water is sucked up inside the vessels as it cools. However, his conclusion is now different. He adduces a supplementary experiment in which the vessel is perforated by a hole above water level, in which case the water no longer rises. The cavities within mountains clearly do not comprise an airtight system, and it cannot therefore be the attractive power of heat that directly effects the raising of water. The lack of a sealed system also causes problems for any theory that envisages the circulation of water as driven by the inner heat of the earth, which he discusses elsewhere in the Codex. On the same page, he undertakes a form of mental modelling as he speculates about the possibility that water is distilled within the mountains, condensing on the roofs of high caverns and running out of fissures in the form of springs.

The method of modelling the physical basis of a natural phenomenon through an experimental 'demonstration' is very characteristic of Leonardo. Sometimes the experiment is at some remove from the actual phenomenon, serving to demonstrate the cause rather than the effect. On 4v, he recounts a game played with counters or chequers, in which the lowest disk in a vertical stack is rapped out of the way with a sharp blow, leaving the remainder of the stack to settle down in their vertical array. This is cited to suggest what happens when the base of a wave is suddenly removed. When Leonardo wishes to demonstrate the difference between a 'push', caused by continuously falling water, and a 'blow' resulting from the sudden fall of a body of water, he thinks about two men standing at either end of a seesaw (f. 8r). If one attempts to jump upwards, his effort will be futile unless the other man strikes downwards with his feet on the opposite end. It would be nice to think that Leonardo staged such a game in his workshop with his assistants, but it is more likely that the experiment is designed to clarify the conceptual difference he is wishing to establish. Similarly, the 'impetus' game he envisages with pea-shooters and soft pellets exists in a realm somewhere between thought experiments and physical ones. It may be that his description of what happens when balls of mud are thrown against walls is based upon his having done so to see what happens.

The most important piece of experimental apparatus that he describes in the Codex is his glass-sided tank. In the late 1490s, on folio 109r of MS I, Leonardo had already mentioned the making of an experimental tank with four glass sides, and on folio 115r of the same manuscript he talks about an experimental channel with one glass side to observe water flow. In the Codex, a comparable tank makes a number of appearances, not only so that he can observe underwater currents, but also so that he can test the effect of wind blown from an aperture across its surface. On 9v, he plans to ask a specialist in ceramics to construct an experimental tank consisting of a terracotta trough with glass sides, measuring about four feet in length and about one foot in width. What may be the same tank is sketched in the margin of 29v, where he envisages using a coating of dust to register the pattern of motion on the sidewalls. On the next page, when he is discussing how to measure the speed of a ship, he tells us that he intends to make the sidewalls of a tank sticky, so that dust will adhere to the glass in order to record its height accurately.

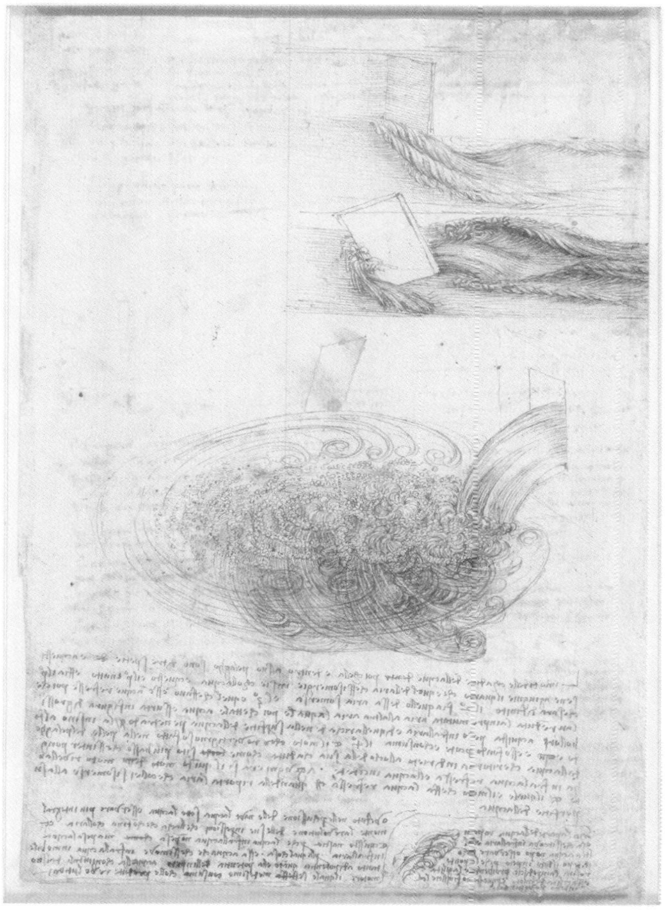

FIG. 10
Studies of water passing a rectangular obstacles and pouring into a pool from a rectangular outlet. Royal Library, Windsor 12660v.

In the case of the experimental tank, his account of the apparatus is sufficiently precise and the various references are sufficiently consistent to encourage us to think that it was actually constructed. He variously describes using seeds, such as millet or grains of panic grass (*panicum vergatum*), and papyrus paste, or more generally 'dust', so that the underwater course of the whirling currents can be witnessed more clearly.

The series of spectacular drawings of water falling from a rectangular outlet into a body of still water at Windsor may well be the result of observations made using his experimental tank.[63] If so, they show him intellectually separating out the component motions witnessed in water and of air in water before combining them in a great synthetic representation of the internal anatomy of the very complex phenomenon (Figure 10).

In MS F, which we have already noticed as overlapping the Codex in date and content, he talks of plans to place sand and fine gravel on the bottom of his tank to understand what happens to the bed of a river under various conditions (f. 34v). In the Codex on folio 15r, he informs us that 'what happens with the beds at the mouths where the two rivers join can be demonstrated with a simple experiment using sand'. This innovatory 'laboratory' modelling of water interacting with the bed is taken to a higher level of complexity in the Codex Atlanticus on folio 227v (Figure 11), where he puts forward the remarkable idea of making a proportionally scaled experiment [*la sperienza nelle minute dimostrazione*] of the gulfs and seas in the Mediterranean with the major rivers that feed it, in order to test his reconstruction of the

63 Windsor 12659r, 12660v, 12661, 12662, and 12663.

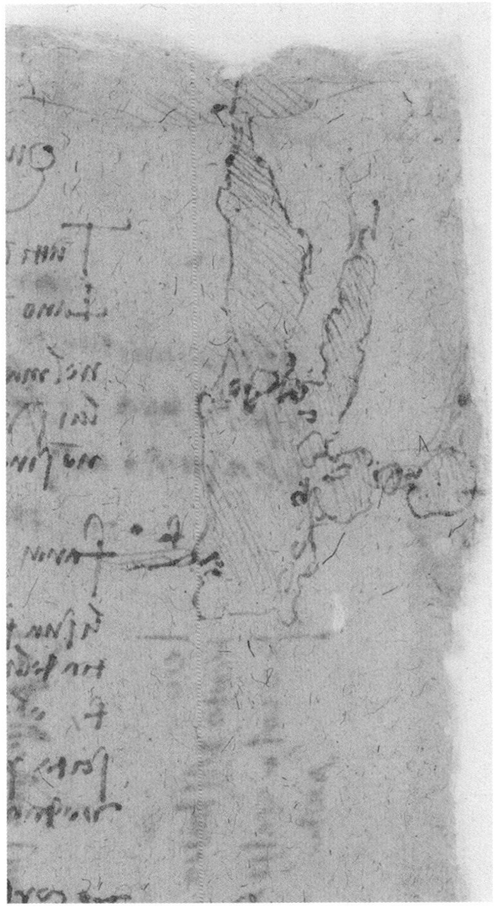

FIG. 11
Sketch-map of the Mediterranean (west at top). Biblioteca Ambrosiana, Milan, Codex Atlanticus, f. 227v (detail).

ancient geological processes. He says he will 'give the science through the experiment of the courses of the waters'. This is accompanied by a small sketch-map of the Mediterranean orientated with the Strait of Gibraltar at the top.[64] This planned experiment embodies the principles behind the phenomenon within a scale replication of a geological process over a compressed scale of time.

The experiments that he dedicates to the replication of processes in the body of the earth include those on 10r and 15r that are intended to determine to what extent water expands when it is vaporized by heat. Leonardo well knew that the expansion of water in steam could exercise a fearsome degree of force and had speculated about a steam-driven canon of the kind Archimedes was said to have made. Hot springs and medicinally beneficial 'volcanoes of mud' were the result of such steam-driven effusions from the fiery innards of the body of the earth.

More generally, Leonardo is concerned to demonstrate the nature of water pressure, which was directly relevant to how water emerged at different levels from springs and from higher and lower outlets in canals, envisaging a kind of pump with a spout on 11r, and a very ingenious tank with a parchment side on 6r. When full of water, the parchment wall bulges outwards, but it is tugged back by a series of horizontal strips of wood that are pulled inwards by suspended weights. He envisages that the weights necessary to pull each strip inwards increase in steady and proportional increments as they act progressively lower on the tank. Again we many wonder if he had built this apparatus or whether it was

64 See also his sketch-map of Europe and West Asia on CA, f. 1006r, which does not include the shore of North Africa.

sufficient to envisage it as a demonstration of the mathematics of increase of pressure at additional depths. The latter seems more likely.

Some of the experiments or observational set-ups can be simple and even homely. His lovely account on 23v of a drip running along the lower edge of a horizontal pipe on 23v, stretching itself downwards before finally separating itself as a falling droplet, is probably a chance observation rather than a deliberate set-up. It becomes an experiment through the act of observation. Equally simple are the two deliberate experiments on 36r that are designed to demonstrate how particles of moisture in the air are responsible for its blue colour when seen against a dark background. The first involves creating moist smoke from green wood on a fire, while the second requires the experimenter to disperse water in fine droplets by blowing it vigorously into the air, presumably out of a thin tube, or perhaps directly from the mouth. In both cases, the particles or 'atoms' catch and scatter the light in such a way that they create a blue tinge, just as happens when we look at the dark heavens through the moist air. In fact, one of the more remarkable sub-themes in the Codex is the particulate nature of water itself, not just of the fine sediments that it transports.

The Codex 5: Atoms and Particles

By the time that Leonardo compiled the Codex, he was increasingly conceiving of the elements and their behaviour as particulate (comprised of small particles) or as sharing the properties of an aggregate of particles. He had undoubtedly come to know something of ancient atomic theory, which was familiar through the didactic poem *On the Nature of Things* written in the 1st century BCE by Lucretius. The Roman philosopher, whose ideas were exercising an increasing impact the Renaissance, not least on Machiavelli, was expounding a set of ideas associated with Epicurus (4th–3rd centuries BCE) and Democritus (5th–4th centuries BCE), whose writings survived only in fragmentary form.[65] The central notion of Democritean physics was that the universe was composed from tiny 'atoms'. That Democritus was known in Leonardo's Milanese orbit is shown by the fresco by Bramante of the weeping Heraclitus and smiling Democritus (Milan, Brera).

One of the pieces of evidence regularly adduced for atomic theory was the appearance of particles dancing in the air when a beam of light passes into a shady interior. This observation was cited by the Islamic philosopher Ibn al-Haytham (Alhazen), of whom Leonardo acquired some detailed knowledge. It was also exploited poetically by Dante in a vision of a luminous cross in his *Paradiso* (Canto XIV, ll. 112–17).

Leonardo himself uses the term *attimi* on folio 4r when discussing the cause of the blue colour of air, which he also analyses on 36r, where he notes that 'the water blown in the form of atoms into a dark place into which the orb of the sun passes makes the sunbeam blue'. He speaks of the 'atoms' of both dust and smoke, and unusually differentiates between them: 'the difference between the atoms of dust and the atoms of smoke can be seen when the rays of the sun pass through small apertures in a wall into dark places, that the one ray seems to be ash-grey and the other of the fine smoke seems to be of the most beautiful azure' (f. 4r). The difference clearly lies in particle size.

65 Brown 2010.

There are many places in the Codex where he discusses phenomena that rely upon the dispersal or evaporation of water into fine particles, and sometimes on its subsequent condensing into liquid. Dispersal and condensation are as essential as flow to the way that water exercises its power in the world. Leonardo notes that water is dispersed by both evaporation when it is heated and by scattering into particles when it is propelled violently into space. The latter is accomplished in a variety of ways, as when fine spray is produced by the percussion of waves or when we use our pursed lips to blow water forcefully into the air. He does not appear to distinguish between evaporation and dispersal. He also thinks that water that is completely dispersed or evaporated assumes the nature of the medium into which it is dispersed. Thus evaporated water is somehow transformed to assume the nature of air.

The force exercised by water as it changes from a fluid to gaseous state comes from its huge increase in volume. We have noted that Leonardo set up at least two experiments to quantify the extent of this expansion, which he regarded as more or less irresistible. It comes into play to within the fiery earth, resulting in effusions that could assume explosive violence. It could drive water into the high and cold caverns of mountains, where it would condense in the manner of a still. It also condenses in clouds as they rise and cool. The resulting decrease in volume in the clouds generates winds. In their fiercest form as 'twisters' or tornados—described so vividly in the Codex—the winds generated by condensation cause whirling destruction of a terrible nature. Condensed droplets of rain may amalgamate into larger bodies of water, or on other occasions the action of friction within the air can result in such fine dispersal that the water does not reach the ground at all. The foundation for such ideas is Aristotle's *Meteorology*, a treatise of which Leonardo became well aware. However, Leonardo concentrates on mechanical causes, such as erosion, attraction between particles and friction rather than the qualitative explanations characteristic of the Aristotelian tradition. His alertness to the role of friction in physical processes is a prominent feature in his late science, and works against the simple proportional rules he earlier favoured.

The notion of particles is crucial to his analysis of radiating ripples, whether he is thinking that the particles do not move laterally but vibrate up and down, or whether he is envisaging that particles do move laterally as internal reflections within the ripples. He also regards the 'tenacity and cohesion' of water as the result of particles exercising mutual attraction, like a 'magnet' (f. 23v). The smaller the individual particle of water the more its cohesiveness results in a perfectly spherical form.

In the Codex, his concept of the particulate nature of water is itself dispersed across a number of analyses of varied phenomena. There is no consolidated statement about it. It is in MS F that we find his most extensive account:

> If you consider well each minute particle of this surface you will find it surrounded by other similar particles that are at an equal distance from the centre of the earth, and at the same distance from this centre is that particle surrounded by them. Therefore, according to the third conception, that particle will not move of itself because it is surrounded by boundaries of equal height. Thus every circle formed from such particles makes itself a vessel for the particles enclosed within this circle and this vessel has its circle formed by boundaries of equal height. In this way this particle resembles all the other similar ones of which the surface of the sphere of water is composed. Of necessity it will be without motion in itself, and consequently each is at equal height from the centre of the world, necessity makes the surface spherical. But it is not necessary that they should be spherical below, as reason and experience show.... So having proved by the testimony of the spheres that the flexible elements are spherical, it is

my purpose to delve into nature in its universality and in the particularities of its elements, the first of fire, then air, then water.[66]

This analysis places particle theory at the heart of the basic concept of the spheres of the elements and their remorseless tendency to seek their true level.

It is implicit that air is also particulate in a very fine manner, since he stresses that its behaviour is essentially like that of water, allowing for the fact that air is compressible. When he mentions that air exercises friction on particles of water falling from a cloud, the rubbing of the elements presupposes that they are both particulate. On folio 19r he suggests an experiment: 'yet if you vaporize with pursed lips a known weight of water into the air, and you let it fall on a sheet, the weight of which is known to you, you will find when you weigh the sheet that little water descended on it and the remaining evaporated into the air'. He cites his earlier record of this in Libro A. The general resistance of air to falling water particles also makes sense in if air has a particulate nature.

In the case of earth as a substance, the situation looks rather different at first sight. Earth has a demonstrably solid nature. However, as is clear throughout the Codex, it assumes granular form in a systematically graded manner. The progression is from large rocks to larger and smaller stones, coarse and finer gravels, sand and soil, or silt (the finest variety of which contains organic fragments). The process of erosion of earth by water, which is a central feature of his geology, moves inexorably in this direction, producing ever-smaller granules. Leonardo is also aware that sedimentary rocks result from the conglomeration of smaller granules, and recognizes a congealing 'virtue' or 'humour' in the earth that manifests itself most evidently when it petrifies shells and other things that result from a kind of sculptural casting.

Even we stop short of seeing Leonardo as fully embracing of all the tenets of atomic theory as expounded by Lucretius, there are clear indications that the concept of the particulate nature of the element of water—and less explicitly of air and earth—had insinuated itself very widely into his thinking. Leonardo does not clearly announce the theory in the form of a joined-up statement of principle, and he may not even have been aware of how ubiquitous it was becoming in his thought. It remains one of the incomplete revolutions in his late thought.

His hesitation to designate atomism as a founding principle may reflect that fact the evidence for the *attimi* comes largely from their effects, that is to say second-hand, rather than having the first-hand tangibility or the level of visual reality that Leonardo valued so highly. If the planets were too far away to be readily available to Leonardo's science, the finest atoms were too small to be conclusively visible. Given the huge range and quantity of material that Leonardo aspired to cover in his book 'On Water' it is probably an advantage that he did not extend its scope too far into the immensely distant and invisibly small.

After the Codex

The Codex is in many ways exceptional in his manuscript legacy. There is no other example where seventy-two pages (or forty-four if we only count the 'inner set') are devoted to a central topic—albeit

66 MS F, f. 26v.

with characteristic Leonardo ramifications. But the texts it contained were very much not the last words, as their questing and provisional tone suggests. There are numerous occasions in other manuscripts and drawings from *c.*1508 onwards when he returns to topics discussed in the Codex, and there are clear signs of some radical revisions to his thinking about the 'circulation' of water in the body of the earth.

We find, inevitably, more incomplete hints about his scheme for 'books' on the various subjects that involve water. In MS F, he talks about producing a sustained account of swimming, stressing the application of each proposition, books on 'hollows and mounds' that comprise the beds of bodies of water, and on the entrances and exits of water from such a body.[67] He mentions a book on a water drop, and books numbered from 9 to 13 on eddies, percussions and types of leaping water.[68] The continued proliferation of topics is signalled on folio 35r, where a discourse on clouds is numbered as book 42. He must have complied alternative lists of books at various times, each accorded a number. The task for a comprehensive treatise on water seemed to be expanding unmanageably.

Swimming seems to have occupied increasing amounts of his attention, not least since he was bringing it into line with his researches on flight: 'in order to give the true science of the motion of birds through the air it is necessary first to give the science of the winds, which will be proved by means of the motion of water within itself'.[69] And, on a fascinating page that deals with bird flight and swimming fishes, he reminds himself to 'write of swimming under water and you will have the flight of the bird through the air'[70]—noting that a place near the Ponte Rubaconte on the Arno will be a suitable place to observe underwater swimming.[71]

It is Manuscript F, which probably overlaps in date with the later stages of the compilation of the Codex, which is closest to it in its lists of topics and indications that he is striving towards some kind of possible order.

There is the same recourse to staged experiments.

> For this experiment [*sperimento*] you will make a square glass vessel, keeping your eye at about the centre of one of these walls. With slow motions drop a few grains of panic grass into the boiling water, because by the motion of these grains you can readily know the motion of the water that carries them. And from this experiment, you will be able to proceed to investigate the many beautiful motions that result from one element penetrating into another.[72]

There is a strong parallel here with the researches that followed his dissection of the old man in the winter of 1507–8. Undertaking detailed investigations into the heart, most notably its valves, he announces his plan to make a model of the neck of an aorta in glass within a clay mould (Figure 12).[73]

67 MS F, ff. 2v and 4v.

68 MS F, f. 66v.

69 MS E, f. 54r.

70 CA, f. 571ar.

71 CA, f. 571ar.

72 MS F, f. 34v.

73 Windsor 19082r.

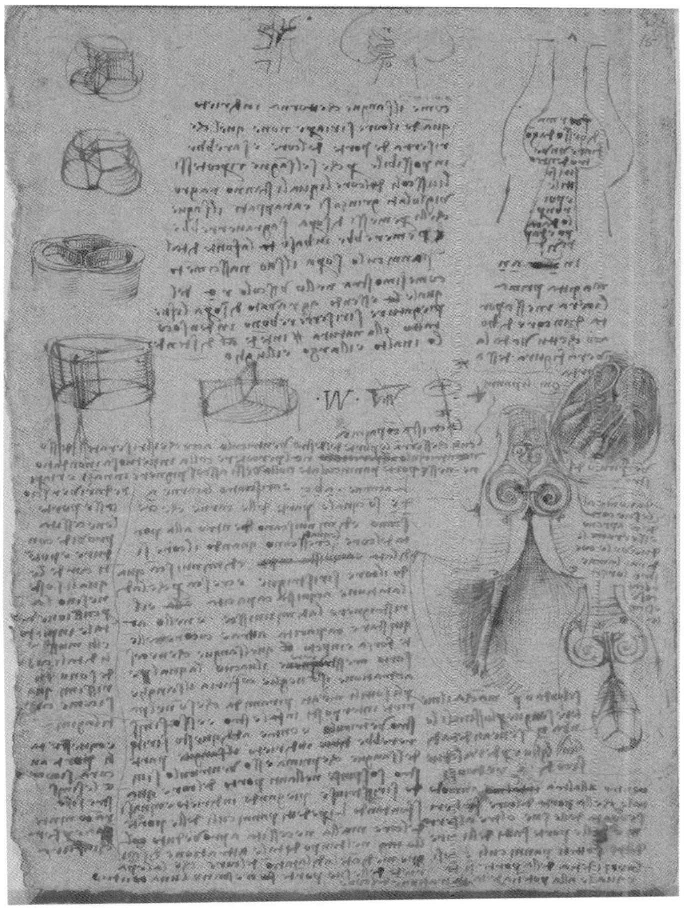

FIG. 12
Studies of an aortic valve and a glass model of the neck of the aorta. Royal Library, Windsor 19082r.

This was to be preceded by the casting of the inside of the valve in a bull's heart to ascertain its true shape. On Windsor 19078v, he refers to his investigations as exploring 'the geography of the heart'. He was hoping to observe the vortex formation in the constricted neck to understand how the revolving vortices filled the three cusps of the valve through haemodynamic rather than muscular action. He also announces that he intends to add panic grass seeds to the revolving water to make the currents more readily visible.[74] The resulting patterns of blood surging through the valves relate closely to his studies of water vortices passing through sluices in weirs and narrow inlets.

Again, some of the experiments are simple and homely. On folio 13r in MS F, he notes that 'when the hand is turned in a circular motion in a bowl half filled with water, it causes an accidental eddy which will expose the bottom of this bowl to the air'. This motion persists until 'the end of its impetus'. Considerably later, he is still puzzling about the remarkable phenomenon that water accelerates towards the centre of a vortex, with a hollow at its centre. In a sheet in the Codex Atlanticus from his years in France (1513–16), he explains that,

> The helical or spiralling motion of every liquid is so much faster as it is closer to the centre of its revolution. This we put forward, as a case worthy of note, in that the motion of a circular wheel is so much slower as we move closer to the centre around which it turns. But in the case of the water we have the same speed and distance in each complete revolution of the water, equally in the greatest

74 Windsor 19116v.

> circumference of its largest circle and in its smallest. The angle of the lesser circle is tighter than that of the larger circle to the extent that the curve of the greater circle is less tight than the lesser. Thus the water is of equal motion in all its revolutions. If this were not so, the cavity [at the centre] would immediately be filled again, but because the lateral weight in such spiralling motion is doubled, the cavity does not have any enduring motion. Of this doubled weight, the first part arises from the rotating motion of the water, and the second is generated at the sides of the cavity, where the water supports itself, and it finally collapses over the air that has filled up the cavity.[75]

He is grappling in a not wholly consistent manner with the paradox that revolving fluids turn faster towards the centre, unlike a turning wheel, and in opposition to the rule that a moving body gradually loses its impetus. It seems he is saying that a whirling current of water acts as a continuous substance, maintaining a constant impetus throughout, and that as the circle of revolution becomes tighter so the water has to revolve faster. However, he notes that the weight of the water at the sides of the cavity is doubled by its being suspended, so that it will eventually fall, and the cavity of air will disappear.

Inevitably the major topic of 'circulation' in the body of the earth looms large in the sections on water in MS F. The question is posed with increasing focus:

> If the water that pours forth from the high summits of the mountains comes from the sea, the weight of which drives it up there so that it is higher than these mountains, why has this portion of water the capacity of raising itself to so great a height, and penetrating the earth with such difficulty and over such a time while it has not been granted to the rest of the element of water to do the same, although this borders on the air, which would not be able to resist it and so prevent the whole from rising to the same height as the previously mentioned part?
>
> You who have found such an invention must return again to the study of natural things, because you will be found lacking in relevant knowledge. And for this you have powerfully fortified yourself by means of the volume of the friar of which you possess.[76]

The cover the manuscript gives a clue as to what the friar owned. He notes, 'Albertus, *De Coelo et Mundo*, from Fra Bernardino'. An edition of Albertus Magnus's treatise on heaven and earth had been printed in Venice in 1490. The fact that his study of 'natural things' is to be articulated through a study of one of the great Aristotelian philosophers of the Middle Ages reflects the increased 'bookishness' of Leonardo's researches at this time, as he seeks to define ever more sophisticated principles behind observed effects.

That he is still asking the basic question about how water ascends to high springs indicates that the 'circulation' theory based on the macrocosm–microcosm analogy is proving to be a less than wholly satisfactory answer. We find a number of nagging doubts being expressed. On the first page of MS F, he notes that 'the concavities of the hollows of the veins of the earth are enlarged by the prolonged and continuous passage of water'. This is opposite to what happens with human blood vessels, which become silted up and pass progressively less blood, as he ascertained when dissecting the old man.

75 CA, f. 813r.

76 MS F, f. 72v.

Later, on a page of anatomical studies at Windsor dating from around 1510, he declares conclusively that 'the origin of the sea is contrary to the origin of the blood, because the sea receives into itself all the rivers that are only caused by water vapours raised into the air'.[77] And on a late folio in the Codex Atlanticus, where he lays out a series of 'propositions' about water and the earth, he adds assertively in the margin: 'the water of rivers does not have its origins from the sea but from the clouds'.[78] After all his agitated debates about the internal circulation of the 'veins of water' as the cause of high springs, he settles on the simplest explanation. Rivers, however high their sources, are caused by precipitation.

These laconic notes are not part of a sustained discourse at this time on the complex issue of the 'veins' in the body of the earth and human body, but they do seem decisive in their rejection of the microcosmic analogy as an automatic explanation of how water can spout from high fissures in mountains. The analogy was not destroyed as such, since the human body and the body of the earth comparably expressed the rule of 'necessity' in shared natural laws. However, the implication is that each physical effect and its relationship to underlying causes now needs to be thought through in its own right rather than being subject to an overarching explanation.

An example of his developed mode of analysis during this later stage in his researches is his extended and remarkable analysis in MS G of the conservation of salt in the bodies of both earth and man. On folio 48v, written around 1510, he begins by refuting the explanation given by Pliny for the saltiness of seawater, which involves the evaporation of water from the sea. He then declares that,

> The saltiness of the sea arises from the many veins of water which encounter mines of salt when penetrating the earth, and these mines dissolve in part and the veins carry salt with them to the Ocean and the other seas, from where the clouds, which feed the rivers, can never lift up the salt.

Again, he is clear that it is rain from the clouds that generates rivers, not the blood pressure of a macrocosmic 'circulation'.

On the next folio he envisages a kind of salt conservation in which the human body plays a key role:

> Salt can be dug out wherever pigs have pissed, and the sea breezes are salty. . . .
>
> But to express it better, given that the world is eternal, it is necessary that its peoples are also eternal. Thus eternally the human species has been and will be consumers of salt. . . . Either the species of salt will be eternal together with the world, or it dies and is born again like the men who devour it. But experience teaches us that salt does not die, as is shown by fire, which does not consume it, and by water that becomes salty to the extent that salt is dissolved in it. . . . Salt passes through the bodies of men either as urine or sweat or other excretions to be returned [to the world]. And every year much salt is carried into towns.
>
> We therefore say that the rain that penetrates the earth is that which is found under the foundations of the cities and populations, and this is what restores the saltiness of the seas via passages within the earth. The transformations of the seas, which once were over the mountains, caused the salt to be deposited there in the mines that are discovered in these mountains etc.

77 Windsor 19003r.

78 CA, f. 433r.

> The third and last reason is that salt is in all created things, and this we learn from water passed through ashes and cinders of burnt things and the urine of every animal, and the excretions of their bodies, and from the soil into which all things are converted by corruption.[79]

The reasoning is rather tortuous and is not entirely resolved, but the idea that the excretions of humans and animals play a key role in the recycling and conservation of salt in the integrated system of the world and its inhabitants is very remarkable.

This discussion is based on the premise that 'the world is eternal . . . [and] its peoples will also be eternal', which is a different picture from his normal one of an earth that is born, lives, changes and will die. It seems that once he had loosened the framework that the earth always acted in ways that are precisely body-like, he was able to float ideas that should be considered, without necessarily solidifying them into definitive theories. This passage in MS G is rather isolated, but does appear to indicate innovatory if inconclusive directions in his late thought. There is a strong sense of a chaotic radicalism without a coordinated centre of new theory.

Alongside the intellectual earthquakes that were disrupting his cherished theories, he continued to work on major canal designs for his employers. The French authorities in Milan were no less keen than the Sforza family they had ousted on improving the Lombard canal system.

The most polished of his designs is the highly finished drawing of three double outlets of water from the 'Naviglio di S. Cristoforo in Milan': that is to say on the Naviglio Grande where it ran past the church of S. Cristoforo. The inscription at the top records that it was 'made on the 3rd of May 1509' (Figure 13).[80] It is drawn in almost non-convergent perspective, so that the dimensions of features at different depths into the depiction remains more or less constant. It may seem that the drawing was made at this date as a record of a pre-existing structure, or that he submitted his own design at that time—or even that the work had been completed according to his specification. However, the design departs from the actual arrangements of sluices in the canal itself, which followed standard practice.[81] What Leonardo is presenting seems to be an ideal design of how the outlets or *bocce* ('mouths' or *bocchelli*) *ought* to be arranged in contrast to what was actually done. The three outlets each consist of a pair of rectangular *bocche* that spout water into irrigation channels. The vertical width of the upper orifice in each pair appears to be half the lower one. The rightmost one emits water from the lower, larger mouth, the central one from the upper, and that on the left from both. In the cross section he characteristically shows the interpenetration of the streams of water as they emerge with different trajectories according to the water pressure at each level in the canal, and plots their rebounds from the bed on which they fall. He had made a number of studies of the effect of water pressure on the spouts of water from a tank perforated at various levels, including one in the Codex Leicester.[82] He became aware that the flow of water from any orifice was proportional to its depth in the tank and the size of the aperture. The drawing is an overt demonstration of how the 'science of water' relates to its 'applications' in actual engineering practice. The 'science' provides the predictive tools to moderate the flow so

79 MS G, f. 49r.

80 CA, f. 1907r. See Maffioli 2016a, b. Maffioli provides a perceptive review of the project, with comprehensive references to earlier studies.

81 Maffioli 2016a, b.

82 Leicester, f. 27r. Compare MS F, f. 53r for a vertical array of five spouts.

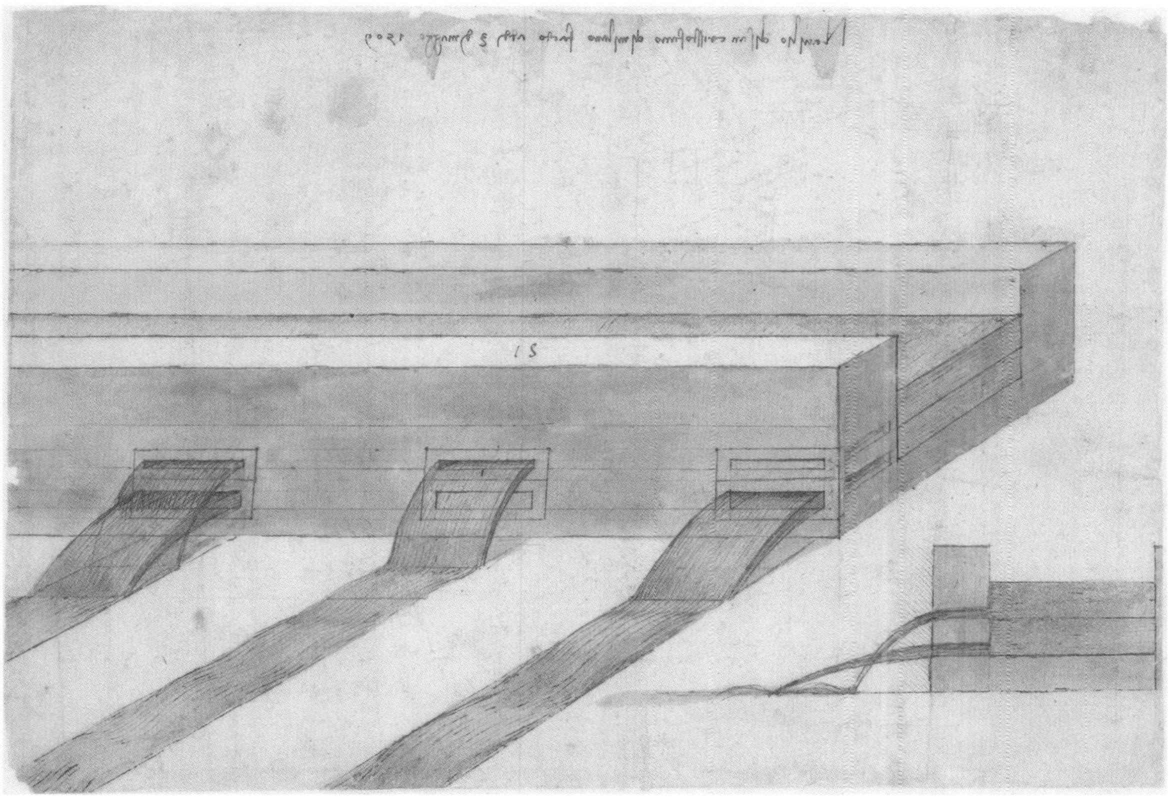

FIG. 13
Sectional studies of the Canal of San Cristoforo with outlets for the release of water, inscribed on 3rd of May 1509. Biblioteca Ambrosiana, Milan, Codex Atlanticus, f. 1907r.

that erosion and deposition could be managed, and so that definable quantities of water could be drawn off, measured in 'ounces'.

We may recall that he had been granted a water concession for this very canal by King Louis XII, and he therefore had a more than passing interest in how its 'mouths' operated. Indeed, in spring 1508 he drafted a long letter to the king's vice-chancellor in Milan in which he complained that he had been unable to benefit from the concession because of drought and the poor regulation of the water outlets (*bocchegli*).[83] Leonardo left the concession in his will to his servant.

In the main canal system to the east of Milan, there remained a substantial obstacle; the Adda passes through very restricted and non-navigable cascades at Tre Corni—named after three massive upright rock formations. Goods had to be transported laboriously overland from Trezzo to the start of the Martesana canal at Vaprio, downstream from Tre Corni. At some point, when Leonardo was staying in the Villa of Count Melzi at Vaprio d'Adda with his pupil and companion Francesco Melzi, the Count's son, he made three small and exquisitely energetic drawings of the Adda, one of which shows a chain ferry close to the family villa, while the other two show the Martesana canal running beside the non-navigable river. The illustrated example shows the rocks of Tre Corni above and to the right of the house and bridge (Figure 14).[84] He may well have been residing in the villa when he was sketching his grand plans for a massive dam and lock at Tre Corni.

83 CA, ff. 872r and 1037v; Villata 1999, pp. 221–3.

84 Windsor 12398, 12399, and 912400.

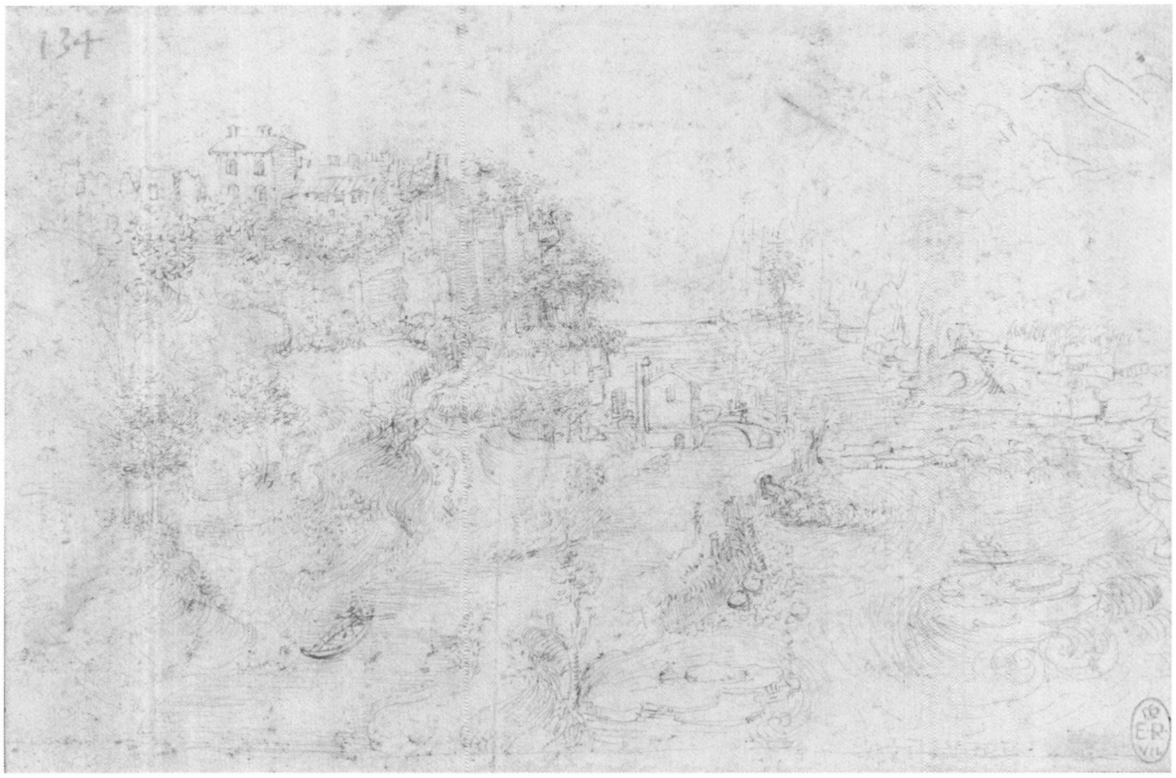

FIG. 14
View of the Adda and the Martesana Canal with the Rocks of Tre Corni, Royal Library, Windsor 12399.

On Codex Atlanticus 911r (Figure 15) Leonardo drew a notably careful map of the River Adda from the basin of the 'Port of Trezzo' near the top, flowing downwards through the rocky obstacles at Tre Corni.[85] Below this we can see the Martesana canal diverging to the west, pursuing a different course from the Adda. His projected new canal begins just below Trezzo at *b* and passes beside Tre Corni to join the existing canal of the Martesana at the 'Mill of Travaglia' at *a*. This looks straightforward enough, but the difference in the height of the water above and below Tre Corni and the hostile topography along the reach with the three massive rocks presented great problems. He spends a good deal of time recording distances and water levels, and makes calculations of costs, which total 18,000 ducats. He also looked more widely at the relative levels of the lakes in the region to understand the direction and speed of any flow between them.[86]

An idea of how he intended to overcome the problems at Tre Corni can be gained from a series of designs and notes on folio 388v in the Codex Atlanticus (Figure 16).[87] He writes:

> From the beginning of the canal to the mill.
>
> From the beginning of the canal of Brivio to the mill of Travaglia is 2794 *trabochi*, that is 11176 *braccia* [almost 4 modern miles], which is more than three miles and two thirds; and here the canal is 57 *braccia* [about 100 feet] above the surface of the water of the Adda, giving a fall of two inches in every hundred *trabochi*; and at that spot we propose to take the mouth of our canal.

85 For previous discussions, see Beltrami 1902; Zammattio 1981; Marani 2005; and Maffioli 2016a, b.

86 Codex Atlanticus, f. 740r.

87 CA, f. 388r contains diagrams of the distance involved with the various waterways.

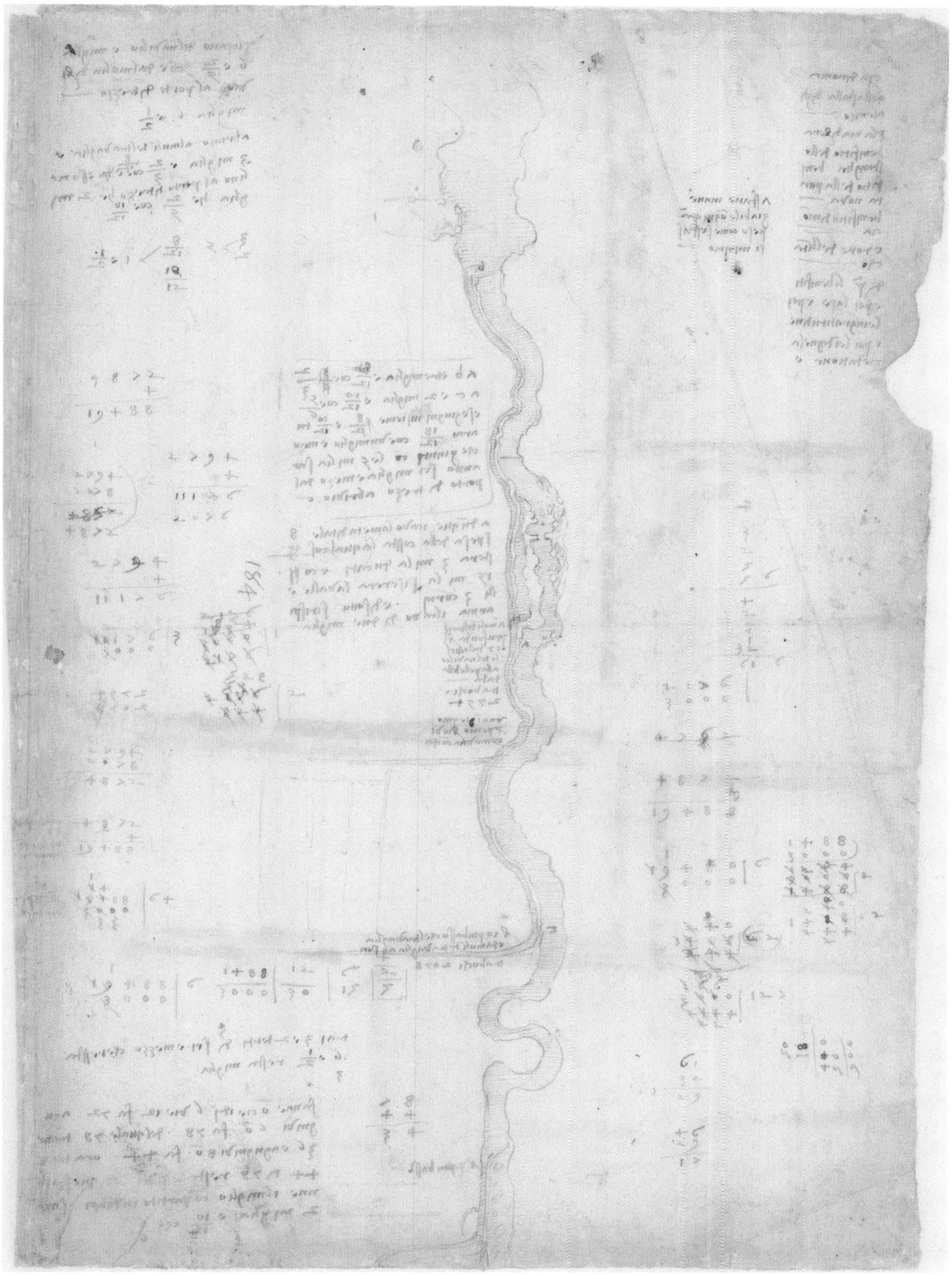

FIG. 15
Map of the River Adda and Martesana Canal with calculations. Biblioteca Ambrosiana, Milan, Codex Atlanticus, f. 911r.

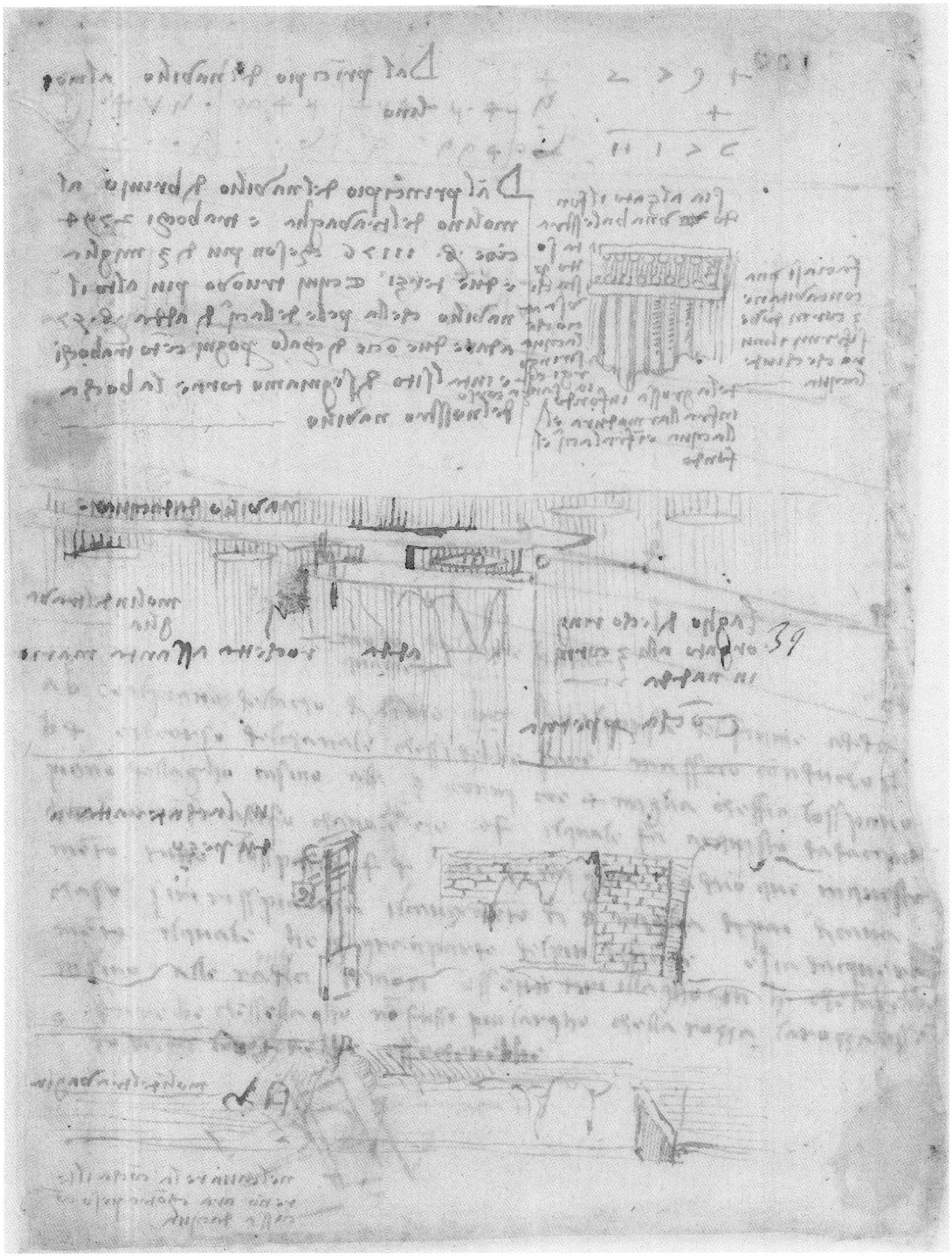

FIG. 16
Plan and other studies of the proposed dam, lock, and canal at Tre Corni. Biblioteca Ambrosiana, Milan, Codex Atlanticus, f. 388v.

> Let the bed below this dam be raised as far as an arrow-shot so that the water can be dammed up and left without any current. . . .
>
> Make a hollow in the Tre Corni to attach the wall which dams up the water.

The drawings are labelled:

> canal for irrigation;
>
> the Mill of Travaglia;
>
> the small rock of Santa Maria;
>
> the River Adda;
>
> the lake of Lecco dammed up at Tre Corni, in Adda;
>
> permanent lock;
>
> here the flood gate is all one piece;
>
> Mill of Travaglia;
>
> in excavating the sluice the soil will have as a counterweight a tank of water.

The drawings and notes are not easy to interpret, given their speculative quality. In the plan in the centre of the sheet the Adda is to the left and the dammed up waters coming from Lake Lecco (a branch of Lake Como) are to the right. They are separated by a massive dam that holds up the waters from the 'lake of Lecco'. A single boat is in the lock and two boats are to the left, while two other boats enter the high canal at the top of the plan. The elevation drawing of the lateral wall of the lock is shown to incorporate vertical rock formations into the masonry. Above the plan, Leonardo draws an open rectangular wall constructed from closely packed piles. The lowest diagram indicates that the high channel is supported by the rock and masonry wall, beside which is the lock that spans the high levels to the right and low levels to the left. He also roughly indicates a counter-weighted lifting mechanism, probably to remove soil when 'excavating the sluice'.

On Codex Atlanticus, f. 949v (Figure 17), a rough sketch added 'upside down' in the corner of a sheet of optical studies (from *c.*1508) indicates that the narrow canal above the lock is that of the 'commissioners' (the consortium that will enjoy the financial gains from tolls and drawing off water). The lock flanks this canal, as in the previous plan. Boats approaching from the right have a choice of continuing in the high canal or entering the lock to be lowered to the level of the Adda to the left. A sketch above in elevation (with elements that seem to be in plan) appears to show an extremely deep lock with steps down which water could cascade safely. Such a steep change in the height of water on either side of the dam presents huge constructional and hydraulic problems, and the sketchy drawings leave many questions unanswered.

The lock designs are not dated, but can be no earlier than 1508. There is no indication that Leonardo was directly commissioned to undertake the project alongside the other Milanese engineers who were involved, but his ideas may have fed into what was a contentious debate about how to accomplish what was an extremely difficult task. In 1516, a year after succeeding Louis XII, Francis I granted the considerable sum of 10,000 ducats to Milan, half of which was to be devoted to canalization, of which the Tre Corni project was the most pressing. It is possible that Francis was encouraged in this ambition by Leonardo, who was by then working for the king in Amboise.

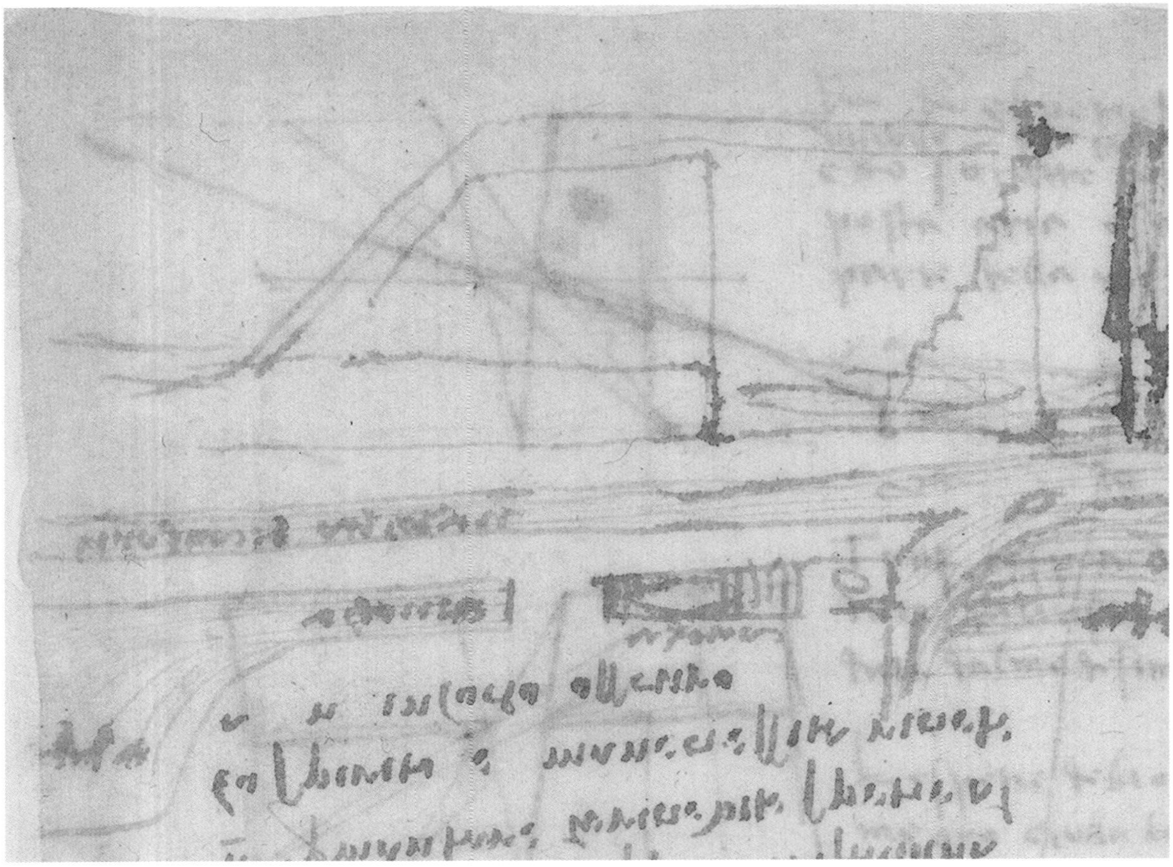

FIG. 17
Scheme for a lock at Tre Corni. Biblioteca Ambrosiana, Milan, Codex Atlanticus, f. 949v.

A report was produced by a commission headed by the engineer Carlo Pagnani in 1519, and it was hoped that the new lock, dam, and canal would actually be built. In the event, the bypassing of the river at Tre Corni was not to be realized until the later years of the eighteenth century.

Having moved to Rome in 1513 under the patronage of '*Il Magnifico*' Giuliano de Medici, Leonardo was involved in a scheme to drain the costal marshes south of Rome, reclaiming land for agricultural use and reducing the grip of malaria. Giuliano had been granted rights by his brother, Pope Leo X, in a *motu proprio* of 14 December 1514, to 'each and every territory which the putrid and stagnant waters flood'. Unsurprisingly, this ran into opposition from those who claimed to have existing titles to the lands and fishing rights.[88] Leonardo's proposals for drainage channels are illustrated in a magnificent coloured map, labelled neatly by Francesco Melzi (Figure 18).[89] He also made sketches of pumps and excavation equipment.[90] Work was begun under the supervision of the surveyor Fra Giovanni Scotti of Como, but Giuliano's premature death in March 1516 curtailed his financial investment in the project. How directly Leonardo was involved in the actual work before his departure for France is unclear.

When he moved to France in 1516, Leonardo was involved in water management in somewhat less taxing terrain. Resident in the grand manor house of Clos Lucé on the outskirts of Amboise, he studied

88 Bambach 2013; Linoli 2005, pp. 27–56.

89 Windsor 12684.

90 MS E, ff. 15r and 75v.

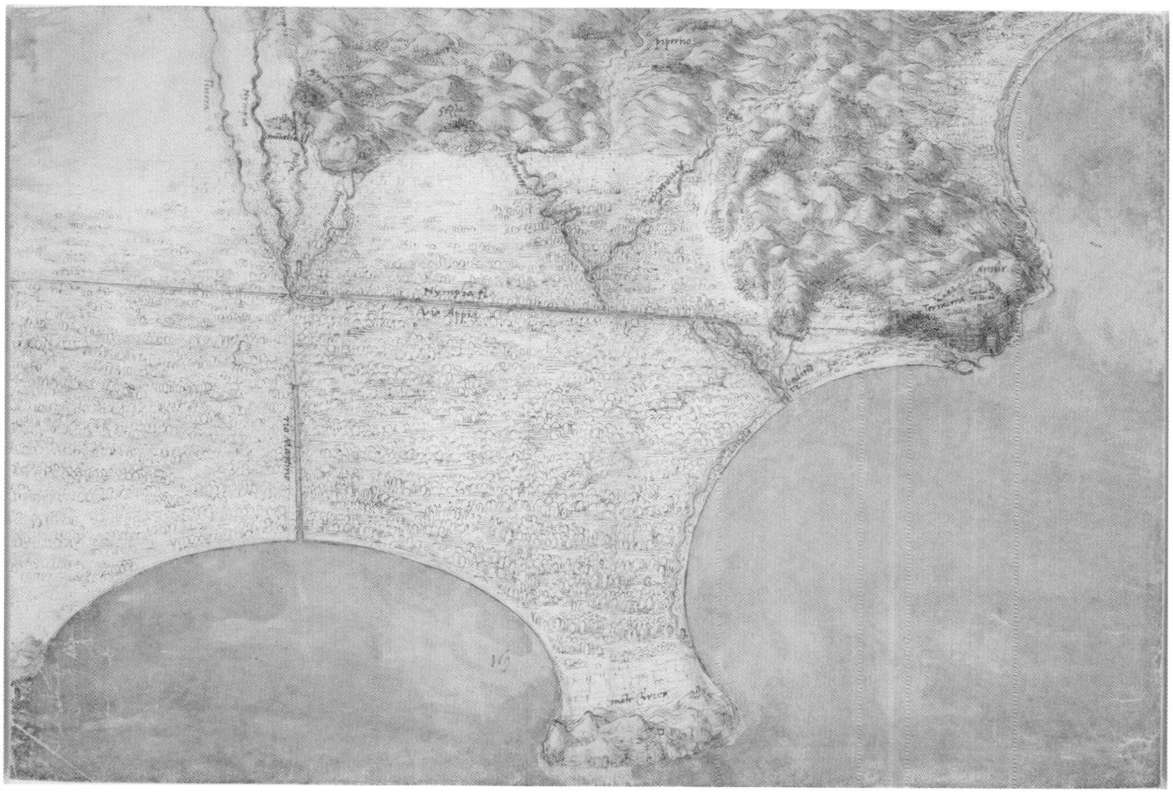

FIG. 18
Scheme for the draining of the Pontine Marshes. Royal Library, Windsor 912684.

problematic aspects of the flow of the River Loire. In a map in the Codex Arundel (Figure 19) he observes that:

> The Loire River which passes through Amboise passes through *abcd*, and after having passed the bridge *cde* it turns around going upstream through the canal *debf* along the bank *db*, which is interposed between the two contrary motions of the said river, that is *abcd* and *debf*; afterwards it turns around again, going down through the canal *flghnm* to re-join the main current of the river.[91]

It is clear from his account and from the conflicts of currents and banks in the drawing that the flow was far from optimum, but we do not have any studies that show how he would rectify these specific problems. On folio 270r, when he was considering general improvements to the flow of the Loire and its tributaries, including the 'river of Romorantin', he reminds himself of work he had earlier undertaken: 'Let the sluice be mobile like the one I arranged in Friuli, where the water, on the channel being opened, excavated the bed'. We see such a mobile weir on folio 22v in the Codex Leicester.

His major project for King Francis was a massive palace or double palace at Romorantin on the River Saudre. A drawing in the Codex Arundel shows that significant amounts of canalization were involved.[92] We can see a great double palace (one for the king and the other for the queen or dowager queen?) divided by a broad central canal which is flanked by large rectangular pools and surrounded by narrower peripheral canals running around the perimeter of the rectangular site. It seems as if the palace scheme was mutating into a small new town. Related drawings show various configurations of secondary

91 Arundel, f. 269r. Sammer 2019, chapter XVIII, for a revised chronology of the Romorantin project.

92 Arundel, f. 270v.

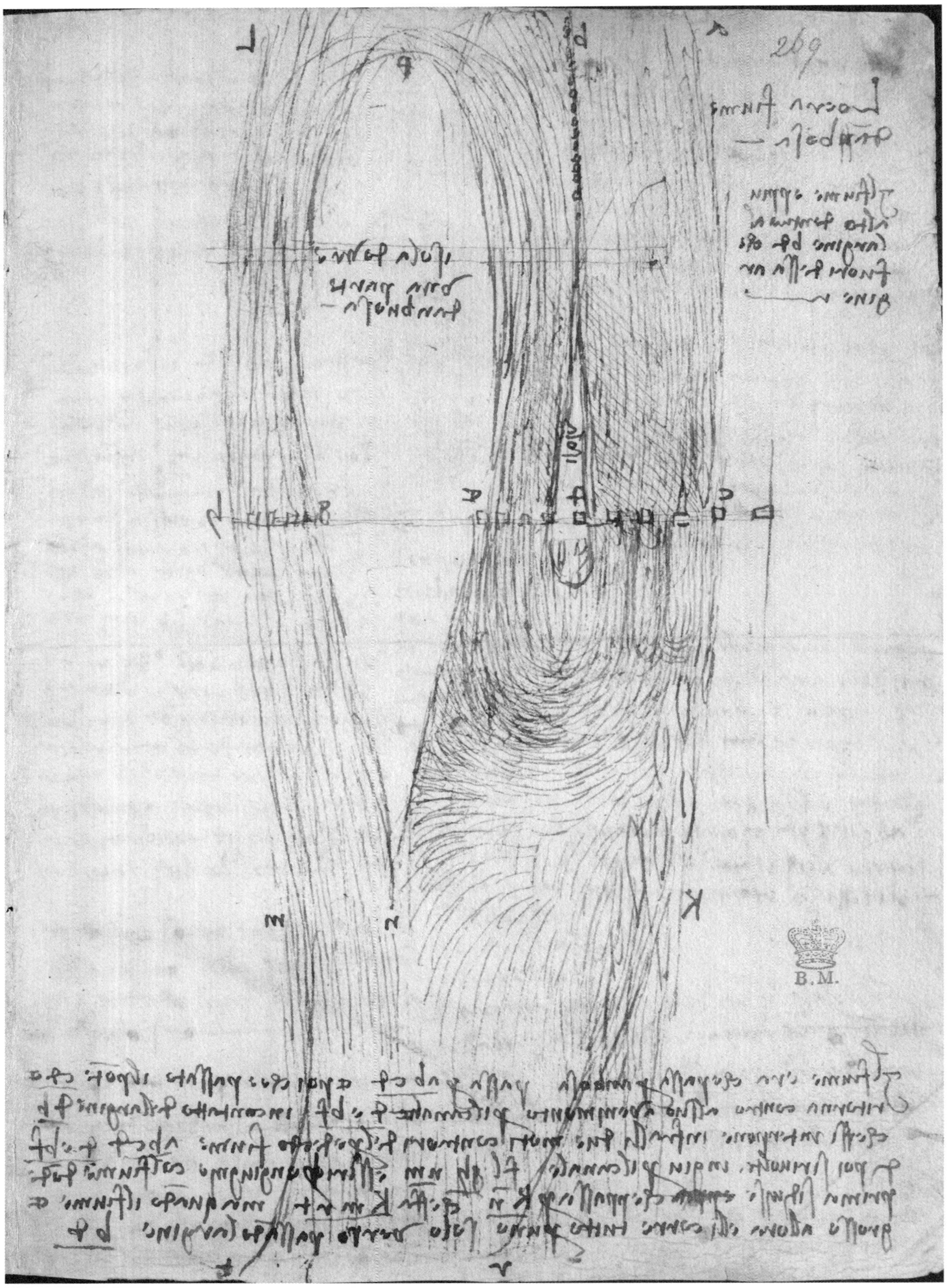

FIG. 19
Map of the Amboise Island on the Loire. British Museum, London, Codex Arundel, f. 269r.

channels within the site. These would have required carefully calculated engineering to ensure non-turbulent and non-turbid flow. As he says, with a characteristic concern for the economics of the scheme,

> The river in the middle must not receive turbid water, but the water should pass through ditches on the outside of the town, with four mills at the entrance and four mills at the exit, and this can be done by damming up the water at Romorantin.

The issue of water supply was also key. On Codex Atlanticus, folio 920r, Leonardo makes calculations about the relative water levels in the 'river of Villefranche' and the 'river of Ramorantin', so that water for the latter might be drawn off from the former.

Not only would the canals have facilitated the irrigation of the planned gardens of trimmed bushes and sweet flowers, but they would also have provided the opportunity for water features, most notably fountains adorned with sculpture. He reminds himself that there should be 'fountains on every piazza'. On a double sheet in the Codex Arundel, which contains a beautiful design for a spiral staircase, probably made for his projected palace, he draws a pump driven by the narrow vanes of a watermill of precisely the kind that could have fed fountains for the palace gardens.[93] He was also designing an automated water clock in which an animated figure would have struck a central bell on the hour.[94]

We know that Leonardo was on site working on the plans in January and June 1518, but the project was not to come to fruition. In less than a year after his last recorded visit to Romorantin he died in his chateau of Clos Lucé.

The Drawn and Painted Earth: Some Landmarks

Leonardo's first securely datable drawing is what would later be called a 'landscape'. It represents a plunging view across a deep tract of land viewed from a high viewpoint between two mountainous promontories, and is inscribed: 'day of Saint Mary of the Snows day of 5 August 1473'—that is to say, when the draftsman was twenty-one years old (Figure 20). It is a pioneering drawing with respect to its subject matter, and is unique (or is a unique survival) as a precisely dated drawing of a landscape. The inscription identifies the day as that on which a miraculous snowfall in 352 guided Pope Librerius in defining the ground plan of the basilica of S. Maria Maggiore in Rome. Whether Leonardo is simply referring to the traditional association of the day or making some kind of conscious allusion to the miracle is unclear. The precise view has not been convincingly identified, and it is likely, given illogicalities in the spatial array of the foreground ledge and flanking promontories, that it is confected from the artist's imagination. In any event, what it presents is a highly novel vision of the living earth, suffused by implicit and explicit motion. Not least, a waterfall plunges from a fissure to fall over the cliff face on the right. The surprising emergence of water from such high places was to become one of the major bones of contention in the Codex.

Landscape thereafter plays an unusually active role in Leonardo's art, compared to the works by most of his contemporaries. Let us look at the main landmarks across his career. Already in his earliest years as an independent painter in Florence, he cultivated an exceptionally subtle understanding of the visual

93 Arundel, ff. 264r–269v.

94 Windsor 12688 and 12716; Kemp 2006b, pp. 187–8.

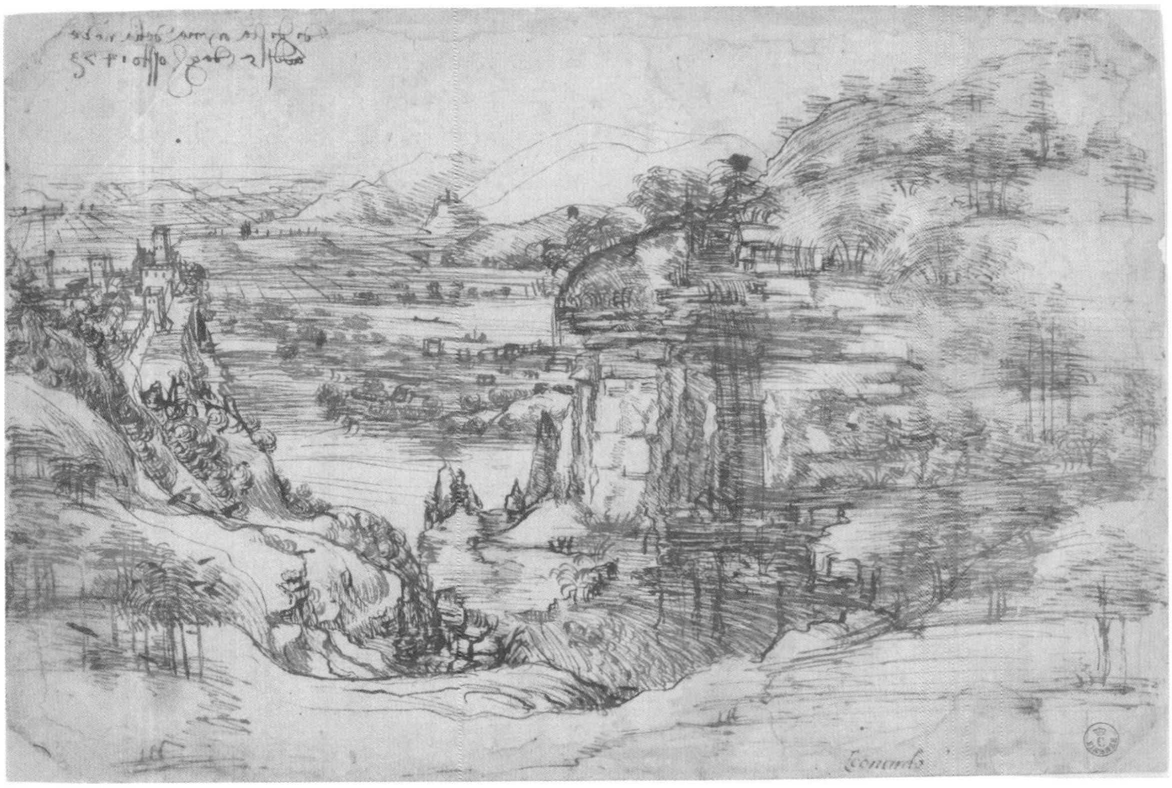

FIG. 20
View from an elevated position of cliffs, waterfall, castle, plain and distant mountains. Uffizi, Florence, f. 1453.

qualities and emotional suggestiveness of landscape. The enclosed garden in his early *Annunciation* (*c.*1475) is adorned with a lively carpet of wriggling plants, while the background fades into an atmospheric vision of a port with boats and blue mountains. His portrait of *Ginevra de' Benci* from about 1478, developed on the basis of Netherlandish precedents, sets her symbolic juniper (*ginepro*) against a shimmering watery scene of delicately dematerialized trees and towers.

The first commission that he undertook in Milan was to provide the painted adornments for the large architectural and sculptural altarpiece for the Confraternity of the Immaculate Conception in S. Francesco Grande, commissioned in 1483 from himself and two Milanese brothers, Ambrogio and Evangelista da Predis. The central painting provided by Leonardo was the *Virgin of the Rocks*, known in two versions. The earlier of the two, in the Louvre, never seems to have reached the confraternity's altar; while the painting in the National Gallery, London, was the one that he actually supplied. No Madonna and Child composition had ever granted the setting such an assertive role. The meeting in the 'wilderness' between Mary, Jesus, and the infant St John (accompanied by the Archangel Uriel) occurs immediately in front of a mysterious grotto through which we glimpse a notable series of upright rocky formations. In the foreground there appears to be a pool.

The geologist Ann Pizzorusso had undertaken an effective modern analysis of the rocky structures in the first version:

> The grotto is made of weathered sandstone which has been dissected by a layer of harder rock. At the top of the grotto are rounded weathered mounds (spherical weathering) of sandstone, a sedimentary rock. Above the Virgin's head is a rock with vertical relief. This is diabase, an igneous rock. The diabase was injected as a molten liquid, forming a band (sill) several feet high over the Virgin's head. Since it

was a molten liquid when it spread over the sandstone, it contracted as it cooled, forming vertical or columnar joints. Directly above the Virgin's head is a horizontal crack in the rocks, called a basal or bottom contact, which is the seam between the diabase above and the sandstone below. The column of diabase extends upward until it meets another horizontal contact surface and the rock formation changes to sandstone at the top of the grotto.

The rocks below the horizontal line (basal contact), near the Virgin's head, and extending down to the foreground are sandstone. This is the same stone that appears at the top of the grotto. The texture and rounded weathering pattern are the same below the basal contact as in the rocks at the top of the grotto. In the foreground, the layered (bedded) sandstone has not been heavily weathered, thereby retaining its highly defined horizontal layered structure. The rocks in the center, forming the diabase sill, are harder and more weather resistant. They are therefore, less prone to erosion, have sharper edges and more vertical relief than the sandstone.

The jagged rocks in the background, rising from a blue-gray mist, are remnants of erosional processes which stripped away the overlying softer rock and left the harder rock remaining. These formations have been subtly, yet accurately depicted, consistent with Leonardo's unwavering commitment to geological realism.

What is especially intriguing is Leonardo's placement of vegetation in the picture, not simply for aesthetic effect, but rather according to the location where such plants would grow. At the top of the grotto, the sandstone would have decomposed sufficiently to allow roots to take hold. This is true for the plants growing in the foreground and in the background. No plants are growing out of the diabase, however, since it is too hard and resistant to erosion to provide a suitable habitat for plant growth.[95]

Leonardo would not have known about diabase and its volcanic origins in our modern terms, since he was only aware of sedimentary processes. Even in the late eighteenth century, granite was assumed to have sedimentary origins. However, he was alert to the difference between rounded and columnar rocks, the different erosion of soft and hard rocks, and their varied stratification, as shown in a drawing of a rocky landscape in Windsor, dating from the same period as the *Virgin of the Rocks*.[96] In themselves such observations are not unique to Leonardo. Jan van Eyck, Andrea Mantegna, and other fifteenth-century painters differentiated between layered, rounded, and columnar rocks, sometimes portraying them in even more detail than Leonardo. However, the other painters displayed these rocks in a bravura manner as isolated elements, often in the foreground. It may be that the rounded rocks, still covered by soil and accompanying plants, are supposed by Leonardo to be younger than columnar rocks, which were weathered by waters in more distant times. For him the geological features serve as integral parts of a coherent landscape, which, from the foreground to the background, acts as a unitary organism that has lived over long periods of time.

In any event, Leonardo's painting does not present us with a literal portrait of an actual place. Tre Corni has been unconvincingly claimed as its specific location. As is his customary way of proceeding, he is confecting a consistent and dramatic setting based upon much hard looking and thinking about geological features and the processes that gave rise to them. He is re-making nature rather than reproducing the appearance of a specific location.

95 Pizzorusso 1996.

96 Windsor 12395.

At the time of the first version of the *Virgin of the Rocks* in the 1480s, his sense of geological process is as much instinctive as analytical, to be supplanted by a more analytical standpoint in the Codex Leicester, where he describes systematic accumulations of sedimentary material in defined strata. However, he was accumulating observations from an early date. On a sheet in the Codex Arundel datable to *c.*1481 he states that, 'on account of the two layers of shells it is necessary to say that the earth was indignantly submerged under the sea and made the first layer, and the deluge made the second'.[97] This is a precocious instance of his arguments about the Biblical Deluge that feature so prominently in the Codex Leicester.

Where he definitely uses conscious analysis in the *Virgin of the Rocks*—very definitely in the second version and less prominently in the first—is in his exploitation of edge contrast when he juxtaposes the dark borders of rocks against the misty radiance of the atmosphere (Plate D). He knows that a dark border looks darkest against a bright background and *vice versa*. As he says in his *Treatise on Panting*, 'colours of equal whiteness will look brightest against the darkest background, and black will display itself at its darkest against a background of the greatest whiteness'.[98] In the second version he deliberately plays up the contrast at the edges of the light sky and dark rocks. This is precisely the subjective effect he uses to account for the varied apparent radiance of light across the shaded side of the moon on folio 2r in the Codex.

The reason why the air assumes a blue colour is also analysed in the Codex on folios 4r and 36r. On the former, he explains how

> The blue of the air is not its natural colour but it is caused by the warm moistness within the air which is comprised of very small and invisible atoms of evaporated water. These are percussed by light from behind, becoming luminous against the dark region of the fire above the air, which covers it.

The overlying of a particulate layer of white on a black surface produces a form of optical blue. The phenomenon is indeed the result of the particulate matter of the air. We know that the blue is the result of the scattering of sunlight off the molecules of the atmosphere, predominantly at the blue end of the spectrum. It may well be that Leonardo had already arrived at his explanation at the time he painted the first version of the *Virgin of the Rocks*, since his paintings testify that he has been keenly interested in the effect of 'bluing' as early as the mid 1470s.

By the time he painted the *Mona Lisa*, which was begun in 1503 but seems to have been executed over a protracted period, we know for certain that he was thinking about processes in the formation of the body of the earth.[99] Indeed, it was commenced at the time when the Arno canal was very much on his mind, and was probably still being painted when he was beginning to assemble the Codex Leicester. The landscape is full of implied processes, not only ones that have occurred but also ones that are occurring and will occur. In particular, we can see two lakes, one high, the other low, like those he supposes to have occupied Tuscany in years gone by. As he writes on folio 9r of the Codex:

> The rock [Gonfolina] in ancient times was joined with Montalbano in the form of a very high bank; this dammed up that river [Arno] in such a way that, before empting itself into the sea, which was afterward at the foot of that rock, it composed 2 large lakes, the first of which is where now the city of

97 Arundel, f. 156v.

98 Cod. Urb. Lat., f. 75v.

99 Kemp and Pallanti 2017.

> Florence is seen flourishing together with Prato and Pistoia; and Montalbano continued the remaining part of the bank until where Serravalle is now located; in the upper part of the Valdarno, above here as far as Arezzo, a second lake was generated, which emptied its waters into the above-mentioned lake, enclosed about where we now see Girone; and it occupied all the mentioned upper valley for a space of 40 miles in length

On the right a high lake is dammed up behind a rock barrier, while on the left a lower lake seems to seep its waters behind Lisa and across to the mid-ground on the right. Alternatively, there may be an outlet in the rock barrier of the upper lake that feeds the lower waters on the right. Either way, a river is formed, running under a three- or four-arched bridge of a traditional kind. By contrast, on the left, below the lake, is a meandering road that appears to be a dried-up riverbed. There are three possible explanations: (i) the dry bed may have resulted from ancient processes when that portion of land was once emerging from the waters; (ii) its course may have dried up as the river on the right opened up; (iii) it may have been deliberately dammed up, perhaps to divert water along the river on the left. Since no dam is visible, the third alternative is the least likely. The second seems most plausible.

This is not to suggest that he has literally reconstructed this prehistoric landscape of high and low lakes in the *Mona Lisa*, or that he has portrayed an actual location that demonstrates comparable topography.[100] Rather he has synthesized his dynamic vision of the body of the ancient earth, not least as a macrocosmic analogy to the microcosmic body of the woman (the 'lesser world'). Lisa del Giocondo's cascading hair and fluent draperies underscore the deeper philosophical echoes between the two bodies.[101]

By the time of he painted his *Virgin, Child, St Anne and a Lamb* in the Louvre, probably from 1508 on, he was rendering the elusive beauties of a distant landscape shaped by water and veiled in atmosphere in an increasingly elusive manner. It stands as an imaginative construction founded on the inner understanding of the body of the earth that he had achieved in the Codex. In the more focused foreground, he delights in the fine-bedded sedimentary rocks and in a series of rounded pebbles whose eroded faces display beguiling patterns of mineral veins. Such stones disclose the congealing action of the 'petrifying virtue' that produces fossils and conglomerates, while their curved contours tell of their rolling over by impetuous currents of water in earlier times. The crumbling ledge and vertical face of the rocky plateau on which the figures are located speak of the continuous action of mobile water of the kind he discusses so insistently in the Codex. The distinction between the geological formations is now less intuitive and more conscious than in the *Virgin of the Rocks*. Relatively newer, partially solidified sedimentary rocks, with grained or layered shapes, are visible in the foreground, while older columnar rocks, weathered by sustained erosion, characterize the high peaks of the mountains on the background.

As Leonardo stressed in his extended notes on the role of water, the action of mobile water—and indeed of mobile air—the fluid elements brought both life and death. The raw power of the elements running amok is depicted with expressive fury in his series of late *Deluge Drawings* at Windsor (Figure 21).[102] On one hand, there is the human tragedy:

100 Of the many attempts to recognize the landscape as a specific location, the most considerable is by Albini 2015.

101 Kemp and Pallanti 2017, p. 192.

102 There are ten drawings in the 'set' of deluges: Windsor 12377–12386, with related drawings on 12376, 12387, and 12388.

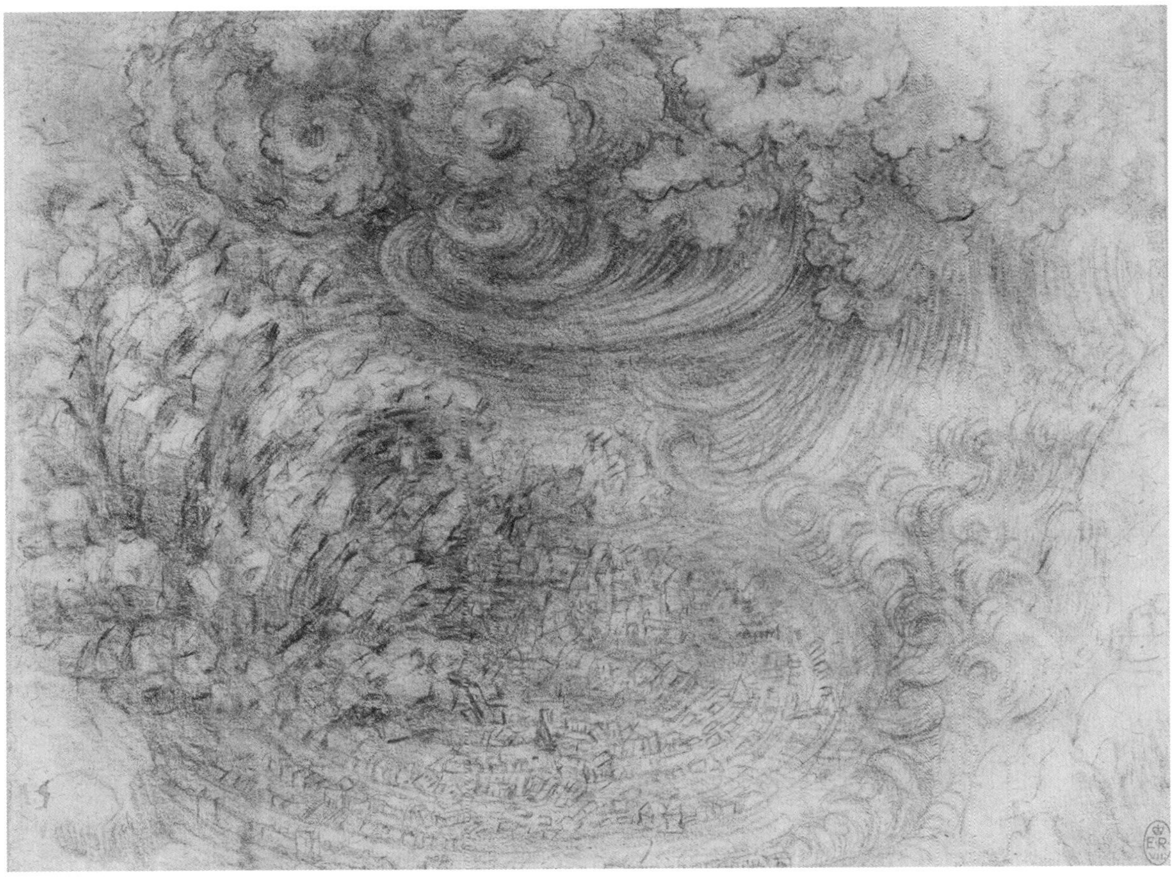

FIG. 21
Cataclysmic storm overwhelming a mountain and town. Royal Library, Windsor.

> O how many might be seen closing their eyes with their hands to block out the immense rumbles made in the lowering air by the fury of the winds mingled with rain, the thunder of the heavens and the fury of the thunderbolts; others, finding it is not enough to shut their eyes, laid their hands one over the other to cover them more closely so as not to see the cruel slaughter made on the human species by the wrath of god. O how many lamentations and how many flung themselves from the rocks in terror![103]

On the other hand, there is the geometry of impetus, described with dynamic exactitude:

> The swollen waters gyrate within the lake which contains them, and with eddying vortices percussively strike diverse objects, and leap into the air with muddy spume, and then falling back and making rebounds in the air with the percussed water. And the circular waves which fly from the place of percussion march with transverse impetus against the motion of other circular waves which move in opposition to them, and after making their percussion they leap up into the air without being separated from their bases. . . . If the waves encounter various objects, then they return in direct opposition to the blowing of the other winds, observing the same increase in curvature which they would have originally acquired in the observation of their principal motion.[104]

We can therefore, with only a selection of some of the most notable expressions of the body of the earth in Leonardo's pictorial works, see how his depictions and his written analyses are part of the same

103 Windsor 126665v.

104 Windsor 12665r.

developing enterprise across the whole span of his career. The human narratives—what he called 'the fictions that signify great things'—manifest a substratum of science, while the science is never devoid of humane and expressive dimensions.

A Long Journey

No themes are more persistent across Leonardo's career than the behaviour of water and its role on the body of the earth. His recurrent efforts to master the 'look' of water covers more or less the whole span of his career, and his attempts to master its theory and practice persisted across at least three decades of heroic scientific and technical endeavour. Nowhere does his mathematical theory, his intense powers of observation, the staging of experiments, and the practice of engineering work together in such remarkable concord. We know how complex water is, even for modern science, and many of the deductions that Leonardo made when analysing water in motion and planning its engineering belong in the realm of chaos theory and unpredictability. The most rounded expression of his range of thinking about water is the Codex Leicester, and it bids fair to be the most remarkable of his scientific manuscripts. Its ideas are startlingly radical, and fed into the revolution in geology, as we will see. The fact that his writings on water were never drawn to a conclusion is unsurprising and does nothing to reduce the sustained beauty of his thinking. Indeed the infinitely fluid but shapely complexities of water in motion serve as a mirror of his mind.

Of the Earth in Itself 4

Leonardo's 'Geology,' or Towards a History and Anatomy of the Earth

DOMENICO LAURENZA

WHILE LEONARDO WAS compiling the Codex Leicester (*c.*1505–8), the news of a New World—the *Mundus Novus*—was spreading across Europe. There was much more dry land than previously supposed. The 1501–2 travels in South America of the Florentine navigator Amerigo Vespucci demonstrated that the land encountered by Columbus was not already known (the Indies or Asia), but an unknown and huge land mass extending into the Southern hemisphere, previously supposed to be, at least according to a widespread theory, largely covered by the ocean. Probably printed in 1507, the map *Universalis cosmographia* by the German cartographer Martin Waldseemüller clearly showed this new land for the first time, labelling it '*America*'.[1]

Leonardo's age was the age of the enlargement of the geographical, or *spatial*, limits of the earth. In the very same period, the Codex Leicester contains signs of a *historical* vision of the earth's existence. Leonardo understood that the appearance of the earth exhibited a *history* comprising local and global changes and, even in a more ambiguous way, that these changes had occurred over eras of exceedingly long duration. The first part of this chapter is dedicated to the reconstruction of the *historical* side of Leonardo's study of the earth. We will also consider the possibility that Leonardo based his conclusions, at least in part, on the news concerning the new geographical discoveries.

The other aspect we will consider in the second part of this chapter, Leonardo's study of the *anatomy* of the earth, is conceptually connected to the former. The earth portrayed in the Codex Leicester, unlike the *eternalism* of the Aristotelian tradition, has a dynamic *history* with a beginning, an end, and, in between, incessant changes. It is a body which, just like the bodies of man and animals, undergoes processes of generation and corruption and needs to be studied in its *anatomical*, solid, and liquid constituents: rocks, stones, soil, and waters.

1 Brotton 2012, pp. 146ff.; Museo Galileo website, https://exhibits.museogalileo.it/waldseemuller/index.html.

This *historical* and *anatomical* vision of the earth is the most specific feature of Leonardo's approach to such a subject.

The world and the earth, according to ancient Greek philosopher Aristotle (384/383–322 BCE), were eternal, with no beginning and no end and therefore with *no history* of its own. All the changes happening in the earth were cyclical, secondary, and merely local, superficial phenomena. The opposite, atomistic theory, known to us above all through Lucretius (*De Rerum natura*, esp. book III), considers an origin and an end of 'our' world, but within purely atomistic physics (the roughest atoms gave rise to the earth, the smoothest ones to water and other elements) and within a predominantly philosophical and cosmological discourse, devoid of 'geological' and 'geographic' data, which are typical of the Aristotelian tradition. The latter also informs Seneca's thought (*Quaestiones naturales*, esp. books III, IV, and VI) but only in his description of the current state of the earth; for example, it is used to explain the presence of water springs in the mountains or earthquakes. Seneca's approach lacks a general historical sense and is further neutralized by the Stoic conception of the continuous birth, death, and rebirth of a self-same world; it also assigns ample space not to the creation, but rather to the end of the world through fire (*ecpyrosis*), or, according to Seneca, in the Deluge.[2]

Judeo-Christian thought, on the other hand, imagined a unidirectional *history*, with a creation and an end of the world, but this was a divine and compact history, assumed to be around 6000–7000 years and implying no substantial changes in the basic structure of the earth since God's creation; the Deluge, thought to be of divine origin, was the only great event.

During the Middle Ages the natural philosopher Jean Buridan took up Aristotle's idea of the eternity of the world and/or a span of time much longer than the few thousand years supposed by the theologians. Like Aristotle, Buridan hinted at geological changes and an inversion between land and sea. However, the eternalist vision, without a beginning or an end, removed the potential of any historical implications of those changes.

Regarding Leonardo's *anatomical* vision, according to a vision established by Aristotle in his *Meteorology*, the earth was mostly shaped and changed by two fluid *exhalations* generated by water (wet exhalation) and earth (dry exhalation) by a force placed outside the earth, far into the cosmos and in the sky: the heat of the sun. The cyclic inversions of water and lands were also the consequence of alternate dry and humid macro-seasons affected by the sun. In the Middle Ages, the astrological influence of the stars was added as a development of this vision dominated by an external/celestial force shaping the earth. By contrast, Leonardo tends to study the earth as a self-sufficient *body*, in its intrinsic constituents, regardless of the influence of the sun. He even denies the influence of the moon on the tides of the seas. Leonardo's study is one 'of the earth in itself', to use the title of a set of geological studies from the Codex Leicester (f. 36r).

Leonardo is one of the most famous victims of that schizophrenia that long divided historians of science between supporters of the genial *forerunners* of modern science and supporters of *continuism*, or 'everything was achieved earlier'. In the following pages we will try to escape these rigid alternatives and understand the specificity of Leonardo within the context of his own age.

2 Two classic studies on historical versus cyclic theories of the earth are Gould 1987 and Oldroyd 2006. On 'geology' in antiquity, see Gohau 2010, pp. 14–27; Ellenberger 1988, pp. 13–63.

Thus, although on a single and particular occasion Leonardo considered the eternalist hypothesis, his prevailing vision of the earth was unidirectional and historical, and this, along with his study of the earth *in itself*, in its more solid or anatomical structures and mechanical processes (deposition of strata, erosion, repair, etc.), seems to give his research aspects that are more characteristic of later periods in the history of geology.

But, at the same time, we will see how his *anatomical* and *historical* vision owes much to the ancient and mediaeval tradition of studies, from Strabo—an underestimated source of Leonardo's research—to Buridan and Albert of Saxony, and how it owes much to its age, from the 'anatomical revolution', shared by many artists like Leonardo, which started the slow shifting from 'fluid' and humoral to an anatomical and solidistic medicine, to the increasing interaction between empirical and scientific knowledge. This, for example, led Leonardo to use his skills in casting bronze sculptures as a conceptual model to explain what we call today *fossils*.[3] Such knowledge was set alongside travels and geographic explorations and the humanist rediscovery of ancient scientists.

Modern scientific theories have often been anticipated in the past. But what matters is the vision that dominated in a given historical period. In the case of Leonardo, what matters and fascinates is *the way* in which he elaborated his theories, which were more or less new, more or less 'anticipatory', forged out of an interaction between empirical-artistic and theoretical knowledge.

Leonardo's vision, though marked by new aspects, is actually a theory of the earth painstakingly and unsystematically developed within the framework of an Aristotelian cosmology and meteorology, dominated by the physics of the four elements in a geocentric universe. The four constituent elements of the sub-lunar world (fire, air, water, and land) form concentric spheres, from the most external and lighter, the sphere of fire, to the innermost and heaviest, the irregular earth, situated at the lower and more central point of the four spheres and, therefore, of the cosmos. Each element that finds itself outside of its sphere naturally tends to re-join it. This natural desire of every element to return to its sphere is a fundamental dynamic force and, in the case of heavier elements, water and earth, constitutes their 'gravity'. The geological theory developed in the Codex Leicester is triggered by, and in a certain sense consists of, a study of the 'gravitational' relationship between earth and water, the two heavier elements. *Gravity*, in its pre-Newtonian sense, is the most recurrent theoretical notion in the Codex Leicester and a unifying concept of its three main areas of study: astronomy, water, and 'geology'. The 'gravity' of the moon is the central concept in Leonardo's astronomy as well as in hydrology: the mountain springs are cases of apparent contradiction of gravity, while, in geology, continents emerge for a gravitational rebalancing of weights. This and other aspects connect Leonardo to mediaeval cosmology and mechanics.

In talking of the history of geology, we should bear in mind that the word *geo-logy*, in its modern sense of study of the earth in itself, was introduced at the end of the eighteenth century as a counterpart to *cosmo-logy* as the study of the earth as a part of the cosmos that also included the sky.[4] As we have seen, Leonardo works within a cosmology and meteorology dominated by the four elements and spheres. But, at the same time, he also tends to study earth and water as an interconnected whole, sharing the shifting from the two-spheres theory in favour of the new concept of terraqueous globe, which, as has

3 See Chapter 5 in this volume for ancient and modern meaning of the word 'fossil'.

4 Rudwick 2005, pp. 134–5 n. 3. For an alternative hypothesis, see Vai and Cavazza 2003.

been noted, 'was formulated outside the scholastic tradition', following the new geographical discoveries achieved between the end of the fifteenth and the beginning of the sixteenth centuries.[5]

'Of the water' (*Dell'acqua*) is a title used in the Codex (e.g. ff. 5r and 15v) to designate hydrological studies. 'Of the earth in itself' (*Della terra in sé*) is the title of what is the most important page of geological content in the Codex Leicester (f. 36r). The Codex Leicester is the most significant effort made by Leonardo to connect these two areas of study, which, in modern terms, we refer to as hydrology and geology. Leonardo explicitly claims it in a note added later in a sheet (f. 5r) of hydrological studies within a section of the Codex, the 'outer set' dominated by cosmological and geological studies:

> These books contain, in the first ones, 'On the Nature of the Water in itself, in its Motions'; the others contain 'On the Things Made by its Courses, which Change the World in its Centre and Shape'.

Since the beginning of the last century, there has been no comprehensive and up-to-date study of Leonardo's geological vision, above all as far as it concerns the history of the earth.[6] The same is at least in part true for classical, mediaeval, and Renaissance geology.[7] What follows is therefore only the first and necessarily provisional attempt at a systematic reconstruction of Leonardo's ideas in this field.

The History of the Earth

The Codex Leicester contains striking and laborious demonstrations of how the earth possessed a history comprising local and global changes. In very ancient times, the Mediterranean Sea covered lands that are now mountains, and similar changes affected the Black Sea, the Adriatic Sea, the Red Sea, and the surrounding areas. On a global scale, the dry land or what Leonardo calls 'our hemisphere' had not existed for ever, but emerged from the waters because of catastrophic events in a remote era, and then continued to change in a less catastrophic and slower way.

One dominant vision in antiquity was that of an earth that was eternal and therefore had no history of its own. No geological event exercised a formative and irreversible role that would explain the form of the earth as a whole. All geological events were cyclical, regularly repeated and reversed, and therefore with no impact on the general structure of the earth. Aristotle explains this vision above all in his *Meteorology*. The atomistic and philosophical-cosmological limitations of the Lucretian construction of the 'history' of the beginning and end of 'our' world, Seneca's cyclical, and thus ahistorical, conception, and his emphasis on the Deluge as the final event, make these ancient visions less interesting for the purposes of a more scientific reconstruction of Leonardo's geological thought, which goes beyond the 'catastrophism' of the late Deluge drawings series. What is more important in order to better understand Leonardo is above all the Aristotelian meteorological tradition and in particular, within this tradition, Strabo's geographic vision.

We will often cite Aristotle's vision, not only because many aspects of his geological model were widely accepted from antiquity until the seventeenth–eighteenth centuries, but also because Leonardo quotes more than once from Aristotle's *Meteorologica*, and likely gained access to a fourtheenth-century Italian

5 Grant 1994, pp. 635–6; Wootton 2016, pp. 110ff.

6 Baratta 1903; Duhem 1909–13, esp. vol. 2, 1909, pp. 281–357; Duhem, 1913–59, esp. vols 8 and 9; De Lorenzo 1920. For the bibliography on fossils, see Chapter 5 in this volume.

7 Duhem 1913–59, esp. vols 8 and 9; Ellenberger 1988; Grant 1984, 1994, ch. 20; Wootton 2016, pp. 110 and ff. More up-to-date studies on the ancient and mediaeval conception concerning the antiquity of the earth, eternalism, etc. are available: Bianchi 1987; Dales 1990; Nothaft 2014; Dal Prete, 2014.

translation, including excerpts from Aristotle's text and from Albert the Great and Saint Thomas's commentaries. A copy of this translation almost certainly was amongst the books in his library.[8]

Aristotle describes changes in the earth that involve the rising of land previously under the seas and conversely the flooding of dry land, but these are regarded as events that did not radically affect the global and essential structure of the earth. They act only locally and above all according a cyclic and balanced sequence: if an area undergoes a drying-up process, another area of the earth undergoes floods and slowly becomes a wet area. Both will later return to their former conditions, before changing again. These local and cyclic changes are driven by temperature and *meteorological* events. The processes of drying-up and flooding are more or less directly attributed to the greater or lesser influence of the sun, which governs their cyclic succession. In a given region over long periods of time, there are macro-seasons dominated by cold and heavy rains (floods), which give way to macro-seasons dominated by heat (drying up or desertification).[9]

These are 'accidental' (i.e. secondary) phenomena and they do not generate or affect the global structure of the earth, which has no beginning and does not change in its nature over time. As Aristotle declares, 'it is absurd to argue that the whole is in process of change because of small changes of brief duration'.[10] It is not by chance that such a statement occurs in *Meteorologica*, a work that deals with *meteora*, that is, changing and irregular phenomena. The fact is that, even if the land is home to the continuous change of the elements, animals, and plants, from a cosmological point of view the world and the earth, as a whole, are unaffected by these changes.[11]

The vision of Leonardo is very different, but it did not come from nowhere. Some authors already in antiquity gave more attention than Aristotle to local geological changes, though they continued to consider them as basically irrelevant for the comprehension of the form of the earth as a whole. Later on, at least for some mediaeval philosophers, attention was directed towards global changes in the earth, but within a rather synthetic and theoretical vision, which involved at most a limited study of real and local events. The ancient and mediaeval traditions seem to converge and complement each other in Leonardo's geological vision of the global history of the earth. The historical vision of the classical authors loses its local limitations; that of the mediaeval philosophers loses its theoretical limits.

Local changes (Mediterranean and other minor seas): Strabo and Leonardo's scientific humanism

Among the more spectacular passages of the Codex Leicester are those in which Leonardo describes the past, present, and future history of the Mediterranean Sea.

In the beginning the Mediterranean was a sea completely closed, with no connection to the ocean.[12] Surrounded, like a mighty lake, by a continuous ring of lands it was fed by the great rivers of Europe,

8 Madrid II, f. 3r (*meteura d'aristotile*); CA. f. 611ra (*Metaura d' Aristotile vulgare*); MS F. first cover verso (*Meteura*); Codex Arundel, ff. 190v, 191r. Some texts on the origin of the winds in CA, f. 461v, present similarities with this fourteenth-century Italian translation. Solmi 1976, pp. 44–6, 76; Librandi 1995, pp. 40–3.

9 Aristotle, *Meteorologica*, Bk I, ch. 14.

10 Aristotle, *Meteorologica*, Bk I, ch. 14, 352a 25. See also Aristotle, *On Generation and Corruption*, Bk II, ch. 10, 338a, discussed in Oldroyd 2006, p. 12.

11 On this distinction see also Albert of Saxony, *De caelo*, quoted in Grant 1994, pp. 618–19.

12 Except, starting at some point of its history, it had a communication with the Red Sea. See below and Leicester, f. 31r.

Asia, and Africa. In this ancient period its level was higher and it covered Italian lands including most of the Apennines, with only the higher summits emerging like islands. In Egypt, Memphis was on the sea coast and in North Africa the sea reached the Atlas Mountains:

> In the bosom of the Mediterranean, which, as a large body of water, are received the regal waters of Africa, Asia and Europe that were directed towards it and with their waters came up to the bulwarks of the mountains which surrounded it and formed its bank; and the peaks of the Apennines stood in this sea in the form of islands surrounded by salt waters; and Africa, within its Mount Atlas, did not yet reveal to the sky its uncovered earth with its great plains about 3000 miles long; and Memphis stood on the shore of this sea; and above the plains, in Italy, where today birds fly in flocks, fishes used to move in great shoals.[13]

Subsequently, a more or less sudden and catastrophic event occurred, although it had long been prepared by the lasting action of weather and erosion of the waters. The rocks at Gibraltar ruptured, and the waters of the Mediterranean began to stream into the ocean:

> [T]hen Mount Calpe was cut in the west, 3 thousand miles remote from this place and was detached from Mount Abyla; and this incision was made very low in the plains that lay between Abyla and the ocean [. . .]; and Hercules came to open the sea to the west, and then the marine waters began to pour into the western ocean.[14]

As a consequence of this opening at Gibraltar, the level of the Mediterranean Sea was lowered, disclosing lands that were previously submerged. We will see how this is one of the two causes (the *local* cause) adduced by Leonardo in order to explain marine fossils in mountains.

A drawing in the Codex Atlanticus (f. 901r, Figure 22) represents the first part of this history: the Mediterranean Sea, indicated by the tinted wash, covers wide areas of North Africa, together with Piedmont, Lombardy, and Tuscany and part of the Roman area in Italy.[15]

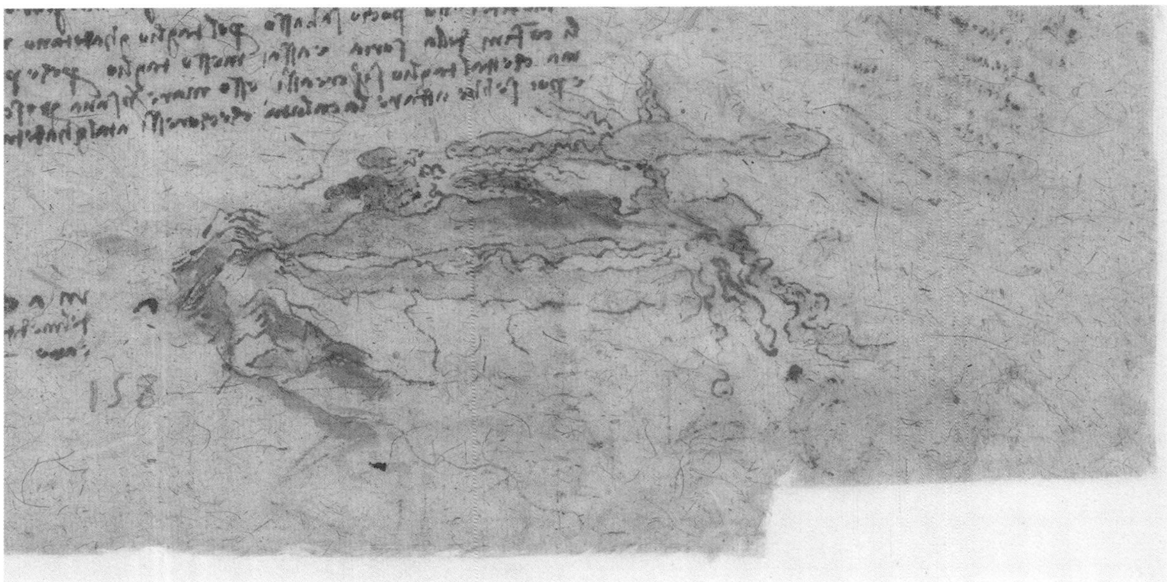

FIG. 22
Geological map of the Mediterranean: the sea covers inland areas of Italy where Leonardo had found marine fossils. Biblioteca Ambrosiana, Milan, Codex Atlanticus, f. 901r (detail).

13 Leicester, f. 10v.

14 Leicester, f. 31r; see also CA, f. 901r.

15 The drawing thus likely dates no earlier than *c.*1515, when he studied the marine fossils of Monte Mario in Rome. See also Pedretti 1978–9, vol. 2, p. 169.

The wash covers inland areas of Italy where Leonardo had found marine fossils, and the map is therefore based at least in part on direct observations. On the left, after Gibraltar (marked *a*) a precipice indicates the higher level of the Mediterranean Sea in comparison to the Ocean. The image is in a sense duplicitous in that events that are only imaginable are represented in a realistic drawing as if actually witnessed. On the other hand we will see how the Codex Leicester (f. 36r) contains images of the past of the earth that are of a less ambiguous status, that is to say of an overtly schematic kind, comparable to those used by modern geologists.

Leonardo also outlines the ultimate end of the history of the Mediterranean. Eventually the rivers that created it with their waters will also cause its end: by unloading sediments in it, they will cause an advancement of the coastline and a progressive silting up of the sea:

> How the soil, carried into the Mediterranean seas by the rivers that pour into them, chases the Mediterranean away from its place.[16]

> The marine shores move continuously toward the middle of the sea and drive it away from its original place.[17]

Eventually the bed of the Mediterranean will be completely exposed to the air and all that will remain of its waters will be those of the greatly extended Nile and its tributaries, which will open directly into the ocean.[18]

The Codex Leicester contains similar reconstructions of the history of the Black Sea, the Danube valley (f. 1v), the Adriatic Sea with the River Po (ff. 10r, 27v), the Red Sea (f. 31r), and the Persian Sea (f. 31r). The description of the changes that had occurred in the Danube valley, once covered by the Black Sea as far as Austria, contains very precise geographical names, following Ptolemy (100–175 CE), as punctiliously reconstructed by Gerolamo Calvi at the beginning of the last century.[19] But Ptolemy's *Geography* is not enough to fully understand Leonardo's historical account of the Black Sea. Ptolemy's book is basically a list of place names with indications of their correct representation in a map. It is neither descriptive nor historical geography. These are instead the main features of another masterpiece of classical science, the *Geography* by Strabo (58 BCE–*c.*25 CE). The humanists rediscovered both works in the Renaissance. The Latin translation of the *Geography* by Ptolemy played a central role in the renewal of geography, including the trans-oceanic explorations. Ptolemy described an extension of dry land greater than that supposed during the Middle Ages, and this was also Columbus's assumption.

In general, the birth of modern science in the Renaissance drew upon humanistic roots. The rediscovery of the great classical scientific works provided models to be emulated and even surpassed by modern scientists. The new geography benefitted from the rediscovery of Ptolemy, just as the new anatomy and geometry benefitted respectively from the rediscovery of the classical anatomist, Galen, and of the classical mathematician, Euclid. The role played by the humanistic rediscoveries of ancient Greek and Roman authors in the history of geography, medicine, and other sciences is by now a well-studied phenomenon, along with an enhanced appreciation of the heritage of Islamic science.

16 Leicester, f. 17v.

17 Leicester, f. 10r.

18 Leicester, ff. 20r, 10r; CA, f. 227v.

19 Calvi 1909, pp. xix–xxii.

By contrast, the specific role played by the humanistic rediscovery of Strabo for the birth of modern geology and his influence on Renaissance geography has not been fully elucidated,[20] and this explains why the importance of Strabo for Leonardo's geological vision has been underestimated in previous studies.[21]

Strabo's *Geography* was translated from Greek into Latin for the first time toward the middle of the fifteenth century by the humanist Guarino da Verona (1370–1460). This translation was printed in several editions, starting at least in 1469, and the reception of Strabo was particularly strong in Tuscany and papal Rome.[22] New geographical works were produced, sometimes in Italian (like the *Septe giornate della geografia—Seven days of geography*—by the Florentine Francesco Berlinghieri, 1440–1501), in which the purely geographic model by Ptolemy was combined with the more descriptive model by Strabo. Such information as the practices and history of people living in a given region were included.

The changes of the earth in Strabo's Geography

Strabo's treatise, especially Book 1, contains an extensive description of changes that occurred in two geographical regions: the Mediterranean Sea and the Black Sea. The starting point for these descriptions are the marine fossils found far from the sea, in inland Egypt and in the inland areas to the South and East of the Black Sea (approximately modern Turkey and Armenia). Strabo lists fossils of various shells, oysters, and bivalves, in addition to describing the remains of boats (*Geography*, 1.3.4). Correctly understanding fossils as evidence of ancient marine animals, his observations led to the conclusion that in the past these regions were covered by the Mediterranean Sea and by the Black Sea.

One thing has to be underlined: Strabo considers these changes as specifically local, with no wider implications for the general history of the earth. He writes: 'such changes [. . .] do not in any particular alter the earth as a whole; changes so insignificant are lost in great bodies' (*Geography*, 1.3.3). The general vision is therefore still that of Aristotle. Notwithstanding his reservations, the space given by Strabo to these 'insignificant' local changes is much greater than in Aristotle's *Meteorologica*.

Strabo, besides presenting his own theories, quotes and examines critically those by previous authors (Eratosthenes, Strato, and Xanthus of Lydia) and, so doing, he reports a series of possible explanations for the lowering of the Mediterranean and the Black Seas. According to the theory expounded by Strato (*c.*335–269 BCE) and shared by Eratosthenes (276–194 BCE), there was once no communication at the Bosphorus between the Black Sea and the Mediterranean, and no connection at Gibraltar between the Mediterranean Sea and the ocean. It was the long-term pressure of the waters unloaded by the rivers into the Black Sea and into the Mediterranean that were to rupture the rocky barriers at those two points, enabling the waters of the Black Sea and of the Mediterranean to surge outwards, leaving uncovered extensive regions in Turkey and Egypt (*Geography*, 1.3.4).

20 Gautier Dalché 2007, pp. 311 n. 175: 'The history of the Italian humanist' confrontation of Ptolemy's and Strabo's opinions has not yet been written; there are no real studies of the reception of Strabo.'

21 Short mentions in Solmi 1976 [1908], p. 267; Calvi 1909, p. xxi and n. 1; Duhem 1906–13, vol. 2, 1909, pp. 293–4; and Zubov 1968, p. 229. Strabo has been generically included among previous authors who advocated the animal origin of marine fossils. See Laurenza forthcoming a.

22 *Io. An. episcopi Aleriensis ad Paulum 2. Venetum pont. max. in recognitionem Strabonis epistola. Geographiam*, Rome, 1469. See Fryde 1983, pp. 55–82; Jones 1917, pp. xli–xlii; Gautier Dalché 2007, 2009, pp. 246–7 and 403; Aujac 1993; Larner 1998; Lazzi 2008, p. 94; Dueck 2017.

Strato believed that the cause was the different height of the seas, the Black Sea being higher than the Mediterranean, which in turn was higher than the ocean. The waters of these seas were like the water of a river, inclined to flow toward the ocean: 'Strato, [. . .] assumes this, believing that what happens in the case of rivers occurs also in the case of the sea, namely, that the flow is away from the high place' (*Geography*, 1.3.5). The differences in heights were in turn caused by the silting up by the rivers that unloaded sediments into the seas, which occurred to a greater extent in the Black Sea than in the Mediterranean. According to Strato, the Black Sea in the future would have been completely filled with soil: 'Strato further says it is his opinion that the whole Euxine Sea will be silted up at some future period, if such inpourings continue' (*Geography*, 1.3.4).

Eratosthenes and Posidonius (135–50 BCE) were even more radical. The former believed that even within the same sea there could be different heights of water, thus explaining the tides, that is the flow and ebb of waters in the Straits of Messina in Sicily and elsewhere in the Mediterranean (*Geography*, 1.3.11).

Strabo, who quotes their theories, which we only know thanks to him, criticises both the theory of the different levels of the seas and that of their silting up by rivers (*Geography*, 1.3.7, 1.3.9, 1.3.11). According to Strabo, silting up is a cyclic and reversible event and happens above all close to the coast (*Geography*, 1.3.8). The sediments unloaded into the sea by the rivers are mainly pushed back onto the coast by the sea, in an endless movement of coming and going, which is like the 'breath' of the sea. Since the bottoms of the high sea are basically not affected by these events, the seas are not inclined and since they are at rest and in communication among them their surfaces, for the law by Archimedes on floating bodies, will be spherical and equidistant from their centre, with no difference in height (*Geography*, 1.3.7, 1.3.11).

How can we explain marine fossils inland? Strabo tends toward an explanation close to Aristotle's cyclic vision. On one hand, he agrees with Xanthus of Lydia's theory of a long period of heat as the cause of the withdrawal of the Black Sea from inland Turkey and Armenia (*Geography*, 1.3.4); on the other, in a rather unclear way, he attributes the fossils to cyclic, sudden, and reversible floods caused by earthquakes (*Geography*, 1.3.10, 1.3.16–20). The bottom of the sea rose not because of slow and unidirectional processes of silting up, but in a sudden and reversible way because of earthquakes generated by the pressure of hot air circulating in the subterranean bowels of the earth. Water, after flooding the land and before withdrawing, sustained marine animals that were then left stranded on the land.

Strabo explains that hot air, tending to rise and burst forth, ruptures and uplifts the earth's crust. This happens especially along the coast, where earthquakes determine collapse of isthmuses and therefore the creation of islands, as in the case of Sicily or of the islands along the coast of Naples (*Geography*, 5.4.6 and 6.1.6). Floods might follow these movements of the earth, as Strabo says he has observed directly at Alexandria in Egypt (*Geography*, 1.3.17). Only in high seas would small islands be generated because of earthquakes and uplifting, like the islands of Thira and Therasia (present-day Santorini; *Geography*, 1.3.16). The formation of volcanoes, almost always along the coast, as in the cases of Vesuvius and Etna, diminishes the tension in the earth's crust, allowing the venting of the hot air trapped within the earth. Strabo claims to prefer this volcanic explanation because it is more directly observable than the slow and long sedimentary processes by Eratosthenes and Strato. He is basically juxtaposing an Aristotelian explanation that looks towards the dynamics of heat to the more mechanical explanation by Strato. Thus Gibraltar was opened not by the pressure of the waters but because of a collapse caused by an earthquake (*Geography*, 1.3.17).

Strabo as a model for Leonardo

We do not necessarily have to assume that Leonardo read in detail all the Latin text of Strabo's *Geography*. It is possible that a learned friend discussed Strabo with him or that he had access to a paraphrase of the classical geographer's general ideas, especially those outlined at the beginning of his long treatise (Book 1). It is known that Leonardo studied the ancient classical mathematician Euclid in this way, with the help of his friend and mathematician Luca Pacioli. However, by the time of the Codex, if not before, he could probably struggle his way through Latin if he had to, and there are many signs here and elsewhere that he drew upon Latin sources that were not translated.

A good number of Tuscan humanists were interested in Strabo. A possible indication of contact with this cultural context is a memorandum from the time of the Codex Leicester in which Leonardo reminds himself to ask 'Maestro Maffeo' about the periodic flood of the River Adige.[23] This might refer to Raffaello Maffei (1451–1522), author of an encyclopaedia published in 1506,[24] which included geographical sections in the descriptive style of Strabo. Maffei was Roman, but was in frequent contact with Tuscany where he died, and he was buried in Volterra.[25]

Strabo's treatise is not amongst Leonardo's surviving book lists, nor is it explicitly quoted in the Codex or elsewhere. Leonardo's booklists seem to signal books specifically in store, and do not necessarily include all the books he possessed. Moreover, the absence of a book from a private library does not necessarily imply ignorance of the text. With few exceptions,[26] Leonardo rarely cites authorities in the Codex. A common characteristic of all the long passages in the Codex Leicester is the absence of direct quotations or the names of authors. He only makes generic references to one or more 'adversaries', while the few names listed in the margins of the pages refer to authors whom he intends to study.

Leonardo's notable use of Strabo, like any act of scientific humanism, simultaneously involves imitating and surpassing. As far as the histories of the Mediterranean and the Black Sea and the theories of their different heights and of their silting up are concerned, Leonardo uses Strabo's text as a starting point, seeming to agree more or less consciously with the more mechanical theories criticized by Strabo. Leonardo's description of the opening of Gibraltar from the pressure of the waters and the consequent lowering of the Mediterranean (ff. 31r, 8v) is not far from that attributed by Strabo to Strato. Even Leonardo's conception of the end of its history, with the Mediterranean almost totally silted up (ff. 10r, 17r, 20r) may be seen as an extension of Strato's theory of the future of the Black Sea. Leonardo's analogy between seas and rivers[27] seems connected with the theories of the slanting of marine waters—the Black Sea to the Aegean to the Mediterranean to the ocean—that are quoted and criticized by Strabo.

Most of folio 31r is a free and original development of topics treated by Strabo in a section of Book 1 of his *Geography* (1.3.11–15). In addition to looking towards the different heights of the sea as an explanation (1.3.11) Strabo reconstructs the history of the Red Sea in relation to the Mediterranean (1.3.13). In origin, the Mediterranean poured into the Red Sea, but later on, after the opening of

23 MS F, first cover verso: 'ask maestro Mafeo why the river Adige rises every 7 years and lowers every 7'; 'da maestro Mafeo perché 7 anni l'Adice alza e 7 abbassa', *c.*1508.

24 *Commentariorum Urbanorum octo et triginta libri*, Roma, 1506.

25 Benedetti 2006; Tolias 2007.

26 e.g. Andrea Cattaneo (f. 1v), Ramon Lull (f. 2r), Theophrastus (f. 16v).

27 CA, f. 901r: 'the main river in our world is the Mediterranean'.

Gibraltar, it was lowered, the communication between the two seas ceased and the Red Sea remained at the same height of the external ocean. As Leonardo writes, 'the Mediterranean water poured for a long time into the Red Sea,' until 'Hercules came to open the sea to the west [at Gibraltar], and then the marine waters began to pour into the western ocean, and, because of this great lowering, the Red Sea remained higher'.[28]

It is also apparent that the lists of marine fossils in Leonardo[29] are close to those provided by Strabo ('mussel-shells, oyster-shells, scallop-shells'—'*concharum & ostreorum & cheramydum multitudo*[30]). Strabo also notes the remains in inland Egypt of ships (*fragmenta navium*), just as Leonardo describes the finding of the remains of ancient ships in inland locations (f. 9v).[31]

Strabo as a model to be superseded

The influence of Strabo is only one side of the coin. The work by Strabo is for Leonardo a model to be developed and corrected in emulation of the *principles* of investigation demonstrated by such ancient authors. Just as Andreas Vesalius, the great reformer of Renaissance anatomy, can be seen as more truly 'Galenic' in spirit than the more literal Renaissance Galenists—subjecting the great classical anatomist to admiration and criticism in equal measure—Leonardo can be said to be more 'Strabonian' than Strabo.

First of all, Leonardo extended the historical reconstruction to geographic areas not considered by the classical authors. Strabo and other ancient writers concentrated on two inland regions where the finding of marine fossils suggested the former presence of seas: North Africa and in particular Egypt, with respect to the Mediterranean, and the regions immediately surrounding the Black Sea (Turkey, Caucasus, Caspian Sea, and the mouth of the Danube). Leonardo directly studied marine fossils in inland Italy (Tuscany, Lombardy, Piedmont, and, in the years after the Codex, in the Roman area). On the basis of information supplied by travellers whom he judged reliable, he seems to have known about the presence of marine fossils along the Danube valley, from Austria to the Black Sea. Then, emulating his classical model, and using such empirical data, Leonardo reconstructed the geological history of these regions. If the fossils found in Tuscany and other areas on the Tyrrhenian side of Italy confirmed the lowering of the Mediterranean as suggested by Strabo, the fossils found in other areas allowed for new hypotheses.

Thus, fossils in Lombardy and Piedmont were evidence for the fact that the Adriatic Sea once touched the Alps. Later on the sea was lowered and left lands uncovered, traversed only by the waters of the River Po (ff. 10r, 8v). In the future the remaining waters of the Adriatic Sea will also completely disappear, because of the processes of silting up due to the River Po: 'How the river Po in short time will cause the Adriatic Sea to dry up, in the same way it dried up a great part of Lombardy.'[32] Leonardo thus foresaw an end to the Adriatic Sea similar to that which he anticipated for the Mediterranean and the River Nile.

28 Leicester, f. 31r.

29 e.g. Leicester, f. 1v: 'oysters and sea snails and turbinated shells and bivalves and the bones of great fishes'—'*l'ostriche e li nichi e bovoli e cappe e ossa di gran pesci*'.

30 Book 1, p. xiiiir in the Latin edition published in 1510. English translation from Jones 1917, p. 181.

31 See, for comparison, Ovid, *Metamorphoses*, Book 15, ll. 262–5: 'Often sea-shells lie far from the beach, and men have found old anchors on mountain-tops.' ('Et procul a pelago conchae iacuere marinae, et vetus inventa est in montibus ancora summis.').

32 Leicester, f. 27v.

He formulated a magnificent vision of the reconstructed history of the Danube valley and the Black Sea on folio 1v of the Codex, a masterly melding of Strabo's and Ptolemy's classical geographies (Figure 23) in the light of new travels and fresh empirical findings of marine fossils in Northern Europe. Leonardo begins with a description, based on Ptolemy, of the long chains of mountains limiting the valley of the Danube, running from Central Europe (Austria) toward the southeast, across the Balkans until its final termination in the Black Sea. To the north the valley is limited by the Carpathian Mountains, to the south by a long mountainous chain that originates in Northern Europe and extends to the south along the Balkans, continuing after the Bosphorus in the Taurus Mountains in Turkey (Figure 23).

Leonardo argues that, in the past, there was only a small break in the mountains at the Bosphorus and that only a small quantity of water found its exit from the Black Sea toward the Aegean and the Mediterranean. This continuous chain of mountains prevented most of the water of the Black Sea from exiting. The consequence in the distant past was that this sea once extended a long way to the north, even as far as Austria. Then, over time, erosion by the water permanently augmented the previously narrow exit, causing more and more water to flow towards the Mediterranean. The Black Sea was thus lowered, uncovering Austria, the Balkans, and the Carpathians. The finding of marine fossils led Leonardo to include these regions among those formerly occupied by the Black Sea. Therefore he radically extended the historical map of the world, substantially increasing the submerged regions as previously defined by classical geographers (Asia Minor, Caucasus, the Don, and the Caspian Sea).

But how did Leonardo know about the presence of marine fossils along the Danube valley? There is no evidence of any travels by him to these areas. His only travel in that direction was to the River Isonzo

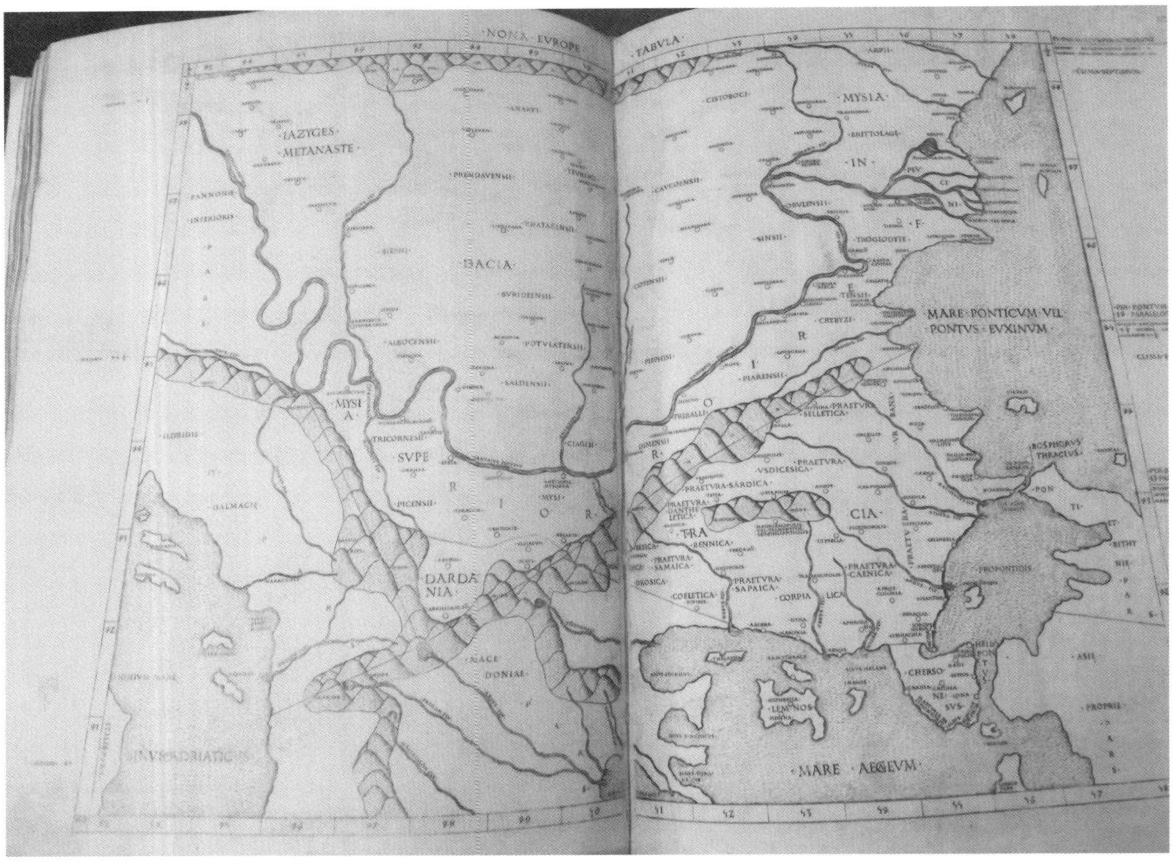

FIG. 23 Map showing the River Danube running from Central Europe toward the Southeast, across the Balkans until its final termination in the Black Sea. From Ptolemy, *Cosmographiae*..., Rome, 1490, 'Nona Europe Tabula'.

(Friuli, north-east Italy, and Slovenia) in order to arrange water defences against the Turks on behalf of the Venetians (*c.*1499–1500, CA, f. 638dv). A possible answer to our question is actually to be found in Venice, above all in its travellers. Leonardo's notebooks contain a number of memoranda about his intention to seek information from people who knew foreign lands. For example, in the Codex (f. 26v) he reminds himself to ask someone from Genoa about the tides. Such first-hand contacts would have vitally supplemented any published records to which he had access.

Some evidence points toward Venice as a source for eyewitness accounts. A note from the same period in the Codex Atlanticus leads us into the world of the Venetian travellers to the Black Sea and more adjacent regions: 'Write to Bartolomeo the Turk as to the flow and ebb of the Black Sea, and whether he is aware if there is such a flow and ebb in the Hyrcanean or Caspian Sea'.[33] This person may be Bartolomeo dalli Sonetti, also known as Bartolomeo Turco, a Venetian traveller and author of a famous *Isolario*, which contains descriptions in verse of the Aegean islands (printed in 1485 possibly in Venice and now very rare).

Venice and its tides are quoted at least three times in the Codex (ff. 9v, 13r, 27v). Perhaps when he describes cisterns of fresh water bounded by sand and clay in order to prevent the penetration of salt water—'as it is seen in the cisterns made in salt waters, which, outside their walls and sand, are surrounded by that finest soil of which vases are made; and the potency of the salt waters can never penetrate it and so the water always remains sweet in the cisterns'[34]—he has in mind the cisterns still visible today in the centre of many ancient squares in Venice and which were built just as Leonardo described. They were still renowned in the eighteenth century, when the scientist and geologist Antonio Vallisneri cites them in his book on the mountain springs.[35] Finally, the memorandum in Manuscript F (first cover verso), which we will regularly quote because of its citations of various authors consulted by Leonardo, contains also the note of 'books from Venice' ('libri da Vinegia'). His stay in the maritime republic lasted no more than six weeks, but it was a time in which fruitful contacts were and much information was gathered.

On the basis of his expanding knowledge of distant countries, Leonardo emulated and extended the classical model of Strabo, devising one the earliest schemes for the history of the earth. But he went much further. As we will see later on, marine fossils, merely listed by Strabo, become for Leonardo the focus of a wide and profound analysis of the genesis and structure of the changing earth. The vision of change in the fabric of the earth in Strabo and other ancient authors addresses specifically local events. In Leonardo's thought this is extended into a much wider vision of global change.

Global changes of oceans and lands: Leonardo and the mediaeval philosophers

The history of the Mediterranean, developed by Leonardo from classical authors, involves the passage of trapped waters towards the ocean after the opening of the Strait of Gibraltar, followed by a lowering of the surface of the Mediterranean and the uncovering of lands previously submerged—along with traces of the marine animals that once lived there. Classical authors, such as Strabo and those whom he quoted, did not venture further than this. Leonardo went radically further in his geological vision. He

33 CA, f. 797r.

34 Leicester, f. 3r.

35 A. Vallisneri, *Lezione accademica intorno l'origine delle fontane . . .*, Venezia, 1733, col. 13a.

transformed the passage of waters and sediments from the Mediterranean into the Ocean from an event with only local implication to become integral to global changes in the earth. These changes involve the whole relationship between the Ocean and dry land in the 'two hemispheres' of the earth. We move from a local account of geological changes to a global vision.

In this respect also, Leonardo's thinking did not come about from nowhere. Some mediaeval authors had elaborated a global vision. Yet, this was mostly theoretical and, in Jean Buridan's case, a cyclic theory that lacked any directional sense of history. But it was in his time the most advanced global explanation available.

Jean Buridan: A global theory of the earth

In mediaeval science, the earth was assumed to be immobile at the centre of the world, and, on the basis of available geographical knowledge, was conceived as having an asymmetrical setting (Figures 24 and 25).

On one side was the habitable hemisphere composed mainly of dry land (mountains or plains) and by 'all the seas that can be crossed';[36] on the other side was a hemisphere completely occupied by oceanic waters and corresponding to the sphere of water. The existence of the latter was demonstrated by the inability to reach land beyond the Strait of Gibraltar.[37] The ocean was part of the sphere of water, which was supposed to be up to ten times more extensive than the earth and surround the irregular sphere of the earth.[38] This two-spheres model was the most common model in the Middle Ages[39] and was the model adopted by Jean Buridan, the philosopher who lived in fourteenth-century Paris (*c.*1295–1363). The earth emerged from the water only in the upper hemisphere, and was made mainly of dry and porous land and mountains. The habitable 'hemisphere' was therefore like a massive island in the ocean-sphere of water. Outside the sphere of water, there was firstly the sphere of air and then the successive spheres of fire and the heavens, all surrounded by the sphere of the fixed stars. Even though irregular in form, the earth as made by God stood at the centre of this world.

Buridan applied to this cosmological conception a series of ideas from the science of statics (*de ponderibus*), involving the behaviour of weights and balance.[40] His theory was developed and broadcast by other authors, such as Albert of Saxony (*c.*1316–90). According to Buridan, there is not only an asymmetrical arrangement of forms across the two hemispheres, with dry land concentrated only in one hemisphere, but also an irregular arrangement of weight. Dry land is lighter in comparison to the oceanic hemisphere for two reasons: on one hand, being uncovered by water, the sun can heat it readily,

36 Jean Buridan, *De caelo* (*Johannis Buridani Quaestiones super libris quattuor De caelo et mundo*), book II, questio VII. The standard critic edition is by E. A. Moody, Cambridge, Mass. 1942. See also the Italian edition by A. Ghisalberti, *Giovanni Buridano, Il Cielo e il Mondo. Commento al trattato 'Del Cielo' di Aristotele*, Milano, 1983 and Clagett 1981, pp. 642–5, Document 10.1 (Book II, *questio* XXII).

37 Grant 1984, p. 27, n. 88.

38 See e.g. Ristoro d'Arezzo, *Della composizione del mondo*, *c.*1282: 'A sapere quanto l'acqua è più che la terra . . .', Book IV, ch. 3 (E. Narducci edn, Rome, 1859, p. 70; see also Morino 2007). In general, see Wootton 2016, p. 115.

39 Wootton 2015, pp. 110–21.

40 On Buridan's geological theories, see Duhem 1913–59, vol. 9, pp. 293–306; Goldstein 1972, esp. pp. 32–3 (pp. 33 and 35 contain short notes on Leonardo); Moody 1975; Ellenberger 1998, pp. 97–107; Grant 1994, ch. 20; Gohau 1990, pp. 27–31, 2010, pp. 27–33; Oldroyd 1996, pp. 26–7, 2006, pp. 17–19; Dal Prete 2014; Wootton 2016, p. 114.

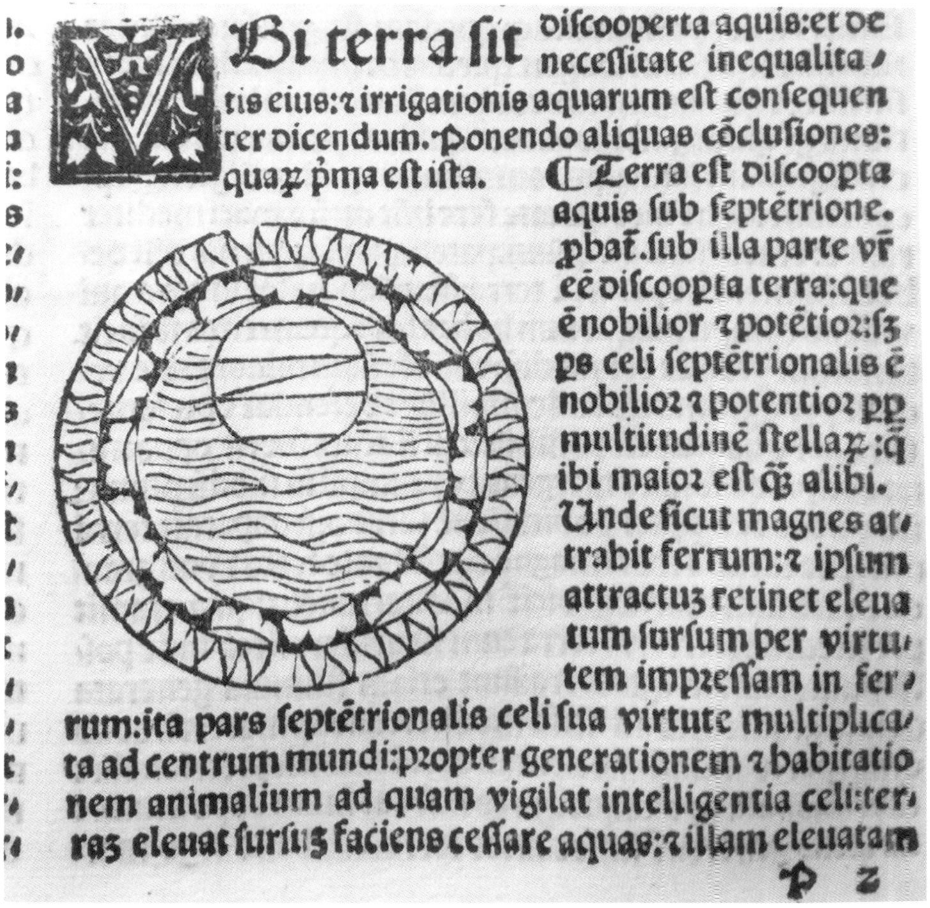
Vbi terra sit discooperta aquis: et de necessitate inequalita- tis eius: & irrigationis aquarum est consequen- ter dicendum. Ponendo aliquas cõclusiones: quarum pma est ista. ¶ Terra est discoopta aquis sub septẽtrione. pbat. sub illa parte vr eẽ discoopta terra: que ẽ nobilior & potẽtior: sz ps celi septẽtrionalis ẽ nobilior & potentior pp multitudinẽ stellarum: q ibi maior est q alibi. Unde sicut magnes at- trahit ferrum: & ipsum attractuz retinet eleua- tum sursum per virtu- tem impressam in fer- rum: ita pars septẽtrionalis celi sua virtute multiplica- ta ad centrum mundi: propter generationem & habitatio- nem animalium ad quam vigilat intelligentia celi: ter- raz eleuat sursuz faciens cessare aquas: & illam eleuatam

p z

FIG. 24
The four spheres model of the world. Xilography from Paolo Veneto, *De Compositione Mundi*, Venice, 1498, f. 112r (detail).

making it 'more subtle, rare and light';[41] and, on the other, because the incessant erosion of the mountains subtracts soil, moving it toward the oceanic hemisphere, the mainly earthly hemisphere is progressively rendered lighter, with the watery hemisphere becoming heavier. Therefore the portion of the earth that forms a mass protruding from the sphere of water comprises the lighter, but bigger,[42] hemisphere, while, the other hemisphere is denser and heavier.

Because of this there are, in such an irregular and asymmetrical set-up, two distinctly definable centres: the centre of *magnitude* or volume or size (*centrum* or *medium magnitudinis*) and the centre of weight or gravity (*centrum* or *medium gravitatis*).[43] The centre of magnitude is the geometric centre of the earth as a volume or form.[44] Given that the earth has an asymmetrical volume, its centre of magnitude (*B* in Figure 25) does not correspond to the centre of the world (*A* in Figure 25), but it will be offset toward the hemisphere that protrudes from the sphere of the watery ocean. On the other hand, the centre of the sphere of water corresponds to the centre of the world (*A* in Figure 25), since it is a regular

41 *Quaestiones super libris* . . ., Book II, [11]: 'subtiliant, rarificant et alleviant'; see Clagett 1981, pp. 640–5).

42 Being that the emerged land is lighter than that covered by the waters, it had necessarily to be bigger in size in order to float.

43 'Modo ultra, quia terra per suam gravitatem tendit ad medium mundi, ideo *centrum gravitatis* terrae est medium mundi, et non *centrum suae magnitudinis*, propter quod terra ex una parte est elevat supra aquam et ex alia parte est tota sub aqua.' *Johannis Buridani super libro secundo de caelo et mundo. Questio Septima*, 206–10 (emphases added). See Moody 1975.

44 Buridan and other mediaeval philosophers variously interpreted the centre of magnitude. See Grant, p. 622 and n. 11.

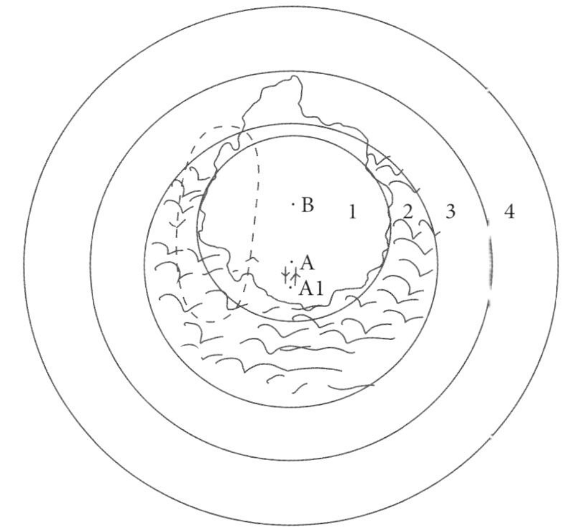

FIG. 25
Diagram illustrating Buridan and Leonardo's theory of the Earth.

(1) Sphere of Earth. (2) Sphere of Water. (3) Sphere of Air. (4) Sphere of Fire.
(A) Centre of the World = Centre of gravity of the Earth = Centre of the Sphere of Water, Air, and Fire
(B) Centre of Magnitude of the Earth
(A1) Centre of Gravity of the Earth shifting up and down in order to be in A.
- - - - -: Cyclic movement of sediments between Northern and Southern Hemisphere.

sphere with its surface equidistant from the centre and its weight uniformly balanced around the centre of the world.[45]

The centre of magnitude of the irregular earthly sphere does not therefore correspond to the centre of the world. Moreover, the earth has a second centre, different from its centre of magnitude. This second centre is that of gravity or of weight (*A1* in Figure 25). In theory, this should be displaced downwards, in the direction of the heavier hemisphere covered by oceanic water, especially because earthy matter or sediments are eroded by the waters in the dry land hemisphere and carried down and deposited at the bottom of the mighty ocean covering the other hemisphere, making the latter heavier, and therefore causing a shifting of the centre of gravity of the earth in this direction. Therefore the centre of gravity, like the centre of magnitude, should not correspond to the centre of the world. But obviously, in a universe with the earth fixed and immobile at the centre of the world, this was impossible.[46]

The conclusion of Buridan is that the centre of gravity of the earth—and the whole earth—does move in small rectilinear shifts: a shift of the centre of gravity out of the centre of the world is counterbalanced by a shift in the opposite direction. These 'little movements' hold the centre of gravity in or towards the centre of the world.

These 'little movements' are the consequence of a geological phenomenon that acts in a direction contrary to the phenomena that contribute to the divergence in weight between the two hemispheres. The compensatory mechanism imagined by Buridan is triggered by the tendency of the shifted centre

45 As later critics, like Copernicus, pointed out, it was difficult, actually impossible, to have the centres of the water, earth, and world (A) overlapping within the two-spheres model, assuming a sphere of the water ten times bigger than the earth. See Wootton 2016, pp. 110ff.

46 *De Caelo*: 'quia, sicut dicis Aristoteles, omens partes tendunt ad medium mundi per suam gravitatem'. See Clagett 1981, p. 645, Document 10.1 [12].

of gravity to coincide with the centre of the world and works as follows: the ocean floor in the 'southern' hemisphere reacts to the arrival of sediments, which make it heavier, with a movement in the direction of the 'northern' hemisphere uncovered by the water. This means that little by little the ocean floor rises towards the uncovered hemisphere and, in so doing, lifts up, reaches, and re-forms the mountains and dry land. In other words, it passes completely through the thickness of the earth and finally emerges in the northern hemisphere, where it continues to form mountains and dry land (Figure 25, dotted line). Just as in a seesaw, the heavier person (corresponding to the submerged lands of the oceanic hemisphere) needs to move toward the fulcrum to achieve a balance, while the lighter one (dry land in the opposite hemisphere) ought to move away from it.[47] There was a continuous, cyclic movement of material between one hemisphere and the other (Figure 25, dotted line). The rising beds of the ocean compensate for the descent of sediments onto them, and this compensation ensures that the centre of the earth continues to be located in the centre of the world, while its centre of size is offset.

Beyond Buridan's ahistorical theory

Leonardo knew Buridan's theories of the earth, perhaps through their propagation by Albert of Saxony. In a memorandum from the same period of the Codex Leicester he writes: 'Alberto de celo e mundo da Fra Bernardino',[48] which is a reference either to Albertus Magnus or Albert of Saxony (both wrote a book with a similar title), although the quotes from the latter, including from the *De coelo*, in Manuscript F, make the second one the more likely candidate.[49] Leonardo's theory of impetus is closely related to that advocated by Buridan, and there is no question that he gained access to the philosopher's ideas, either directly or via intermediaries.

Key aspects of Buridan's and Albert of Saxony's geological thinking helped shape Leonardo's conceptual tools in the Codex Leicester and other manuscripts. He knows about the difference between the centre of gravity and the geometrical centre of the earth (f. 35v: 'centro della gravità' versus 'centro della magnitudine', corresponding to Buridan's 'centrum gravitatis' and 'centrum magnitudinis').[50] Leonardo also deals in detail with the problematic relationship with the centre of the world and the consequent cyclic movement of land between the two hemispheres in order to balance the weight.[51] Leonardo seizes upon the best aspects of this theory, that is say its global vision, extending it in a way that transcends its ahistorical and rather theoretical dimensions.

Apart from the important idea that rivers unload sediments into the ocean, the geological vision by Buridan and Albert of Saxony exhibit mostly a conceptual character. There are no references to detailed

47 Gould 1998, p. 38–9.

48 MS F, first cover verso.

49 See below for one of these direct quotes of Albert of Saxony in Leonardo's MS F, f. 84r Duhem 1906–13, esp. vol. 1, 1906, pp. 1–50 (esp. 29ff.), vol. 2, 1909, pp. 55ff., 332–42; Solmi 1976 [1908], pp. 52–6; Frosini 2005; Nanni 2010; Maffeis 2015.

50 Jean Buridan, *Quaestiones super libro secundo De caelo et mundo*, Questio VII, 206–10 and Jean Buridan, *Quaestiones super libris quattuor De caelo et mundo*, Questio XI; see Clagett 1981, p. 645, Document 10.1 [11].

51 See Leicester, ff. 8v, 34v, 35v, 36r; CA, ff. 280r, 412r. On the centres of the sphere of water and air and their more or less problematic situation with the centre of the world, see MS F, f. 22v (correspondence between the centre of sphere of water and centre of the world); CA, f. 412r and Leicester, f. 17v (not in correspondence: 'How neither the centre of gravity of the water nor the centre of gravity of the earth are not concentric in themselves, nor concentric with the centre of the air and of the fire.'). On the possibility of a plurality of worlds and centres, see MS F, ff. 41v, 56r, 94v; Duhem 1906–13, esp. vol. 2, 1909, pp. 55ff.

observations, and only short and generic notes about the form and arrangement of the rocks. The mediaeval philosophers concentrate on dry land and lesser bodies of water, and some of them make only a few topographical references, mainly to French locations. Buridan's consideration of the three centres occurs within a chapter dealing with a general issue (Quaestio VII), such as the habitability of the world, and is reprised within another *quaestio* dealing with the movement of the earth from a cosmological point of view (Quaestio XXII).

In Leonardo's thinking, these mediaeval theories interact with a more empirical and geographical approach of Strabo and the ancient authors. The weighting of the oceanic hemisphere is now caused not generically by sediments coming from dry land hemisphere, but specifically by erosive and sedimentary processes directly or indirectly studied by Leonardo in actual geographic regions, from the Mediterranean to the Black Sea, from Tuscany to the Alps and the River Po. Leonardo explored the Alps in search of visual evidence of the slow upheaval of lands from the sea. He argued that the lower strata were less petrified because of their later emergence: 'the Northern footings of every Alp are not yet petrified'.[52] Thanks to this new approach, the global vision of Buridan loses its theoretical limit, while the empirical vision of Strabo loses its local limitations.

Another limitation of Buridan's global theory of the earth was its lack of a historical vision. Buridan does not consider the problem of the first origin of the overall asymmetry of the earth. The impression is therefore that the asymmetrical structure has existed forever, in an eternal dimension of time. This Aristotelian idea of eternity denies that of a directional history: a form that has existed forever has no beginning and no end. Buridan considers the 'potential' eternity of motion and time from a more philosophical point of view,[53] and likewise his geological theory, often ambiguously balances between purely speculative hypothesis, knowledge of reason, and faith.

Therefore, he considers Aristotle's idea of the eternity of the world, but, being aware that this ran counter to the Bible, he presented it as a pure theory of reason that occupied a different realm from the truths of revelation.[54] His theory of the eternity of the world was, however, compatible with *cyclic* geologic changes. Cyclicity is in itself an ahistorical concept. The slow processes, through which the oceanic side becomes heavier, raising the side with dry land, are cyclic events[55] that exercise no essential impact on the global structure of the earth over eternity. The asymmetrical earth is only *sustained* by these events, not *generated* by them. Buridan also speculates about more general geological macro-changes (within which the more limited above mentioned circulations of land are set),[56] which in order to act, require very long times—'perhaps a hundred thousand millions of years', he writes. This seems to prefigure the late eighteenth-century discovery of 'deep time' in the history of the earth. The reality is that Buridan's (and Aristotle's) deep time does not carry our kind of historical dimension as a

52 Leicester, f. 10r.

53 Jean Buridan, *Subtilissimae Quaestiones super octo Physicorum libros Aristotelis*, Paris, 1509 (repr. as *Kommentar zur Aristotelischen Physik*, Frankfurt am Main, 1964), Book 8, questio 3 ('Utrum sit aliquis motus eternus'). See Grant 1994, p. 70.

54 'I also suppose that the world is eternal, as Aristotle seems to say, even though this is false from the point of view of our faith'; Jean Buridan, *Questiones super tres primos libros Metheororum et super majorem parti quarti*, Paris, Bibliothèque Nationale, fonds latin, ms. no. 14723; see Duhem 1913–59, vol. 9, p. 299; Gohau 1990, p. 31.

55 See Duhem 1909, p. 329, on cyclicity in Albert of Saxony's development of Buridan's theory; and Oldroyd 2006, pp. 17–18.

56 Buridan speculates about a more general inversion between the two hemispheres: cyclically, the dry land becomes the oceanic hemisphere and vice versa. According to this, not very clear, theory there is a progression of the ocean around the globe attributed to changes in the erosion processes of dry land. See Gohau 1990, p. 30, and 2010, pp. 31–2.

directional narrative. Buridan is setting out a conceptual framework of a cyclic and endless alternation of opposite geological states rather than a sequential transformation that moves in an inexorable course from a set beginning to an ultimate end.[57]

Other mediaeval philosophers, such as Albert of Saxony or Nicole Oresme (1325–1382) qualified the heretical idea of an eternal world (also officially condemned by the Church in 1277)[58] and imagined the initiation of the asymmetry between the two hemispheres, a beginning that is metaphysical and divine. According to Albert of Saxony it was God who preordained asymmetry so that man, animals, and plants could live on dry land. According to Oresme, God with a flick of his finger lifted dry land in one of the two hemispheres. This divinely ordained asymmetry could be maintained by the heat and by the balancing cycles of movements of land imagined by Buridan.[59] The asymmetry could be cyclically inverted by macro-changes across the two hemispheres. Even earlier, in Avicenna's *De Mineralibus* (980–1037), the cycle of formation and disintegration with regard to all mountains 'except those of them which God wills should increase through the petrification of waters upon them'.[60]

In other mediaeval authors, such as Dante (*Quaestio de aqua et terra*) and Ristoro d'Arezzo (*La composizione del mondo*), the original upheaval of the mountains is attributed to astronomical and astrological causes: the fixed stars above the Northern hemisphere attract and lift up dry land from the sea in the manner of a magnet, thus creating the mountains.[61] Ristoro d'Arezzo's *The Composition of the World* is a very useful point of comparison with Leonardo, not least because the treatise was written in Italian, and Ristoro was an artist, a painter, and a goldsmith. While including geological field observations (from the correct interpretation of the animal origin of marine fossils to notes on stratigraphy and erosion by water), Ristoro develops above all the astrological potentialities of the cosmological hierarchy, attributing to the skies, the stars, and the world above the moon a fundamental explanatory role in geological phenomena. Not only, as already mentioned, have the lands emerged from the sphere of water by attraction from the stars, but also the biblical Flood is 'scientifically' explained according to a particular astral conjunction.

Leonardo's vision has aspects different from both the ahistorical vision of Buridan and from the more historical but metaphysical and astrological visions of other mediaeval authors. Transferred from a local dimension to a global one, Leonardo's history of the earth becomes the key theory to explain the general form of the earth as a whole. This form had an origin, has a development, and may well have an end. Within this historical vision, all the more or less particular changes considered by Leonardo, from the erosion of the rocks to the lowering of the Mediterranean until it becomes a mighty river, contribute to the birth of the conception of a global history of the earth.

57 For example after the description of the dynamic geological changes maintaining the asymmetry of the earth he concludes: 'et sic *semper* salvatur elevatio terrae' (Jean Buridan, *Quaestiones super libro secundo de caelo et mundo*, Questio VII, 211–23, emphasis added). See also Moody 1975 [1941], pp. 211–23.

58 Grant 1994, p. 65.

59 Duhem 1913–59, vol. 9, p. 204.

60 Avicenna 1927, p. 29; Duhem 1906–13, vol. 2, 1909, pp. 302–9. See Chapter 5 in this volume on Leonardo's possible acquaintance with this work.

61 Ellenberger 1988, p. 86; Gohau 1990, pp. 20ff; Grant 1994, pp. 618ff.; Dal Prete 2016, 2018. For a useful summary of these theories, see also Kerry Magruder with Brent Purkaple, 'History of Science Online: Nicole Oresme, Jean Buridan, and Theories of the Earth', Janux (2014), http://kvmagruder.net/hsci/08-14thCentury/sources/Oresme-Earth.html.

The history of the earth and its times

In the beginning (with a possible link to the discovery of the New World)

On folio 36r, a page in the outermost sheet, Leonardo outlines his vision of a possible generation of the present form of the earth with its dry land and sea. First of all, he argues that the subterranean world is affected by the same phenomena of aqueous erosion as on the surface of the earth.[62] On folio 36r and almost in the entire facing page, folio 35v (on which Leonardo quotes directly two figures from f. 36r), he strives to demonstrate that within the interior of the earth there is a huge quantity of mobile subterranean water.

We have seen how mediaeval authors had calculated, for various reasons, that the sphere of water was ten times bigger than the sphere of the earth, and how their prevailing idea was that of an asymmetrical distribution, with dry land in the northern hemisphere forming a substantial island in an even larger ocean of water, which occupied all, or almost all, of the southern hemisphere (Figure 24).

Leonardo argues that the visible extent of the ocean, together with evaporated water, is insufficient to explain this huge difference, that is the ratio 10:1. It was therefore necessary that part of the water resided within subterranean cavities and that there was less surface water. If this were so, the actual extent of the exposed land on the surface of the world could be much greater than was normally assumed. In one of the figures of the earth on folio 36r (Figure 26), dry land occupies almost all the circumference of the earth. This is a novelty, since, according to the traditional theory, dry land was almost all concentrated in 'our hemisphere', while the opposite hemisphere was almost completely occupied by the ocean (Figure 24).

As is well known, one of the outcomes of the travels of exploration into the New World was to demonstrate that the extension of dry land was much greater than had been supposed by most of the philosophers in the Middle Ages, and accordingly, there was less marine water. Amerigo Vespucci clarified that the lands discovered by Columbus belonged to a previously unknown continent. It was also evident that the southern hemisphere or antipodes were inhabited, as Leonardo speculates on folio 1r: 'similarly the Arctic and the Antarctic poles, if inhabitants are there'.

FIG. 26
The earth surrounded by dry land. Collection of Bill Gates, Seattle, Codex Leicester, f. 36r (detail).

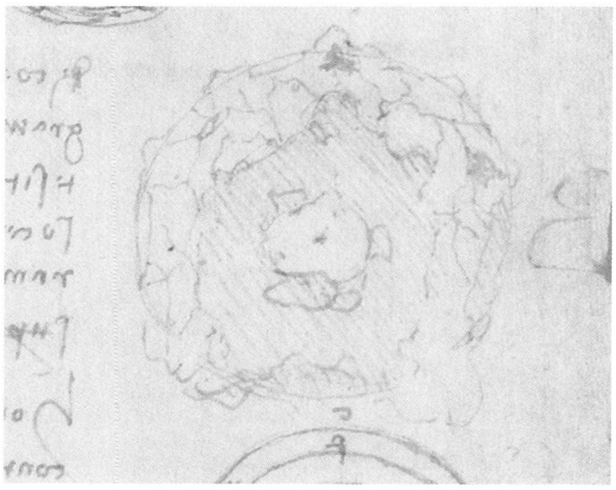

62 See Seneca, *Naturales quaestiones*, Book VI, ch. II, on subterranean erosion. See Thomas H. Corcoran, *Seneca. Natural Questions*, Cambridge, MA, 1971–2.

Is there any chance that Leonardo was aware of what was emerging from the voyages of exploration while he was compiling the Codex? Vespucci's reports about the new continent based on his 1501–2 travels consisted of handwritten letters. They began to be printed from 1504 onwards—including one published by Waldseemüller in 1507. This letter, later printed separately as *Mundus Novus*, was addressed to Lorenzo di Pier Francesco de' Medici in Florence; another, at least in some of its versions, was addressed to another Florentine, the Gonfaloniere, Pier Soderini. All these events took place during the years immediately before and during the compilation of the Codex Leicester, with Florence playing a key role and it has already been proposed that Leonardo's passage dealing with cannibalism practised in some 'islands', which he attributes to the inhabitants of the Canaries (Windsor 19084r), was based on Amerigo's letters.[63]

Some mediaeval philosophers, from Grossatesta to Marsilius of Inghen and others, had claimed, on theoretical ground, that there was less water than earth and had reached a conception close to the later terraqueous globe model. But this was a minority view.[64] However, without excluding a possible influence of these authors—to be demonstrated—evidence in favour of the influence of the new geographical discoveries comes from a comparison between the drawing in the Codex Leicester (*c.*1505–10, see Figure 26) with that in the slightly earlier Manuscript L (*c.*1497–1502; f. 76r, see Figure 27).[65]

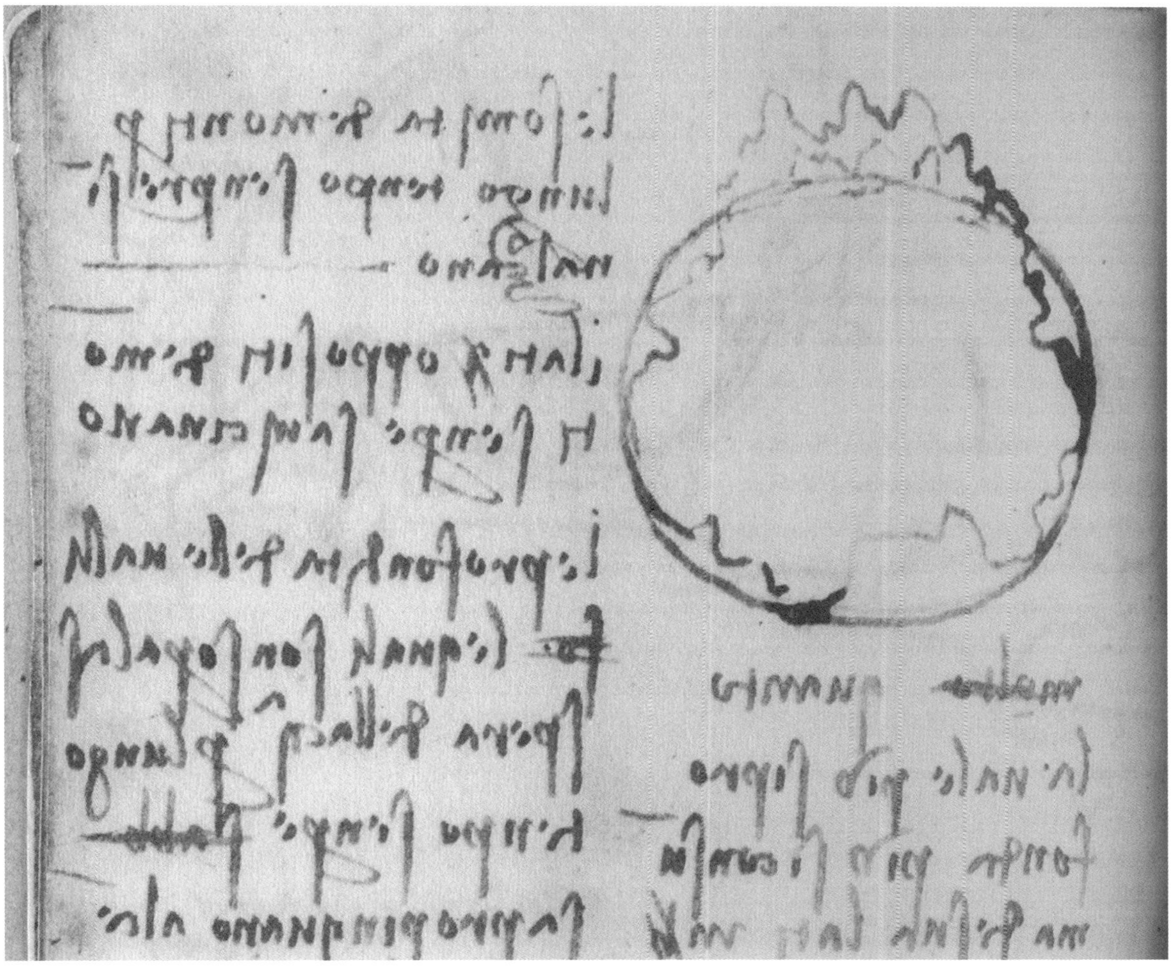

FIG. 27
The Earth. Paris, Institut de France, MS L, f. 76r (detail).

63 Richter 1970a [1883], pp. 130–1, n. 844.

64 Wootton 2016, pp. 117 and ff.

65 See also MS L, f. 13v: 'The surface of the watery sphere is always more remote from the centre of the world [in comparison with dry land in "our hemisphere"].' ('Sempre la superfizie della sfera dell'accqua si fa più remota dal centro del mondo') and compare it with Leicester, f. 35v.

In the earlier image, outside 'our hemisphere' there is only water, and all land lies more or less under the surface of the ocean; while in the later drawing (Figure 26) some mountains are clearly visible above the sea level in the southern hemisphere. Both are rough sketches, but the impression is that in between these two representations of the earth Leonardo became aware of new geographical evidence. The spread of Vespucci's work, *c.*1504, falls in the middle of these two contrasting representations.

Leonardo's figure of the earth, with almost all of its circumference occupied by dry land on folio 36r (Figure 26), may well be one of the first geological representations of the globe to accommodate what was known about the New World. It may also be that Leonardo's geo-historical reconstruction of the origin of dry land was a consequence of the new geographical discoveries. Given the problems about the status of Vespucci's letters and the lack of clarity about their early circulation, adducing their direct impact on Leonardo is difficult to demonstrate conclusively.[66] However, it is unlikely that Leonardo failed to hear about Amerigo's discoveries.

It has been noted that only after the Renaissance explorations of the southern hemisphere, especially along the coast of Brazil (like Vespucci's voyages), and the full rediscovery of Ptolemy's geography, did Ptolemy's *terraqueous globe model*, which had some scarce circulation in the Middle Ages and in the fifteenth century, replace the *two-spheres model* (sphere of water/sphere of earth).[67] Leonardo's figures just examined seem to share this historical shift and in some cases he also tends to consider the centre of gravity of the earth and water as a whole.[68] However, at the same time, he continues to adopt the general framework of the two-spheres model and its static problems (centre of gravity, centre of magnitude, etc.). Indeed, it was by considering the latter issues that he worked out his historical vision of the earth.

Once he had supposed that there was a large amount of water *within* the earth, Leonardo argued that this inner water acted exactly like surface water: being enclosed in irregular subterranean caverns that also included air, it moved, flowed, and cascaded, eroding the inner earth and potentially causing the detachment of subterranean rocks and precipitating massive collapses of subterranean lands. His schematic cross-section of the world suggests that it was a collapse of this nature, but of unusually huge and catastrophic size, that caused the upheaval and origin of dry land at a certain point in the history of the earth, mainly in the northern hemisphere (Figure 29). What happened was that a huge mass of land became detached and collapsed downwards, and, passing through the centre of the earth (occupied by water), reached the lower hemisphere increasing its weight. As a consequence—and in keeping with Buridan's theory about the centre of gravity of the earth, in which its centre of gravity would be sustained at the centre of the world—an equivalent mass of earth emerged in the upper hemisphere and gave origin to dry land and mountains.

66 See, for example, Formisano and Masetti 2007.

67 See Wootton 2016, pp. 110ff. Grant 1994, pp. 635–6, notes that Buridan, *Questiones super libro secundo de caelo et mundo*, Questio VII, 'described an option that is very close to the terraqueous globe, only to reject it'.

68 How the centre of gravity of the water and of the earth joined together is concentric with the centre of the elements. . . . (Leicester, f. 17v)

The centre of the sphere of the water—but not of its weight, because it is not of equal thickness spread over the earth—is always concentric with the centre of the world; and the same happens to the centre of the gravity of the earth and of the waters joined together, but not of [the earth's centre of] gravity nor of its magnitude. (Leicester, f. 35v)

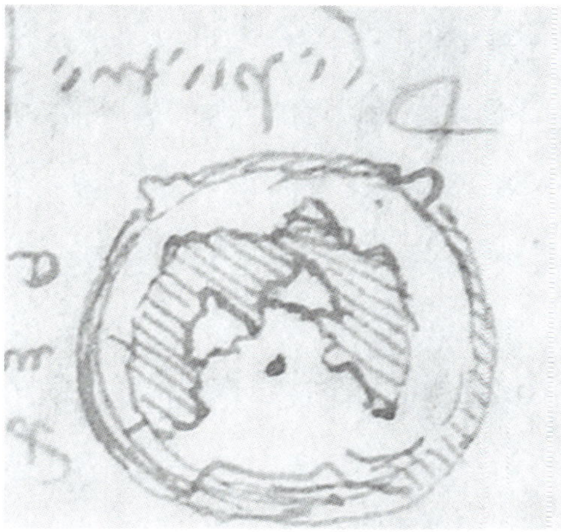

FIG. 28
The Earth.
Collection of
Bill Gates, Seattle,
Codex Leicester,
f. 36r (detail).

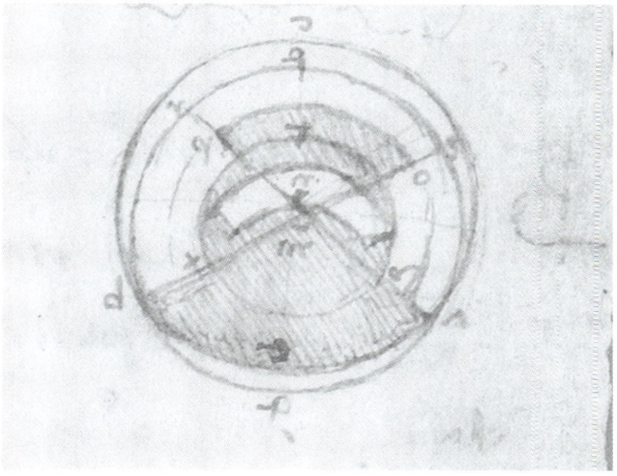

FIG. 29
Schematic figure
of the Earth.
Collection of Bill
Gates, Seattle,
Codex Leicester,
f. 36r (detail).

Most of the figures on f. 36r seem to represent collapses of more limited extent, happening in later eras when dry land already existed. Indeed one of them, the marginal figure right below the upper margin, is accompanied by a note dealing with the hypothesis that the Dead Sea resulted from the collapse of a massive section of the earth's crust. In the figure inserted into the text (Figure 28), we see two triangular masses of land that seem to have recently fallen from the ceiling of a subterranean cavern. They settle around the centre of the earth. Two mountains of approximately the same shape appear to emerge on the surface of the earth.

The first global and catastrophic event, which generated dry land and mountains seems to be represented in the schematic figure in the margin of the same sheet 36r (Figure 29). The figure is labelled with many letters, which are not cited in the text, and therefore the following interpretation is a reconstruction of what he means to say. Perhaps Leonardo copied the figure from another manuscript without the relevant text. The importance of this figure is confirmed by the fact that it is almost the only geological figure by Leonardo to have attracted the attention of historians of geology in recent times.[69]

It differs from other figures on the folio, in that it is schematic rather than pictorial. The schematization is understandable, since Leonardo is representing an event that is not directly visible, as it happened in the very distant past and in the subterranean world. It is the geological counterpart of the famous astronomical image on folio 2r, which, in a similarly schematized manner, tries to reconstruct in optical

69 Gould 1998; Gohau 1990, pp. 34–5; Oldroyd 2006, pp. 19–21.

terms the physical character of the moon, which, with its supposed 'seas', was beyond the immediate access of his direct observations. In both cases they are highly schematized images or 'demonstrations'. In both cases, we now know that his hypotheses are wrong, at least in detail, but they do embody important general truths, namely the earthly and not celestial nature of the moon, and the global transformations of the earth. Like all the later 'theories of the earth' dealing with its very origin, including the modern ones, his was a speculative proposal, but he had striven as far as possible to make it compatible with the available scientific data and theories. He is aware of this when, at the end of the passage, writes, 'and in this case there is nothing which offends rationality'.[70]

Let us look in more detail at the diagram. Leonardo represents the collapse of a huge block of rock in a subterranean cavity. It is represented like an arc with two straight ends, first in a high position (the shaded arc between *o p*), then, after its collapse, around the centre of the earth (in this case not shaded). This downward collapse causes, in accordance with the laws for the balance of weights, the upheaval of land on the surface of 'our' hemisphere of the earth and on the boundary regions between the northern and 'southern' hemispheres. The latter southwards extension might reflect Vespuccian news concerning dry land in the 'southern' hemisphere. The emergent land is represented by the long and curved crescent-shaped area *a c b* along the upper part of the earth. This crescent-shaped area seems to represent, schematically and geometrically, the whole dry land of the upper hemisphere and therefore the subterranean collapse is illustrated as the cause the emergence of dry land.

The image is very subtle in how it embodies theoretical propositions. In addition to the theories of statics by Buridan, Leonardo seems to be applying his geometrical studies of equivalences between rectilinear and curvilinear areas and volumes to these physical and geological phenomena. The visual tools are those typical of his geometrical studies: crescent-shaped areas with curves or straight sides, some shaded, other blank.

In his *Questio de aqua et terra*, Dante describes dry land like a crescent ('*terra emergens habeat figuram qualis est semilunii*'—'dry land has a figure like the crescent'). The schematic figures represent the land more like a spherical cap (Figure 30), while it is the sphere of water outside the earth that is clearly represented as a crescent (Figure 31).

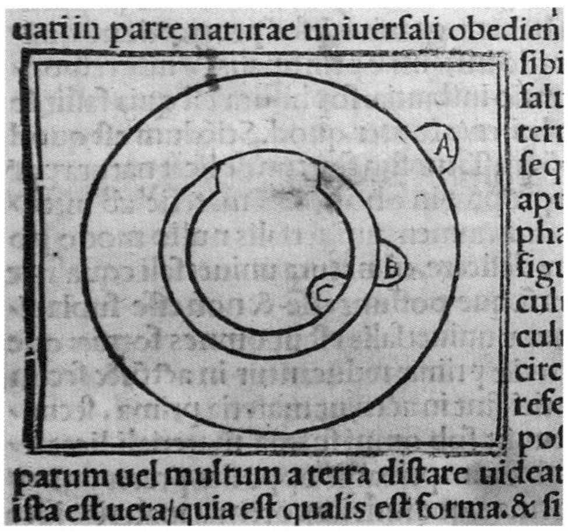

FIG. 30
Schematic image of the spheres of earth and water. From Dante, *Questio de aqua et terra*, Venezia, 1508, f. B(iii)v (detail).

70 Leicester, f. 36r.

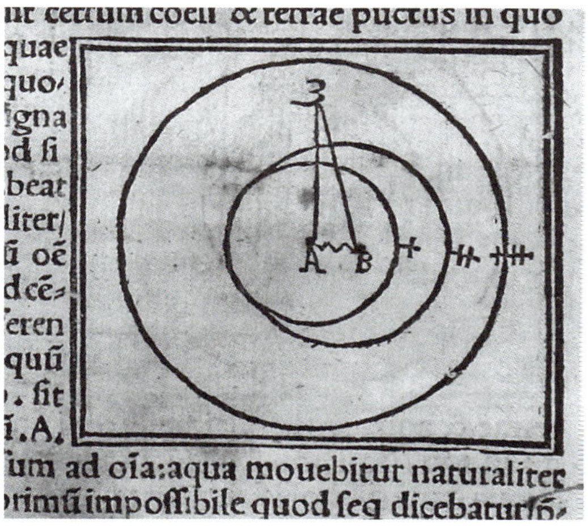

FIG. 31
Schematic image of the spheres of earth and water. From Dante, *Questio de aqua et terra*, Venezia, 1508, f. B(i)r (detail).

This is apparent in the first printed edition of the treatise in 1508, published at the time when Leonardo was compiling the Codex Leicester. The same 1508 memorandum (MS F, first cover verso) that cites Aristotle's *Meteorologica* and the *De coelo et mundo* by Albert of Saxony or Albertus Magnus, includes also the note 'Dante di Niccolò della Croce', which might correspond to the newly published *Questio*.[71] The images in the printed edition of Dante might also have influenced other schematic geological illustrations by Leonardo.[72]

However, even if the figures in the *Questio* by Dante may have influenced Leonardo's geometrical portrayal of the earth, the content of Leonardo's image is very different from Dante's. The geometrical equivalence studied by Leonardo is that between the block that has collapsed within the earth and the mass of dry land that has emerged on its surface. The former, as we have seen, is represented like a blank arc around the centre of the earth. It is symmetrically arranged in relation to this centre because Leonardo supposed that it fell within a cavity filled with water: 'Now this great mass could fall and, given that the centre of the world is within the water, it came to rest with its weight equal and opposite around the centre of the world.'[73] It has only partially fallen into the lower hemisphere. Only the two side portions have crossed the equator (a slight line denoting the diameter of the earth). It is these limited portions that have made the lower hemisphere heavier. The crescent-shaped area *a b c*, representing the dry land that has emerged mainly in the upper hemisphere as a consequence of the subterranean collapse, exhibits an area greater than the two portions. Indeed, as he explains in the text, the upheaval of dry land caused the flow of water toward the lower hemisphere, further weighing it down: 'the earth and the mountains will come out of the sphere of the water on the lightened part, and this will make itself even lighter because of the weight of the water that rested upon it, and it will be elevated so much the more toward the sky'.[74] The crescent-shaped area along the lower margin of the earth (*a d b*) may be a geometrical visualization of the extent of the water that has passed into the lower hemisphere. Consequently, the surface of the upper crescent-shaped area *a c b* is supposed to equal the portions of the arc-block of rocks fallen in the lower hemisphere plus the crescent *a b d* corresponding to the water moved from the upper into the lower hemisphere.

71 See also Solmi 1976 [1908], pp. 128–35.

72 e.g. CA, f. 359r, fourth figure along the left upper margin.

73 Leicester, f. 36r.

74 Leicester, f. 36r.

The meaning of other geometrical areas and equivalences represented in the same figure is not clear. There is a speculative quality to the diagram. Perhaps the large shaded segment in the lower hemisphere represents the extreme depths of the ocean before it became filled in part by sediments generated by the later erosion of the dry land that had emerged in the upper hemisphere. Perhaps the points marked with *n* and *m* represent the dividing of the centre of the world that occurred after the cataclysm: there was at first only one centre of the earth (of gravity and of magnitude), coincident with the centre of the world (point *m*); then, after the subterranean collapse and the emergence of dry land, an asymmetrical situation was generated and the centre of magnitude moved up to point *n*. According to the laws of statics, the centre of gravity *m* must continue to be coincident with the centre of the world. This implies that Leonardo represents here the beginning of the formation of dry land in the upper hemisphere, with the differentiation of the two centres, previously coincident.

He writes that this first upheaval of land may be one of the causes why marine fossils are found on the mountains: 'and this might *yet* be the cause of why the shells and the oysters that are visible on the high mountains, having formerly been in the salt waters, are at present found at such great a height'.[75] The other cause, as we have seen, is more local, namely the lowering of the Mediterranean after the opening of Gibraltar (according to Strabo's classical theory) beside the less catastrophic rising of lands following the passage of sediments and waters from the Mediterranean into the ocean (according to Buridan's theory).

It may be added that according to Leonardo (CA, f. 901r), as a result of the passage of water into the ocean after the opening of Gibraltar (an event described in antiquity by Strabo), the Mediterranean became lighter, and this generated an uplift of lands that contributed to the emergence of the first submerged lands (an event that occurred because of a static-mechanical reason according to the mediaeval theory).[76] The presence of marine fossils in the mountains is therefore not only the consequence of a passive emergence of lands previously submerged (the eighteenth-century 'neptunist' theory), but also the consequence of an active uplift of land on a local scale, in addition to the global, initial, and active emersion of the continents from the sea.

History versus eternalism

The processes proposed by Leonardo do not involve an *eternal cycle*. This leaves open the possibility that the present shape of the earth, with dry land inhabited by man and irregularly emerging from the sphere of water, will have a definite end.

Most of the evidence from his manuscripts indicates that this was his prevailing historical vision, while in only one case does he hint to eternalism, which, by denying any definite beginning and end and any unidirectional development, is the exact opposite of a historical vision.

We have seen how most of folio 36r is devoted to a unique, sudden, and catastrophic event which gave rise to dry land and continents. The schematic drawing representing this event, labelled with letters not cited in the text of 36r, was likely copied from another manuscript devoted to the same subject.

75 Leicester, f. 36r (emphasis added).

76 Discussed above with regard to other aspects. After the opening of the Strait of Gibraltar and the flowing of the Mediterranean waters into the ocean, 'Because of the lack of weight of the diminished Mediterranean waters, the earth was lifted up and changed its centre of gravity' ('Mancato il peso delle diminuite acque mediterrane la terra s'alzò e mutò in sé centro di gravità', CA, f. 901r).

On folio 32v, after a description of the history of the Mediterranean, recounting the phenomenon of erosion and re-emergence of lands, he concludes by claiming that the processes of erosion will prevail over the restorative processes. At a certain moment the erosion of dry land will be so fast and powerful that dry land will no longer be restored by the re-emergence of new lands. The landmass will be flattened and the ocean will submerge it, and in so doing the water will re-appropriate its former location from which it had been expelled:

> [A]nd so it will follow until all the uncovered earth will be carried to the sea by the Nile and by the rivers which pour into it. And so the earth that is located between the rivers that now pour into the Mediterranean will be carried to the ocean by the Nile, together with the turbid waters, which it carries again; *here is how the sea will come back to cover the sites where were formerly the roots and bases of the mountains, and it will cover the earth.*[77]

This global end, affecting the whole earth, is restated in two texts from Manuscript F, dating back to the same years. One text reads: 'It follows that the earth will become spherical and wholly covered by the waters and it will be uninhabitable',[78] the other is a quote from Albert of Saxony.[79]

The end of the history of the earth, at least in its present form, will therefore involve the return of the earth to its previously submerged state, a subject to which he plans to dedicate a whole book within the planned treatise on waters: 'book on how mountains will be flattened and on how the earth in our hemisphere will be covered by water'.[80] This book, if ever written, has not survived. The 'outer set' of sheets of the Codex Leicester represents his clearer surviving exposition of the history of earth and neither it nor the other mentioned texts suggest a new and cyclical re-emergence of dry land.[81]

There is also an early hint, dating back to *c.*1478–80, of the hypothesis of an end of the earth and of man, with the water all enclosed within the earth and the surface of the latter 'left burnt up to cinder and this will be the end of all terrestrial nature'.[82] However, an end by submersion prevailed later on. This had already been considered by Albert of Saxony (and, of course, Seneca). In Manuscript F (f. 84r), Leonardo mentions a Latin passage by Albert of Saxony in this regard and gives an Italian paraphrase.[83] However, Albert of Saxony introduces the hypothesis only to reject it on theological ground: the asymmetry of the earth, with dry land area and mountains on one side was created by God to make the world a living place for animals and plants; and it will therefore remain so over time. The impression is that Leonardo develops, in successive stages and in an original, although not conclusive and unambiguous way, the most advanced and potentially heretical ideas of the mediaeval tradition.

77 Leicester, f. 32v (emphasis added).

78 MS F, f. 52v.

79 MS F, f. 84v. See below.

80 Arundel, f. 35v.

81 On the basis of MS G, f. 49r (see below), Fehrenbach 2015, p. 84, speculates about the possibility that Leonardo envisioned a new uplift of dry lands. See also Dal Prete 2014, p. 302–3, 2018 on Leonardo as an eternalist thinker.

82 Arundel, f. 155v: 'Allora la sua superfice rimarà in riarsa cienere, e questo fia il termine della terestre natura.'

83 Ogni grave attende al basso e le cose alte non resteran in loro altezza, ma col tempo tutte discenderanno; e così col tempo il mondo resterà sperico e per conseguenza fia tutto coperto dell'acqua, e le sue vene sotterranee resteranno immobili. [...] Omne grave tendit deorsum nec perpetuo potest sic sursum sustineri, qua re iam totalis terra esset facta sperica.

See Duhem 1913–59, vol. 9, p. 311, 1991, p. 305; Solmi 1976 [1908], p. 56; Ellenberger 1988, p. 99; Feherenbach 2015.

And this is also true for the eternalist hypothesis, which, being a heretical idea, was presented, for example by Buridan, as a purely speculative and scientific hypothesis.

In Manuscript G, during the course of an elaborate discussion of cycles of salt conservation in the earth, he writes: 'given that the world is eternal, it must be admitted that its population will also be eternal'.[84] This is perhaps the only mention of the eternalist hypothesis in Leonardo's manuscripts and seems specific to the issue considered. In any case, the passage dates back to the last years of his life, when even in his anatomical and embryological research he touched dangerous issues from a theological point of view.[85] It is perhaps no coincidence that next to the famous drawing of the foetus immersed in its bath of fluids in the uterus (Windsor 19102r) is visible a sketch that studies the static equilibrium of a sphere suspended on an inclined plane thanks to an eccentric weight.[86] The old analogy between microcosm and macrocosm, body of man and the body of the earth, contributed to creating the historical side of Leonardo's geological theory, according to which, the earth, just as the body of man, underwent a development with a beginning and an end.[87]

Towards 'deep time': Nature and man

An unexpected consequence of Leonardo's hypothesis of the first and global emersion of dry land from the waters—apparently unrealized by himself—is that fish may have been in existence before the emergence of dry land and the creation of 'Man' (which is consistent with Genesis, according to which fishes were created earlier than man).

When in the Codex Leicester, on folio 9v, Leonardo reports about the finding in inland Tuscany and Lombardy of the remains of ancient ships (actually fossil wood?), besides the fossils of marine animals, he obviously is considering changes to the earth that occurred when man already inhabited it. Local changes, like the lowering of the Mediterranean after the opening of the Strait of Gibraltar, happened later than the first emergence of dry land. These later changes were compatible—within the geological data available to Leonardo and his age—with the finding of these traces of ancient human existence.

However, in another passage Leonardo seems to consider the possibility that marine fossils might also date to eras during which man was not yet present:

> *Since things are much more ancient than letters*, it is no marvel if in our days no written record exists of these seas having occupied so many countries; and even if, moreover, any written record existed, wars, fires, the changes in languages and laws, the floods of water have consumed every ancient thing; but for us there is sufficient testimony in the things generated in salt waters and found again in the high mountains, sometimes far from the seas.[88]

This is an ambiguous statement, since the lack of written documents does not mean the lack of men. Absence of evidence is not evidence of absence. Nevertheless, Leonardo seems aware of a sharp discrepancy between 'human' and 'natural' times, between the relative shortness of the history of man and

84 MS G, f. 49r: 'essendo dato il mondo etterno, egli è necessario che li sua popoli sieno ancora loro etterni'.

85 Laurenza 2004.

86 See Kemp 2000, pp. 20–1.

87 On the ageing processes in man and the earth, see Laurenza 2001, pp. 127–60, esp. 155–60.

88 Leicester, f. 31r (emphasis added).

long duration of the history of the earth. In a passage in the Codex Atlanticus (f. 28or), he generically writes that global changes of the earth occurred 'in many thousands of years'. On the cover of another codex (MS E, first cover verso), a Latin note not in Leonardo's hand—perhaps by his pupil Francesco Melzi—claims that astrology was invented fifty-seven thousand years before the war of Troy. The number, quoted in words, is emphatically repeated in large numbers: '57000'. This duration of time is far longer than that of a few thousand years (in the range of 5000–7000 years), the standard Biblical estimate.[89]

Aristotle had already considered the lack of written documents concerning changes of the earth.[90] During the Middle Ages, some philosophers, most notably Buridan, had proposed hugely extended times for these geological changes, even as many as 'one hundred million years'. But, in all these cases, the hypothesis was neutralized by their ahistorical conception, according to which there were cyclic changes in an eternal time. In the traditional conception of time, the basic nature of the earth was essentially the same from immemorial time, and no passing geological event exercised a conclusive impact on its essential form.

Leonardo, albeit in an unresolved way,[91] seems to have at least considered what will be later called 'deep time', an idea shared by other natural philosophers of his time.[92] On the other hand, there is no doubt about his radical historical vision of the earth.

An Anatomy of the Earth

What we find in the Codex Leicester is a powerful analysis of the solid components of earth (rocks, strata, stones, soil, etc.) and of how they are modified by their interaction with the water. Leonardo studies *the earth in itself*. In Aristotle's *Meteorologica* and in part of the later *meteorological* tradition, the study of the earth was dominated by a different vision, in which the earth was the passive scene of the actions of two fluid exhalations generated from water and earth by the heat of the sun.

Leonardo studies the static and dynamic forms of the two denser elements: earth and water. His studies include: the general form of mountains and seas; the strata of rocks; the form of particulate materials (stones, gravel, sand, silt, etc.); the inner form of the earth (caverns, veins of water or of hot materials); the varied forms of the sandy beds of the rivers and of the sea; the connection between the form of surface water and the configuration of the bottoms; and the form of the banks of the river eroded by water. He is also fully alert to the many different configurations of water in motion, attempting to classify the myriad forms of waves and turbulence.

The ancient analogy between the body of the earth (macrocosm) and the body of man (microcosm) is an integral part of this vision: rocks are like the bones; softer rocks and soil are like cartilage and flesh; subterranean veins of water and of fire are like the veins (blood) and the arteries (spirits and natural

89 On chronologists, see Rudwick 2014, pp. 9ff.

90 Aristotle, *Meteorologica*, Bk I, ch. 14, 351b10: 'But these changes escape our observation because the whole process takes place by slow degrees and over periods of time which are vast compared to the length of our life; and whole peoples are destroyed and perish before they can record the process from beginning to end.'

91 Compare also Zubov 1968, p. 228, on Mona Lisa's landscape evoking 'prehuman geological past', with Zubov 1968, p. 226, on the difference from later geological theories.

92 Dal Prete 2014.

heat) in the animal body.[93] Later on he will cast doubt on some aspects of this analogy, and will note that the veins in the body of the earth are eroded and therefore enlarged by the water while the vessels in the human body are progressively occluded over time.[94] Nevertheless, in the Codex the analogy is still fully accepted and, indeed, is presented in a poetic guise in the famous passage, which includes the presence of a vegetative soul, arising from volcanic heat (f. 34r).

Leonardo's anatomical study of the earth, presented as if it were an animal body, with solid and fluid components, warmed in the earth's case by subterranean volcanic forces, is strictly connected to his study of the history of the earth, because the analogy requires that the body of the earth undergoes processes of generation, growth, and decay, with a beginning and an end, just like animal bodies. The ancient analogy of the microcosm–macrocosm contributed to the formulation of an idea that led to later developments.

Even at the end of eighteenth century, the anatomical dimension was to play a role in the fundamental discovery of the geological cycle of aqueous erosion of mountains and volcanic reconstruction. The chief proponent of the theory, the Scottish physician James Hutton (1726–97) compared the process in the earth to the circulation of blood discovered in the previous century by William Harvey. This circulation had been the subject of Hutton's medical doctorate. Hutton, like Leonardo, was not a professional geologist, since geology, in these early periods, did not exist as an independent and professional science.[95] It is also worth noting that the ideas of Harvey were themselves permeated by micro-macrocosmic analogies, and he adduced canals and locks when discussing the flow of blood through valves.

However, beside the mentioned analogy, Leonardo's deep anatomical research certainly influenced his anatomical study of the earth. In the very same years as the Codex Leicester, Leonardo undertook the dissections of a centenarian and of a child two years old, not least to understand how anatomical forms and functions changed over time (*c.*1507–8). Also his famous embryological studies were initiated in this same period. His studies of the time and change in the body of man, of his generation and decay are locked into his thinking about the life of the body of the earth.[96]

It is not the case that this connection between geology and anatomy is absent from Aristotle's thinking. Aristotle was the first great anatomist, writing fundamental anatomical treatises based on animal dissections. However in his geological thinking, anatomy, in the sense of morphological study of solid structures, plays only a secondary role. This is understandable because the earth and the world according to Aristotle are eternal, and therefore are fundamentally different from the animal body. According to Aristotle, changes in the earth are only local and therefore affect only its *parts* and not its *whole body*.[97] Thus the study of the parts, which gives its name and is the subject of his most important anatomical work (*De partibus animalium—On the Parts of the Animals*), is of secondary importance in his study of the earth.[98]

93 Zubov 1968, pp. 190–5, 226–51; Kemp 2006a, pp. 257–8 and 311–14; Laurenza 2001, pp. 61–4 and 155–60. See also Chapter 3 in this volume.

94 MS G, f. 49v and MS F, f. 1r.

95 Oldroyd 1996; Donovan and Prentiss 1980. On geology and medicine, see Duffin and Gardner-Thorpe 2013.

96 See Laurenza forthcoming c.

97 Aristotle, *Meteorologica*, Bk I, ch. 14, 351a20-30, 351b5, and 352b20.

98 Use of physiological and anatomical analogies in Aristotle's explanation of the earth and cosmos does not imply that he thinks of the earth and cosmos itself as an animal. See Taub 2012, p. 48.

Anatomy was, in Leonardo's age, a vibrant and developing science, whereas the wholesale reform of geology only really began much later—not before the middle of seventeenth century. It is therefore likely that the anatomical model influenced Leonardo's geological vision, not vice versa.

The difference between Leonardo's geological vision and previous conceptions seems not dissimilar to the transition from humoral medicine to anatomical medicine. Humoral medicine was dominated by the fluid and qualitative notion of the four humours, while anatomical medicine relied on the form and function of the organs and hard anatomical parts as determined by dissection, a practice that acquired fundamental importance in the Renaissance.

Aristotle's *meteorological* and geological vision was based on the physics of the qualities (heat, cold, humidity, dryness) variously possessed by the four elements of which natural bodies were composed. It was a vision dominated by the more fluid elements: water and above all air. Thanks to the heat of the sun—a heavenly body—two exhalations were produced from water and earth: a cold and moist exhalation (like vapour) and a warm and dry exhalation (like smoke). The earth, as a planet and as a mass of solid material, was basically the vehicle for these two exhalations and for their cyclic reconversion into water, air, earth, and fire. Stones, metals, and soils were studied less for their forms and more for those qualities that connected their generation with one or another of the two exhalations. For example, dry exhalation prevailed in the generation of stones and soil, while moist exhalation prevailed in the genesis of the metals. Earthquakes and volcanoes were the final outcome of dry exhalation, which rose within the earth, venting outwards to join the sphere of the air. Once it had exited, the part of this exhalation that was most subject to cold, was re-converted into water and fell back and down on the earth as rain. Floods and droughts were the consequences of greater or lesser re-conversions of exhalations in the form of precipitated water.[99]

Even in Avicenna's *De mineralibus*, a text belonging to the *meteorological* tradition and, no doubt, a useful term of comparison for Leonardo's later approach, for example, the attention given to erosive processes, the study of the formation of small and big stones, of mountains and minerals, is dominated by qualitative and elemental notions, such as hot/cold virtues, congealing-petrifying virtues, and so on.[100] Small and big stones, cliffs, heights, and mountains are studied rather independently from one another, and not as interconnected parts of a whole body sharing the same history. Furthermore, consistent with the meteorological tradition, rocky strata in mountains were the production of periodic ebbing of the sea.[101] It is possible that when Leonardo writes, 'Have Avicenna on the remedies translated'[102] he means Avicenna's book *Kitab al-Shifa* (*Book of the Remedy*, or *Book of Healing*), a philosophical encyclopaedia including sections on geology with an influential theory of petrification, which was translated into Latin by Alfred of Sareshel in the twelfth century as *De mineralibus*.[103] However, while

99 See Leicester, ff. 26v [9] and 27r [1] for texts apparently against the notion of dry exhalation.

100 e.g. '. . . mountains have been formed . . . from agglutinative clay which slowly dried and petrified'. The habitable world 'may have petrified beneath the waters by reason of the intense heat confined under the sea' (see Avicenna 1927, p. 28).

101 See Avicenna 1927, p. 31: 'It is possible that each time the land was exposed by the ebbing of the sea a layer was left . . .'. Avicenna's theory also lacks any active uplift of the mountains and besides a very limited role of earthquakes in the formation of cliffs and heights (p. 27) and the assumption that the habitable world was once 'submerged beneath the ocean' (p. 28), the elevation of the mountains was only the passive consequence of erosion: 'their elevation is due to the excavating action of floods and winds' (p. 29).

102 Windsor 19070v: 'Fa tradurre Avicenna de giovamenti'.

103 Avicenna 1927, p. 8; Bertolacci 2013; Solmi 1976 [1908], p. 80; Dal Prete 2018.

Avicenna's theory of petrification was perhaps influential on Leonardo's study of fossils,[104] less so was his basically elemental and qualitative approach to the study of the earth.

In the Codex Leicester, we witness Leonardo's move from this kind of elemental and qualitative geology to an anatomical geology, interested above all in the solid forms of the earth and in their mechanical interaction with water. As already noticed, this parallels the contemporary shift from humoral medicine to anatomical medicine, above all in surgery. In medical practice the humoral conception was to persist long after this period, above all in the practice of physicians. Even in 'physic', the tendency to cure the single sick organ rather than attending to the whole body (a principal mark of modern medicine) has its origin in the Renaissance renewal of anatomy. By contrast, the renewal of geology, based on a historical and anatomical conception, does not really occur on a substantial scale until much later. While in his original anatomical studies Leonardo joined the on-going revolution in anatomical science, his new approach to geology, at least as far as his *historical* and *anatomical* aspects are concerned, is a more isolated event, individual and subject to limits, as Leonardo himself may have been aware.

He shows some awareness, as with his studies of the moon, that visual imagination sometimes has to act as a surrogate for tangible evidence. He writes:

> Here one has *to imagine* the earth *sawn* through the middle and there will be seen the depths of the sea and of the earth and how the veins start from the bottoms of the seas and interweave in the earth and rise to the summit of the mountains and pour again through the rivers and return to the sea.[105]

Leonardo is aware of two things. Firstly, understanding the earth needs a procedure like the dissection of its body in order to discover its inner structure (thus the earth is '*sawn*' through). Secondly, this study, compared to that of human and animal bodies, has real limits, since the deep structure of the earth can only be studied in indirect ways, and has to be 'imagined' (thus, 'here one has *to imagine*'). Even though he starts from empirical observations of those portions of the earth that are directly visible, he necessarily has to pass into more speculative realms—much as Harvey needed to do in his positing of unseen channels that complete the cycle of the circulation of blood. Perhaps Leonardo's awareness of this limitation helps us understand why the Codex Leicester contains a few, mainly schematic images in comparison to his anatomical manuscripts. He could directly portray the veins of the human body, while he could only suggest those in the body of the earth, as he did in the small, sketchy figure that accompanies the quoted passage and represents a section of the earth.

We will now examine Leonardo's anatomical approach to the study of the earth comparing it to Aristotle's different vision in three fields: the general form of the earth; the dynamic processes that change its overall form; and the generation of its component parts.

Within the body of the earth: A self-sufficient system

We have seen how at least two pages of the Codex (ff. 36r and 35v) are almost entirely devoted to the demonstration of the existence within the body of the earth of huge subterranean cavities filled with water. In addition to these caverns, the subterranean structure of the earth includes also conduits con-

104 See Chapter 5 in this volume.

105 Leicester, f. 31r (emphasis added).

veying water and hot material. Leonardo, like the classical philosopher Seneca (4 BCE–69 CE; *Naturales Quaestiones*, III.15), accords these conduits an anatomical name, 'veins'. These veins originate from the sea and, after a subterranean itinerary, open on the mountains giving rise to the rivers, which bring back the water to the sea:

> The body of the earth like the bodies of animals is interwoven with ramifications of veins, which are all joined together and are formed for the nutrition and vivification of this earth and of its creatures; and they depart from the depths of the sea and to them, after many revolutions, have to return through the rivers created by the high rupture of these veins.[106]

This subterranean anatomy is the focus of his discussion of various theories about the origin of the water in mountain springs.[107] He reviews various alternative origins: from rain; the result of the rising up of water through the earth as in a sponge or siphon; the consequence of some areas of the sea rising higher than the earth; internal evaporation and condensation of water as in distillation; active circulation from sea to veins to mountain springs to rivers and to the sea.[108] Almost all these theories are discussed by Aristotle (*Meteorologica*, Bk I, ch. 13) and by Seneca (*Naturales Quaestiones*, Bk III, chs 5–15). Aristotle favours the more thermal and fluid model of the evaporation and condensation of water, which Leonardo equates with the actions of a still. Just as outside the earth, the cold condenses the air into clouds and rain, so the subterranean air condenses in small drops of water, which then unify to form greater amounts of water, finally spouting forth as mountain springs.

Leonardo favours a more mechanistic and overtly anatomical model involving the active circulation of water from sea to rivers via subterranean veins. He writes: 'How all the living veins, ramified and woven into the body of the earth, are joined with the bed of the sea. How by the breaking of the veins, by means of the rivers, the water is given back to the sea.'[109] A drawing on folio 3v represents the macro-circulation of water in the earth: veins originate from the bed of the sea on the left and rise within the mountain to the right, ascending to their final exit. In the small drawing on 31r, showing the section of the earth, the veins are represented within the whole body of the earth.

Later he was to deny the marine origin of mountain springs and therefore the macro-circulation, and favour a meteorological theory: 'The water of the rivers does not originate from the sea but from the clouds.'[110] At the time of the Codex Leicester, he still accepted the idea of the macro-circulation and, independently of whether it was right or wrong, it served to help Leonardo to pass from a fluid and vaporous conception of the body of the earth to a more solid and anatomical one.

Similarly, when the theory of the macro-circulation of water re-emerges toward the middle of the seventeenth century in Descartes[111] and in Athanasius Kircher,[112] it will play a key role in promoting new geological theories.

106 Leicester, f. 33v.

107 Leicester, ff. 3v, 3r, 34r, 32v.

108 Kemp 1982.

109 Leicester, f. 11v; see also Leicester, ff. 21v, 31r, 3v.

110 CA, f. 433r. See also CA, f. 508v; MS F, f. 72v; and MS G, f. 48v.

111 Descartes, *Opera philosophica: Principia philosophiae*, Amsterdam, 1650, pp. 216–17.

112 *Mundus subterraneus*, Roma, 1654–5.

There were of course numerous theories in the ancient world that supposed the existence of large spaces within the earth. Plato, for example, describes an underworld where the souls of dead people are judged.[113] Seneca within his general analogy of what happens within and over the earth,[114] and Lucretius allude to subterranean caverns, veins, and rivers. However, both in Seneca and Lucretius, the general explanatory model still relies predominantly on the fluid elements. Seneca explains the origin of water within the earth and mountain springs through the theory of the transformation of the four elements into each other. Thus earth is turned into water.[115] According to Lucretius, the main subterranean agent is the wind, which explains both earthquakes and volcanic phenomena.[116] In the Middle Ages, Albertus Magnus explains the mountain springs through the subterranean vapours and the heat of the sun.[117]

The inner anatomical structure of the earth provides a fundamental explanatory model for Leonardo not least because it operates as a self-sufficient system. Indeed, according to Leonardo, the mover of the macro-circulation of water is not the heat of the sun, as he had earlier supposed (MS A, f. 56r), but the subterranean and intrinsic heat of the earth, the inner 'fires' which manifest themselves in volcanoes. Indeed a note to the drawing representing the water circulation reads 'sea level exhaled by the fires which are located in the centre or body of the earth'.[118] The Aristotelian tradition located the main motive power outside the earth, namely in the sun, which with its cyclic movements distributed varying degrees of heat across the earth. Geology in this tradition was effectively part of cosmology, and the whole cosmological conception acquired an astrological dimension in the middle Ages.

The shift from an external and cosmological origin for the motion of water to an internal and geological origin, in which heat drives an internal circulation, is perhaps the most remarkable aspect of the limited volcanological aspects of Leonardo's research. His personal opportunities to develop his theories in this respect were hampered by the lack of occasions on which he could have observed volcanic phenomena, other than smaller and local geo-thermal events. On folio 28r, Leonardo's parallel descriptions of how earthquakes are generated within the earth by subterranean 'winds' and of how wind and clouds are generated in the atmosphere still follows the Aristotelian analogy between phenomena within and over the earth. It was only with the volcanic upheaval of Montenuovo near Pozzuoli in 1538 and above all the great eruption of Vesuvius in 1631 that volcanic activity slowly moved to the centre of geological studies.

Processes of erosion and repair

A majority of the forms of earth and water are explained by Leonardo as a consequence of changes generated by their dynamic interaction.[119] To be sure, these changes are important for Aristotle, but they are restricted mostly to the more fluid elements of air and water. Cyclic and dynamic exchanges occur across the four spheres, thanks to which they penetrate each other and permeate them. The

113 Plato, *Phaedo*, ed. D. Gallop, Oxford, 1975, 111a–e and 112b–e.

114 Seneca, *Naturales Quaestiones*, Bk III, chs 15–16, Bk VI, ch. 7.

115 Seneca, *Naturales Quaestiones*, Bk III, chs 9–10.

116 Lucretius, *De rerum natura*, Bk VI, lines 536–705. Similarly in Seneca's *Naturales Quaestiones*, Bk VI, ch. 21.

117 *De Meteoris*, Book II, tract II, ch. XII (see the edition published in Venice, 1488).

118 Leicester, f. 3v.

119 On some aspects of this interaction, see Alexander 1982. See also Chapter 3 in this volume.

sphere of air is composed of hot and dry exhalations above, towards the sphere of fire, and of moist and colder exhalation below, towards the sphere of water.[120] The moist exhalation forms the upper part of the sphere of water (where clouds form), while its lower part is formed by the sea, which contains an earthly material (salt) and is the more earthly part of the sphere of water.[121] Continuous, vertical, and circular movements cause these spheres to interact amongst each other. The water of the sea evaporates and rises because of the heat and is then condensed in clouds and falls as rain because of the following dispersion of the heat.[122] As far the earth is concerned, earthquakes are generated by subterranean hot air that needs to rise and exit. Additionally, the cyclical changes of submerged regions of land and water on the earth's surface are the consequence of the cyclic succession of dry-hot macro-seasons by humid-cold ones. The macro-seasons are thus the consequence of greater and lesser transformations of water within the air.[123]

Leonardo's vision is distinctively different. Geological changes involve the interaction of the two denser elements—earth and water—and therefore involve predominantly mechanical processes. On one side there are erosive processes of destruction caused by 'friction' (*confregazione*) and 'percussion'. Rivers excavate valleys and, in the fullness of time, all the mountains will be eroded. These processes of decay are counterbalanced by opposite processes, again of a mechanical nature. The largest is the upheaval of lands resulting from the re-balancing of the two hemispheres, a process that restores from below what is eroded by the waters above (Figure 25; see also p. 100 n. 76). These processes, in the first instance, are self-sufficient and need only a minimal external cause, in as much as the heat of the sun contributes to the erosion of the upper rocks in the mountains. Erosion is the cause adduced by Leonardo to explain the opening of the Strait of Gibraltar and the subsequent lowering of the Mediterranean (f. 31r). Erosion can of course be directly studied in the present earth, where it is readily visible in the banks of the river, especially when the water 'carries much matter in its fury, such as sand and gravel, which strike, scrape and consume continuously any obstacles to them'.[124] Erosion is also readily visible at the coasts of the sea (f. 35r), and the exposed strata on either side of the cuttings made by rivers vividly attest to erosion by water.

Furthermore, the saltiness of the sea, according to Leonardo, is the result of erosion by the marine waters of salt mines in and under the bed of the sea and not, as thought by Aristotle and Pliny (the latter directly quoted by Leonardo) the result of the exhalation produced by the sun in the earth, which then falls back into the sea as if it were salt rain.[125]

Stones, gravel, sand, silt, clay, tufa. . . .

Leonardo classifies rocks, stones, and sediments according to their form and structure in relation to their origins from the body of the earth. Within the traditional approach, in the so-called lapidaries or

120 Aristotle, *Meteorologica*, Bk. I, ch. 7.

121 Aristotle, *Meteorologica*, Bk. I, ch. 9.

122 Aristotle, *Meteorologica*, Bk. I, ch. 9.

123 Aristotle, *Meteorologica*, Bk. I, ch. 14.

124 Leicester, f. 22r.

125 Leicester, f. 11v: 'Perchè l'acqua è salsa. Dicie Plinio . . . che l'acqua del mare è salata perchè l'ardore del sole secca l'umido. . . . Ancora fu detto che la salsedine del mare è sudore della terra [the latter is Aristotle's theory]'. See also MS G, ff. 49r–v. Cf. Pliny, *Natural History*, Bk II, chs 103, 222 (Pliny, *Natural History with and English Translation in Ten Volumes*. Volume 1: *Praefatio, Libri I, II*, ed. H. Rackham, London and Cambridge, 1959).

petrologic books, including Pliny's sections on stones and minerals in the *Natural History*, they were mainly studied in themselves, independently from their places of origin, and mainly for their qualities (moist/dry/ hot/cold) and for their uses (in masonry, in art, in medicine, in metallurgy).

Leonardo elaborates a granular classification according size and form, although he does not specify actual sizes for the different types. His succession runs in a continuous series from stones (with more or less smooth angles), through gravel and sand to silt (f. 6v). Clay (ff. 3r, 35r) is sometimes seen as synonymous with silt and, like silt is made of sediments not yet petrified or solidified. It is defined as 'fat' ('*grassa*', f. 3r), because it is relatively impermeable to water, just as silt is characterized by greasiness and by its lubricated nature ('*lubricità*', f. 6v). Silt may also contain organic fragments. Tufa ('*tufi*') is a conglomerate midway between clay and rocks with respect to its degree of petrification. Perhaps referring to Pliny, who defined *tufa* as 'tender stones',[126] Leonardo compares them to the 'tender' or malleable anatomical components in an animal body ('*tenerume*'), whereas hard rocks correspond to the bones (f. 34v). Tufa is also associated with conglomerates of fine sand, while stones are associated with conglomerates of thicker gravel (f. 31v).

Apparently less sophisticated than traditional classifications, the originality of his system consists in its association with the places and the mechanisms of origin for the various materials. Weathering, erosion, and transport serve to generate various forms of stones and grains in various stages: stones are found mainly toward the beginning of the river; then comes gravel and, when the river flows into the sea, sand and silt, whose minute particles are mingled with the water:

> The river, coming out from the mountains, deposits a great quantity of coarse stones in its gravel bed, which still retain some part of their angles and sides; and in proceeding along its course, it carries smaller pebbles, with more eroded angles, that is, it renders the big stones smaller; and further on it deposits coarse and then fine gravel, and then follows coarse and then finer sand; and then follows coarse and then more refined silt; and thus proceeding, the water turbid with sand and silt reaches the sea.[127]

Leonardo's stratigraphy

Silt, the final product of erosion, is deposited on the bed of the Mediterranean Sea. When Leonardo writes that more or less thick horizontal strata reflect greater or lesser events of deposition of sediments by the rivers into the sea (e.g. Leicester 10r, 2; see also Chapter 5), it seems to refer to more local depositions and their later unveiling within 'our' hemisphere, both passively as a result of the opening of Gibraltar, or as the result of an active upward motion of lands following the same event (see pp. 79–81 and 100 n. 76). However, the end of a text dealing with the strata seems to imply a cyclical passage of sediments between the two hemispheres, according to Buridan theory: 'whence we conclude that every portion of the surface of the earth was once at the centre of the earth, and vice versa' (Arundel, f. 138r; see also MS E, f. 4v, discussed in Chapter 5).

Because of this upwards movement, the upper strata in the mountains are older than the lower strata (f. 10r). Leonardo undertook 'field research' to confirm this mechanism, and distinguishes three

126 Pliny, *Natural History*, Bk XXXVI, ch 48, 166–7 (Pliny, *Natural History with and English Translation in Ten Volumes. Volume 10: Libri XXXVI–XXXVII*, ed. E. D. Echholz, London and Cambridge, 1962).

127 Leicester, f. 6v. See also Leicester, f. 19v and MS I, ff. 60r and 84r with drawings.

strata or '*falde*': lower strata made of clay not yet petrified; intermediate strata made of tufa with an intermediate degree of solidification; and upper strata composed by rocks completely solidified. Leonardo studied this stratigraphy in Tuscany, the Alps, and elsewhere: 'How the northern footings of every Alp are not yet petrified' and 'are all of a soil suitable to make pots.'[128] In these lower strata marine fossils are often present: 'here are located the shells and this is the white earth used to make jugs' (f. 6v, reaffirmed in folio 36r, paragraph 5).

The complete stratigraphy is summarized in the Codex Arundel (f. 29v): 'In many mountains rocks do not extend to their roots; and, following this same order of the strata [i.e. from above to below], they are replaced by tufa and under the tufa is silt.'

Stones and rocks are no longer studied exclusively in themselves but as part of a wider geological study of the body of the earth. Sediments originally deposited in horizontal strata, of the kind represented in folio 36r (lower left corner; Figure 32), are later on subjected to the huge upward forces of pressure studied in the same sheet, which lift and deform their structure as shown in a famous Windsor drawing (12394, Figure 33).

Georgius Agricola's later stratigraphy[129] strictly served his concerns with the mining and the excavation of metal veins. We have to wait well into the seventeenth century to Nicolas Steno for providing a clear link between stratigraphy, geology, and the history of the earth.[130] Steno, like Leonardo, distinguishes strata that are mainly sandy from strata that are mainly rocky. In his famous schematic visualization of the history of Tuscany (supposed by him to be representative of the whole earth), Steno represents the former with dotted lines, the latter with solid lines (Figure 34).

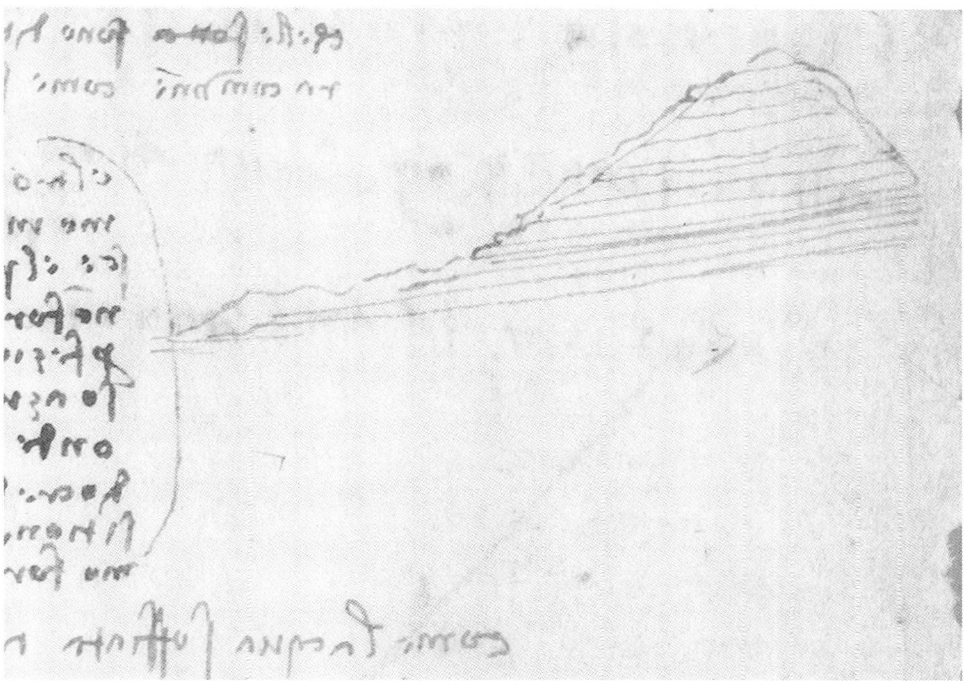

FIG. 32
Stratigraphic study. Collection of Bill Gates, Seattle, Codex Leicester, f. 36r (detail).

128 Leicester, f. 10r. See Cioppi and Dominici 2015; Vai 2003b.

129 *De re metallica*, Basilea, 1556, Book V.

130 Nicolas Steno, *De solido intra solidum naturaliter contento dissertationis prodromus*, Florence, 1669.

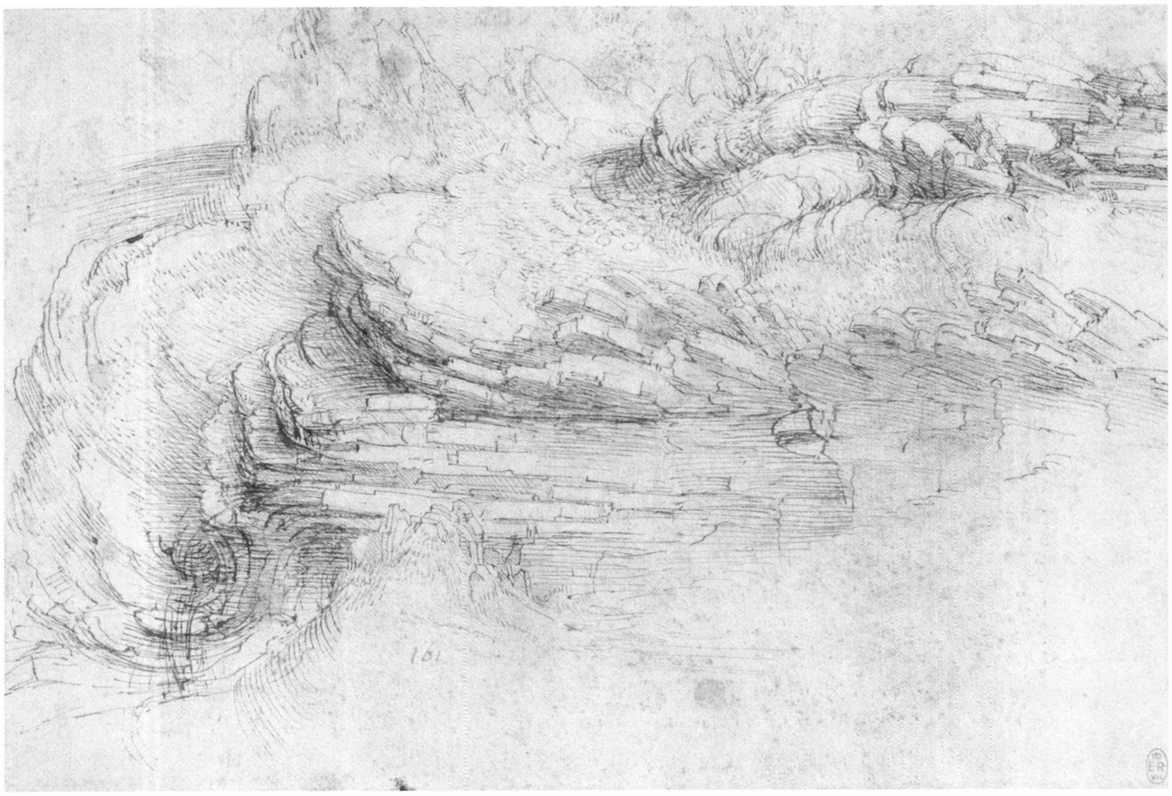

FIG. 33
Stratigraphic study. Royal Library, Windsor 12394.

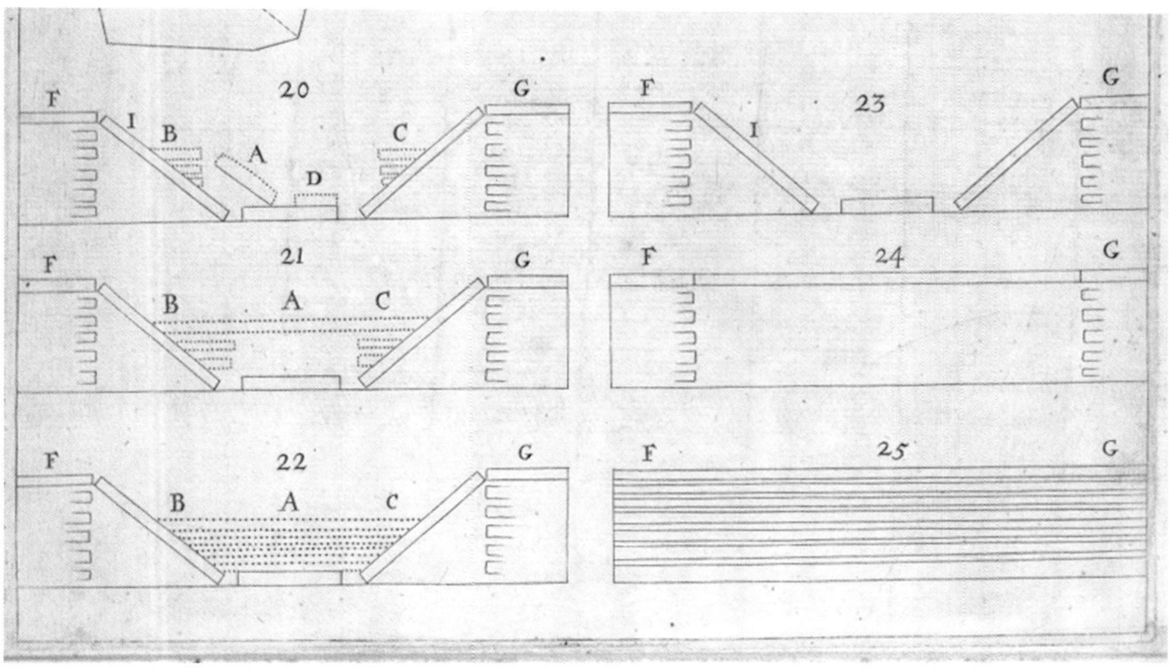

FIG. 34
Stratigraphic study. From Nicolas Steno, *De solido intra solidum naturaliter contento dissertationis prodromus*, Florence, 1669, final plate (detail).

The illustrated plate represents in six figures six successive phases of the earth: the original state of the earth covered with water and the formation of various sedimentary layers (Figure 34, no. 25); underground erosion of these layers (Figure 34, no. 24); their consequent collapses, accompanied by the formation of mountains and valleys (Figure 34, no. 23); a new submersion of mountains and valleys, coinciding with the Biblical Deluge and the subsequent formation of new sedimentary strata

(Figure 34, no. 22); their subterranean erosion (Figure 34, no. 21); and their subsequent collapse with formation of new mountains and valleys (Figure 34, no. 20). The more recent strata are represented as dotted layers.

Unlike Steno, Leonardo did not incorporate the Deluge into the natural causes of the form and history of the earth, and proposes a different succession of strata. According to Steno, an upper layer is necessarily younger than a lower one. Indeed in his case layers are supposed to be the outcome of the unloading of sediments in water within the same hemisphere. The collapses, which uncover the strata, are supposed to occur in the same hemisphere. Whereas according to Leonardo the younger layers are the lower ones in the roots or bases of the mountains. This can imply Buridan's theory or more local upwards movements of lands (Figure 25 and p. 100 n. 76).

At one point Leonardo proposes a theory that is similar to Steno's, that is with respect to the two main phases in the history of the earth, one before and the other later than the Biblical Deluge: 'on account of the two layers of shells it is necessary to say that the earth was indignantly submerged under the sea and made the first, and the deluge made the second'.[131] But this is an early note, probably written around 1481. His later theories do not obviously accommodate the Biblical Deluge.

Leonardo and Steno did agree on one point, namely the absence of marine fossils in higher rocky layers. This does not contradict Steno's theory that upper strata are younger than deeper ones, as sometime asserted.[132] Steno's plate shows clearly enough that, at the end of the history, when the earth has reached its present state, the higher mountains are those formed in the more ancient sedimentary phase (Figure 34: solid lines *F* and *G* in phase 20), at a time when 'all things were covered by a fluid free from plants and animals and other solids'.[133] The underlying observation shared by Leonardo and Steno is just evidence of the accuracy of their fieldwork, and both explored the same area, Tuscany.

More problematic is Leonardo's proposed explanation of how sediments are turned into rocks (in modern terms, diagenesis). In Aristotle's *Meteorologica* (Bk IV, ch. 5) petrification is explained as a drying process under the influence of heat and cold. According to mediaeval philosophers, petrification always requires the presence of water, which acts as an adhesive, of the heat of the sun, and of a 'petrifying virtue' or *virtus lapidificativa* (Albertus Magnus).[134] In both cases, the explanatory model is the fluid and thermal physics of the qualities. Leonardo hints on at least on one occasion at the possibility of explaining petrification through a more mechanical process, as a consequence of the weight or pressure that squeezes out water from the sediments deposited on the bed of the ocean (MS F, f. 79r). But, elsewhere, he prefers to deny that water exercises weight on the bed of the sea (f. 6r).

131 Arundel, f. 156v. See Kemp 2006a, p. 311.

132 Gohau 1990, p. 66.

133 N. Steno, *The Prodromus of Nicolaus Steno's Dissertation*, trans. J. G. Winter, New York, 1916, p. 264 (*Nicolai Stenonis De solido intra solidum naturaliter contento dissertationis prodromus*, Florence, 1669, p. 70).

134 See *De mineralibus* attributed to Avicenna (tenth century) and *De mineralibus* by Albertus Magnus. See Duhem 1909, pp. 302–17 (esp. p. 307 on the heat of the sun and p. 316 on the petrifying force).

More generally, he persists in imagining petrification as the result of an unspecified drying process: 'where in gravelly strata the waters have run off, the materials have, in consequence, dried and been converted into hard stones'.[135] He also alludes in the Codex to the role played by a 'petrifying virtue' which he more concretely defines as 'gum' ('*gomma*', f. 9v).[136]

135 Arundel, f. 138r: 'Dove per li ghiarosi soli l'acque sono scolate e per questo la materia s'è secca e convertita in dura pietra'. See also MS F, f. 11v: 'tali colli comincierebbono a essiccarsi e creare le pietre a falde minori o maggiori . . .'.

136 See also Chapter 5 in this volume.

The Study of Fossils

A Case of Interaction Between Hands-on and Theoretical Knowledge

5

DOMENICO LAURENZA

ANCIENT REMAINS (or marine 'fossils' as they were later called) serve as the 'written documents' that allowed Leonardo to 'read' the history of the earth. The presence of fossils testifies to radical changes that happened in the past history of the earth. Where there once was the sea, now there is dry land. Through the discovery and study of fossils Leonardo was able to map the changes that affected the lands around the Mediterranean, from Tuscany to Lombardy, from the Roman area to the valley of the Danube and the Black Sea.

But, in order to lay out his theories, he had first to solve a fundamental problem, namely to demonstrate the animal origin of the fossils—about which there was no full agreement at this time. Two fields of his multifaceted work and research as an artist-scientist provided operative and conceptual models that he applied to the study of fossils: anatomy and sculpture.

On the one hand, he accomplished his demonstration through a comparative analysis of ancient remains of creatures and the living animals to which they corresponded. As an anatomist he studied also animals and sometime made comparisons between man and animals or among various animals, which provided an important background to his study of fossils.

At the same time, Leonardo's background as a sculptor provided him a conceptual model by which to explain the process of fossilization of animal remains. In particular, the casting techniques developed for the equestrian statue of Francesco Sforza, became *artificial* or technological models through which he explained natural processes such as the transformation of shell animals and fishes into stones.[1]

1 These interactions have never been properly studied. See Laurenza 2018b. For other aspects of Leonardo's studies of fossils, see Duhem 1906–13, esp. vol. 2; Baratta 1903, pp. 221–40; De Lorenzo 1920, esp. pp. 130–32; Ligabue 1977; Vai 1995, 2003a, pp. 230ff, 2003b; Pizzorusso 1996, 1997; Gould 1998; Kemp 2006a, pp. 311–12; Bordonali, pp. 127–52; Cioppi 2006, 2010; Baucon 2010a, b; Cioppi and Dominici 2015.

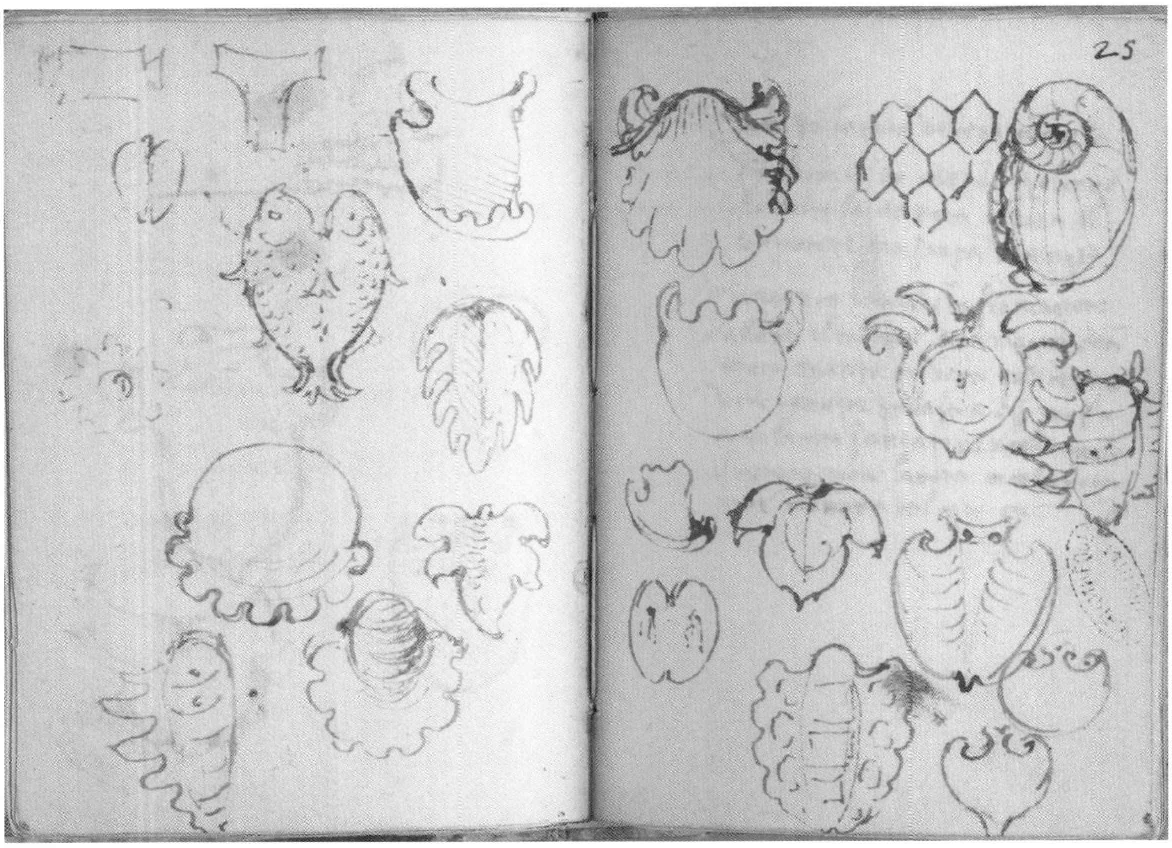

FIG. 35
Fossils and architectural decorative motives. Institut de France, Paris, MS I, ff. 24v–25r.

The Codex Leicester contains his most complete discussion of the meaning of fossils in relation to the history of the earth, whereas the process of their petrification is mainly explained in MS F (ff. 78v, 79r, 79v, 80r), where we also find one of his few surviving illustrations of fossils (f. 80r), in addition to the small sketches in MS I (ff. 24v–25r, Figure 35). It should be stressed that Leonardo does not use the word 'fossils' (*fossilia*), which in his age possessed a different sense from the modern one. It then indicated more generally all stones dug up from underground. For his part he directly refers to the names of the modern equivalents for sea creatures: shells ('*nichi*'), oysters, and so on.

The assertion of the animal origin of fossils and their value as documents of a natural rather than divine history of the earth allows Leonardo to contrast two opposite theories formulated by those he called his 'adversaries': the explanations of marine fossils as the result of the Biblical Deluge or as the result of the influence of the stars:

> Of the foolishness and simple-mindedness of those who require that these animals should be carried by the Deluge to these places, far from the seas. How another sect of ignoramuses asserts that nature or the heavens created them in those places, through celestial influences.[2]

Leonardo rails against these two theories because they were relatively recent and widespread in his age, as attested by the late-mediaeval author Ristoro d'Arezzo and repeated in cosmological books printed at the end of the fifteenth century, such as Paolo Veneto's *De compositione mundi* (Venice, 1498).

2 Leicester, f. 10r.

Ristoro d'Arezzo supports the diluvial origin:

> And we have already found and excavated, almost on the top of a very high mountain, many varieties of fish bones, which we call snails [*chiocciole*] and others call shells [*nicchi*] . . . and this is sign that this mountain was made by the Deluge (*The Composition of the world*, *c.*1282).[3]

The Deluge and the astrological explanations belong to two more general and opposite theories that had been promoted since the classical age: the organic theory, which considered the fossils as petrified remains of animals, and the inorganic theory, which regarded them as stones that had assumed animal forms. These opposite explanations were to continue to dominate debates until the late seventeenth century.

The organic explanation assumed different forms. There was the potentially historical explanation of them as documents of changes that happened in the earth between the sea and dry land, most notably in the ancient Greek authors quoted by Strabo and in other authors such as Xenophanes, Herodotus, and Ovid, and, later on, in mediaeval authors such as Avicenna.[4] An ahistorical theory told how marine fossils were evidence of a subterranean circulation of water from the bed of the sea to the mountains, along which 'petrifying cement' also flowed and formed rocks and fossils.[5] Fossil remains of large vertebrates were interpreted in antiquity as the bones of giants-heroes and mythical monsters once living on the earth, as in the writings of Hesiod, Herodotus, and Suetonius, who quotes the bones of the giants collected by the Emperor Augustus on the island of Capri, an idea revived in the fourteenth century in Boccaccio's *De montibus, silvis, fontibus et maris* and *Filocopo*.[6] The post-classical tradition of the Judaeo-Christian thought defined fossils as the deposits of the Deluge, as we read in treatises by various philosophers from the thirteenth to the fifteenth centuries. The need to accommodate the Biblical Deluge will re-emerge in geological discussion at the end of the seventeenth century and even later.[7] Lastly there was a theory that interpreted marine fossils such as ammonites and sharks' teeth as petrified remains of terrestrial animals, and specifically as ram's horns[8] or headless and rolled up snakes (the ammonites) and as the tongues of serpents or related animals (the sharks' teeth, then called 'glossopetrae', that is 'stony tongues').[9]

Inorganic theories were various and variously connected. Some of them considered fossils as stones whose particular form were impressed by special astrological influences or by an inner plastic force (*vis plastica*). According to the latter theory, sometimes attributed to the ancient Greek philosopher Theophrastus (*c.*371–287 BCE), the material structure (rock) was more meaningful than form in order

3 Morino 2007, II.5.8.

4 Bromehead 1945, esp. pp. 102–6. For Avicenna, *De mineralibus*, see Avicenna 1927, p. 20.

5 Seneca, *Naturales quaestiones*, Bk III, chs 16 and 19: 'this proves that the subterranean regions hides a large amount of water full of fish'; Albertus Magnus, *De causis proprietatum elementorum*, 2.2.5 (in *Alberti Magni Opera Omnia*, ed. P. Hossfeld, Münster, 1980, vol. 2, Part II, p. 101) and (on petrifying force, *vis lapidificativa*, after Avicenna) *De mineralibus*, 1.2.8, quoted in Duhem 1906–13, esp. vol. 2, 1909, pp. 316–17.

6 Mayor 2000.

7 Rudwick 2014, pp. 49ff. and 120ff.; Vai 2003a.

8 See Pliny, *Natural History*, Bk XXXVII, ch. 60, 167: *Hammonis cornu* (Pliny, *Natural History with and English Translation in Ten Volumes*. Volume 10: *Libri XXXVI-XXXVII*, ed. E. D. Echholz, London and Cambridge, 1962).

9 Pliny, *Natural History*, Bk XXXVII, ch. 59, 164 (Pliny, *Natural History*. Volume 10, ed. Echholz). See Leicester, f. 9r: 'teeth of fish, which some call arrows and others serpents' tongues'.

to understand what fossils were. This plastic force was allied to that which generated stalactites and stalagmites. Even at the end of the sixteenth century Ulisse Aldrovandi in his *Museum Metallicum* (published in 1648 after his death) classified animal-shaped fossils according to their material constituents. In the seventeenth century, Martin Lister tried to restate the inorganic nature of them on the base of a link between a certain kind of stone and a certain animal figure particularly associated with it.[10] Aristotle's concept of 'spontaneous generation' (*generatio aequivoca*) from non-living matter provided a further conceptual framework for inorganic theories, while according to another long-lasting theory, it was the rock itself that generated various recognizable forms, because nature delighted in imitating in one of its realms (the mineral) the forms found in two other realms (animals and plants).[11] This was also linked to the more general theory of the correspondences or affinities between all the parts of the universe, including sky and earth, with astrological consequences, and whose most famous example was the correspondence between macrocosm and microcosm.

The organic and inorganic theories both had conceptual and historical limitations. The adducing of the Deluge and/or astrological influence relied upon actions that are matters of faith rather than direct demonstration (which was Leonardo's criticism). The idea that there were races of gigantic men combines anthropocentrism with mythology. More generally, these explanations lacked any real historical vision of global, irreversible, and natural changes in the earth. The Deluge was miraculous, lying outside the scope of natural changes. On the other hand, the absence of any concept of species extinction and the existence of fossils without modern counterparts (so-called 'difficult' fossils such as ammonites or belemnites) justified inorganic theories, which, furthermore, were coherent with the dominant philosophies of the time.

With no academic background and substantially uninvolved with the world of official theology, Leonardo was conceptually freer to review and rework the ideas of his predecessors and to follow the directions in which his field observations seemed to lead him. Furthermore, his artistic background may have drawn his attention to those strange stones in which nature seemed to imitate herself like an artist. His work as a sculptor also provided him unusual explanatory keys from modelling and casting with which to explain fossils in a way that opposed the inorganic theory of nature imitating herself. Later on, in the second half of the seventeenth century, another artist, the Sicilian painter Agostino Scilla, studied fossils 'like a painter' and wrote a treatise, which is considered a milestone in the history of palaeontology.[12] Both Leonardo and Scilla considered only fossils with modern analogues, the so-called 'easy' fossils. This was not a limitation, but the only possible scientific field of action for artists whose art was conceived as an imitation of nature and whose studies of fossils consisted in a visual search for close correspondences between fossils and known animals.

The Biological Model

Detailed comparative analysis of fossils and living animals allowed Leonardo to reject both the Deluge and the astrological theories. His anatomical research, unlike Vesalius's more anthropocentric approach,

10 Ligabue 1977; Rudwick 1985, pp. 39ff.; Vai 2003a, esp. pp. 230–3.

11 Findlen 1990.

12 Agostino Scilla, *La vana speculazione disingannata dal senso* [*Vain Speculation Undeceived by Sense*], Naples, 1670. On Scilla, see Rossi 2003, pp. 39–45; Segala 1996; Findlen 2013; Carpita 2006; Laurenza 2018 and forthcoming b.

included animals. Indeed, his comparisons between man and animals and between various animals were among the most original aspects of his anatomical research, which might have influenced his comparative approach to the study of fossils. In both cases, he was searching for similar features between man and animals on the one side and between fossils and living animals on the other.

Later, at least two key episodes in the history of palaeontology were similarly linked to primarily biological interests: Steno's demonstration of the animal origin of fossil shark's teeth (so-called *glossopetrae* or 'tongue-stones') alongside his dissection of a shark's head in 1666 Florence,[13] and the study of the 'differences' between the fossil bones of a mammoth and those of a living elephant as evidence of animal extinction by George Cuvier (1769–1832),[14] whom many scholars consider the true founding father of modern comparative anatomy. Leonardo was therefore not an exception, and he belongs to a first phase in the history of palaeontology when the agenda concerned the demonstration of the identity between fossils and living animals through the search for 'similarities'. However, unlike the spectacular images that illustrate his studies of comparative anatomy of humans and animals or his physiognomic and psychosomatic search for bestial traits in man,[15] there is no visual representation of the comparisons between fossils and animals by Leonardo. Judging from what has survived, Leonardo only used verbal descriptions. Evidently fossils evoked scenes of life that were not directly observable, and therefore not suitable of compelling visual representations, just like the distant past of the earth and its cataclysms, of which the fossils were an integral part.[16]

The fossils studied by Leonardo are mainly marine animals and plants: bivalve molluscs like oysters, gastropods, corals, small fish bones, fish teeth, squids,[17] and seaweeds. In an early, poetic text he describes the remains of a gigantic fish (Codex Atlanticus 715r and Arundel 156r), but there is no evidence in Leonardo of study of fossils of the great extinct land vertebrates and he does not apparently confront the problem of their not corresponding to living animals. In any event, there was no reliable canon of what animals actually existed in the world.

Although Leonardo's historical vision of nature lacks any extinction or 'evolutionary' base, in that he does not see fossils as the remnants of extinct organisms and he only considered 'easy' fossils, he regarded them as visual documents of the long-term history of the earth. He also demonstrated how the petrified animals once had their own lives and were part of a *biological history*.

The astrological theory implied a sudden and single action of a specific astral conjunction, which catalysed the stone to assume the form of a specific animal. Leonardo emphasizes not only how fossils of different animals are sometimes found in the same level, but also that they are evidently remains of actual animals, which manifestly had lived real lives. He searches fossils for signs indicating the passage of time in ways that correspond to those in living animals. He finds such indications in the growth

13 The 'palaentological' study was published as an appendix to his study *Canis Carchariae dissectum Caput* [*The head of a shark dissected*], in turn printed at the end of his anatomical book *Elementorum Myologiae Specimen . . . cui accedunt Canis Carchariae dissectum Caput . . .*, Florence, 1667.

14 Rudwick 1985, pp. 101ff.

15 See Laurenza 2001.

16 On representing the 'invisible', see Nova 2007.

17 Leicester, f. 9v: '*pesci calamai*'. See the Glossary.

patterns in the bones of the fishes and in the layers of growth in shells, which are compared to the growth rings in the trunks of the trees and in the horns of bulls and related animals:

> [. . .] as if in those places we do not find that the bones of the fishes grew over a long time, and as if in the coats of the shells and sea snails one could not count the years and months of their life, as in the horns of bulls and bullocks and in the ramification of plants that were never pruned in any part.[18]

Just as the geological stratigraphy of the mountains bears witness to the history of the earth so this biological stratigraphy demonstrates the history of the life of the fossils. He also studies the hollow traces ('*tarli*') left by small animals which once lived on the shells of other animals: 'other species of shells attached over them in the form of animals moving over them, and indeed the vestige of their movements still remains on the outside like the wood-worm's mark on the consumed wood'.[19]

Leonardo also looked to similar perforations in corals—if we suppose that this is the sense of his description of the 'porous corals' ('coralli intarlati'). The description is on a folio (9v) adjacent to those containing the study of other holes left by animals in fossil shells (ff. 9r, 10v).[20] Alternatively he may have been describing the intrinsically porous structure of corals.[21]

Also his criticism of the theory that explains the fossils as the outcome of the Biblical Deluge is mainly based on the comparative analysis of fossils and living animals.

With remarkable freedom from Church dogma, Leonardo doubted the reality of a Deluge that covered even the highest mountains and that was caused by a rain lasting forty days and forty nights (ff. 8v, 3r; Codex Atlanticus, f. 418ra). In any event, he argued that the Deluge could not explain marine fossils. He assembled many pieces of evidence against the supposed connection between fossils and the Deluge.

In his stratigraphic studies, Leonardo stated that each separate stratum was the result of a single flood, noting that marine fossils are found in many strata, at different heights and not in one single stratum as implied by a unique Deluge (ff. 8v, 10r). Furthermore, fossils are only absent in the highest strata, close to the tops of the mountains, where we would have expected to find them, given that the Deluge covered the mountains (ff. 10r, 3r, 8v).

If the Deluge were caused by heavy rain, its waters would have flowed towards the sea and not vice versa, in such a way that the transport of the fossils from the sea to the mountains would be impossible (f. 10v). It was also impossible that these heavy and slow-moving animals could have ascended actively from the bed of the sea. They would not have been able to attain their natural place along the new coastline caused by the risen waters at the top of the mountains. Marine snails, Leonardo notes, do not swim and only move very slowly through their furrows in the sand, at a speed of not more than 8 feet a day, certainly not enough to traverse many miles during the forty days' duration of the Deluge (f. 8v). Equally, the theories of the passive transport of the shells by the rising of the sea or by waves moving towards the land are wrong, because shells are too heavy to be carried upwards by water. Moreover, the

18 Leicester, f. 10r.

19 Leicester, f. 9r.

20 According to a recent interpretation, the drawing of a web with hexagonal cells in MS I (f. 25r; Figure 35) represents *Paleodyctyon*, which are the fossilized traces excavated in the mud by an animal according to modern ichnology. See Cioppi 2006, 2010; Baucon 2010a, b; and Codex Leicester, f. 10v: 'How in the strata, between one and the other, are still found the tracks of the earthworms, which proceeded through them when they were not yet dried out.'

21 See Volume IV, Glossary ('Intarlato').

final ebb of the waters from the land towards sea would have brought them back from the shores (ff. 8v, 9r, 3r, 9v).

During the course of these discussions, he introduces remarkable accounts of the fossils as 'snapshots' of arrested life in the very same place where they were later to be found. His comparison between fossils and the corresponding living animals shows that the former are found in precisely the same locations where they must have spent their lives. This means that they were covered by sediments while still alive in their natural habitats. Leonardo quotes two examples: benthic organisms, that is animals that live more or less permanently secured to the bed of the sea or to rocks along the coast; and gregarious animals, that is marine animals that tend to associate in groups or shoals with others of the same kind.

Among benthic organisms he includes corals and oysters (ff. 9v, 10v). Oysters, subject to a meticulous analysis in their living state (f. 9v), live secured to the bed, because they eat the small animals that swim around them, which means that they do not need extensive locomotion. Only rarely they are found separated from their original places on the beach, and, in these instances, they are always found as isolated specimens. The finding of fossilized oyster colonies means that they are still in the same place where they once lived. The same is true for fossils of gregarious marine animals, such as squids (f. 9v): they are congregated in orderly groups, incompatible with their passive and violent transportation by the Deluge.

He also observed the fossils of bivalves revealed that some of them were already dead (shells broken or wrenched open, sometimes filled with sands and broken pieces of other shells, f. 10r), while others were deposited whilst still alive (with closed valves). He knew that the distribution of shells manifested their natural locations, in that dead shells were scattered on beaches while those that were alive were in the sea. Accordingly, broken shells were found in disorderly arrays across the depth of the strata, while closed shells were only found at the borders of low strata. The living shells had evidently been covered by new sediments when still alive and while they were tethered to the bed of the sea that in turn had been formed earlier by the sediments of an earlier flood (f. 10r).

The Casting Model

Around 1505–9, in a sheet of the Codex Leicester (f. 9v) devoted to the criticism of the Deluge and astrological influences as explanatory causes of marine bodies in mountains, Leonardo mentions shells and coral fossils he had found on the mountains of Parma and Piacenza in North Italy and remembers how, while he was sculpting a huge horse, some peasants brought him a large sack of fossils of marine animals, some of which were very well preserved ('of which [fossils], when I was making the great horse of Milan, certain peasants brought to me in my workshop a large sack, which were found in that place, among which there were plenty of those preserved in their original state'). This may refer at this time to the planned equestrian monument to Gian Giacomo Trivulzio, the Milanese general who served the French king, although it is perhaps better read as a retrospective reference to the earlier project from the 1490s for a bronze equestrian memorial to Francesco Sforza. This mnemonic association between study of fossils and work for a statue to be cast in bronze is not coincidental, as suggested, on the one hand, by Leonardo's explanation of fossilization and, on the other hand, by his use of the process of casting to explain the natural generation of fossils.

Two sheets from Manuscript F, filled in the same years of the Codex Leicester (*c.*1508), form a diptych in which Leonardo distinguishes two kinds of fossils of marine animals: fossils of animals 'which have their bones outside' (f. 79r), namely closed inside a shell or exoskeleton, and fossils of animals 'which have the bones inside their skin' (f. 79v), such as fishes. Leonardo distinguishes, with regard to these two different types of fossils (molluscs provided with outer shell and marine vertebrates), two different processes of fossilization.

According to Leonardo, in the case of molluscs with outer shells (MS F, f. 79r), the fossil consists of three parts: the shell (1), the external impression imprinted by the outer side of shell in the material surrounding it (2), the internal impression imprinted on the surface of the material that filled the space inside the shell and that maintained the form of this space (3) (Figure 36). The animals, after having been covered, on the seabed, by the weight of the 'mud' and devoid of ailment, died. Their soft parts dissolved and the internal void was filled by the mud penetrating through the openings of the shells, while other mud surrounded the shell. Afterward, 'over time, lowering the sea', the mud, both inside and outside of the shell, turned into stone, while the shell, especially in the case of old animals with very thick and non-porous shell, was not penetrated by the 'petrifying' matter and preserved its 'natural' composition, retaining not only the shape but also the original substance of the shell. The fossil turned out then to be of three parts, the 'natural' (i.e. preserved) shell (1) included 'between the two stones': the outer rock with the negative imprint of the outer surface of the shell (2) and the stone that had filled the space inside of the shell and therefore bears the imprint of the inner surface of the latter (3) (Figure 36). Given that in the gastropods and bivalves the external ornamentation of the shell is generally not replicated on its inner surface, the result is that the inner petrified cast reproduces the general shape of the shell, but not its external ornamentation. The passage added by Leonardo in the margin of the main text in the sheet refers to this: 'On the imprinted and petrified shells, which have not an external shape corresponding to the original shell' ('De nichi inprontati e petrificati che non han la figura superfiziale che avea la loro scorza').

The formative process and the structure of the fossils of fish or marine animals with 'internal skeleton', is different and is described in the following sheet of the same manuscript (MS F, f. 79v) (Figure 37). In this case a 'viscous and petriyfing humor', acquired from the mud, fills the cavity left empty by the animal's soft parts consumption and slowly dried up forming a solid cast.[22]

It is clear that Leonardo's sophisticated accounts of the casting of fossils were deeply informed by his experience of casting sculpture in clay and bronze. He was trained in the workshop of the greatest master of bronze casting of the later fifteenth century, Andrea Verrocchio. He was also involved, on his own account, in various metal technologies, including the casting of guns. And he spent years on the design and technological planning of the massive equestrian memorial to Francesco Sforza, his patron's father.

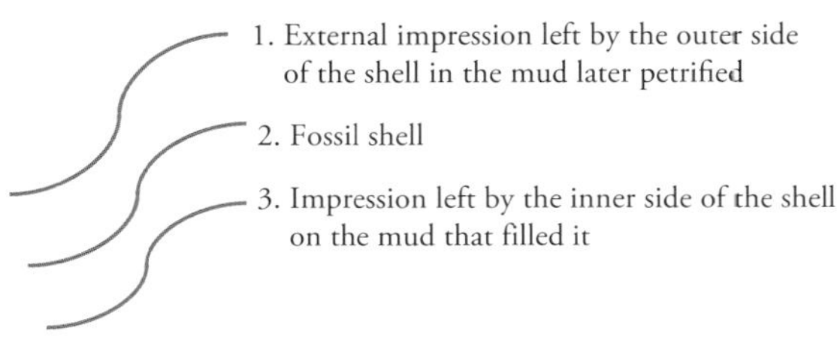

FIG. 36
Diagram illustrating Leonardo's theory of the generation of the fossils of shells.

22 See below for more details.

On his return to Milan after 1507, when actively compiling the Codex Leicester, he was engaged on a second project for a bronze equestrian memorial, this time for Gian Giacomo Trivulzio. He was therefore well aware of the making of three-dimensional models from which an outer mould would be made. The mould might be made from two halves or from smaller part-moulds. It was within the assembled mould that the sculpture would be cast from liquid plaster or molten bronze. The equivalent for fossils is the making of an external mould from the body of the animal and the casting of the interior space as the animal decayed.

The simplest and most obviously relevant processes for the casting of fossils were direct casts in plaster and small-scale bronze and silver items. In this process, fusible material is poured into the mould through an opening. Once the material has solidified by drying or cooling, the mould is stripped away, leaving a replica of the original model. The most straightforward and practical form of this type of casting in the Renaissance was the making of plaster models of heads and faces as life or death masks, which either served as portraits in their own right or could be used as the prototype for a carved portrait bust. This was a practice inspired by ancient Rome. The most famous example, of which Leonardo would have been well aware, was the death mask of Filippo Brunelleschi, the great architect, piously preserved in the workshop of the Cathedral in Florence and still extant. It served as the model for the marble memorial bust of Brunelleschi on the south wall of the Duomo. There was also a widespread practice of making casts from living human bodies that could facilitate the study of anatomy in the artist's studio. Vasari testifies that Verrocchio was a keen advocate of this practice, retaining casts of many 'hands, feet, knees, legs, arms and torsos' ('*mani, piedi, ginocchia, gambe, braccia e torsi*') in his workshop.[23]

Direct, solid casts are analogous to the way that fossils of fish are formed (Figure 37). In Manuscript F, f. 79v, he considers those marine animals with 'internal skeletons'. After having been enclosed in mud that solidifies, the soft parts of a fish were consumed leaving an empty cavity, while the solid and bony parts fell to the bottom. The cavity was progressively filled by the 'viscous and petrifying humour' ('*umore vischioso petrificante*'), acquired from the mud, which, as a result of geological transformations, was exposed to the air and 'became dry of the aqueous humour' ('*s'è risecco dell'umido acquoso*'). This 'viscous humour' filled the entire cavity, driving out the air, which was released through 'minute pores of the earth' ('*minute porosità della terra*'), but only sideways, because the outlets at the bottom and top were blocked by the humour that was filling the cavity. In time 'the humour, once dried up, became a stone without granules and retained the shape of the animal, and the bones remained within it' ('*risecco, tale omore si fa pietra sanza granosità, e riserva la medesima forma dell'animale che quivi s'improntò, e dentro a lui restan l'ossa*'). The fossil fish is a simple, solid cast, arising from a single original animal mould, which records the original shape of the encased fish.

The direct, solid casting techniques could be used for small bronze items, but the favoured method by the end of the fifteenth century was the lost-wax process. There were direct and indirect methods. In the standard method of direct casting, the core of the model will be made in clay and coated with a layer of wax from which an external mould of the original model would be made, recording all the

23 Vasari 1967–87, vol. 3, p. 543–4:

> Andrea used to make moulds of various parts of the body—hands, feet, knees, legs, arms, torsos—so that later, having them before him, he could imitate them with greater ease. Later during Andrea's lifetime, people began to fashion inexpensive death masks [from this material], and over the fireplaces, doors, windows, and cornices of every home in Florence, one can see countless portraits of this kind, which are so well made and lifelike that they seem alive.

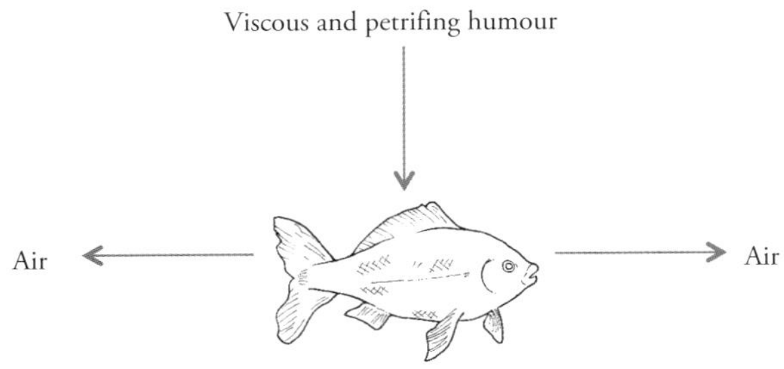

FIG. 37
Diagram illustrating Leonardo's theory of the generation of the fossils of fish.

surface features of the model in negative. The gap between the mould and the clay core would be secured with rods and ties, so that the wax could be safely melted out. Molten bronze at very high temperatures would be poured into the space vacated by the wax. On cooling, the outer mould would be stripped away and the inner fire clay core chipped out, leaving a hollow layer of bronze. This would require a good deal of tooling and polishing to create the final bronze statue or statuette.

Leonardo, for his huge Sforza horse sculpture in the 1490s, was considering an indirect casting method, akin to that used by casters of large canons.[24] In this case, piece-moulds would be made from the original model as before. The disassembled and baked moulds would be lined with wax or friable potter's clay to the desired thickness. Leonardo was favouring the latter for the giant horse. The interior of the piece moulds and their clay lining would be filled with a fire-clay core, which would be separately baked. The whole would be disassembled and the layer of potter's clay between the mould and the core would be removed. The piece moulds and the core would then be reassembled with rods and ties to ensure that the gap left by the potter's clay was secure. The whole would be placed in a casting pit and buried in the earth. The molten bronze would be poured via inlets into the gap between the mould and core as before, while the air rushed out through outlets. He designed the conjoined piece moulds in such a way that they could be levered open along their median axis like a bivalve shell after the bronze had cooled. The whole procedure was fraught with fiery danger. The hope was that an intact sculpture in hollow bronze would emerge, as with the lost-wax method. Not the least advantage of the indirect process was that the original model survived (Figure 38).

There is a morphological parallel here with the formation of a fossil of a shelled organism. The outer mould corresponds to congealing mud outside the shell; the bronze corresponds to the shell; and the inner clay core corresponds to the congealed mud that is cast inside the shell. However, there are differences. In the case of a bronze statue, it is the hollow layer of bronze that is cast, whereas the hard shell of the organism persists throughout. And the inner core of the bronze is quite rough and only approximately assumes the shape to be cast.

Reading Leonardo's accounts of the casting of fossils and sculpture, we find that the terms he uses overlap significantly. The outer mould 'imprinted' with the shape of the horse, is called a '*cappa*', the name he often uses to indicate one of several fossils of bivalves he studied. The term, in its zoological sense, appears on the same sheet of folio 79r in Manuscript F on which Leonardo describes the threefold structure of these fossils: 'On animals which have their bones outside, as shells [nichi], snails, oysters,

24 For a full account of the casting method that Leonardo was considering for the horse, see Kirwin and Rush 1995. See also Kemp 1995; Bernardoni 2007, esp. pp. 30ff.

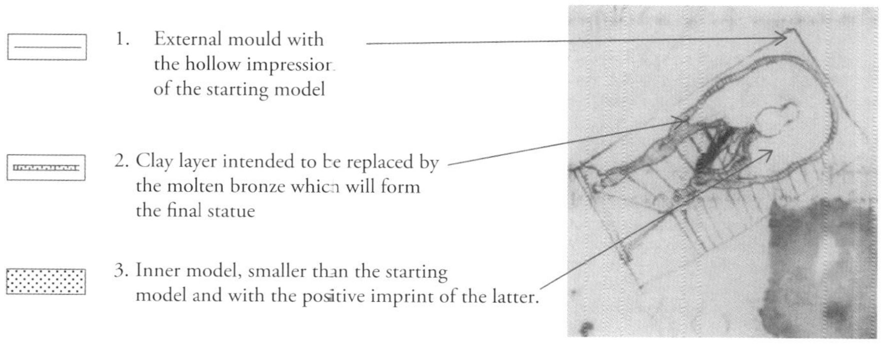

FIG. 38
Diagram illustrating Leonardo's cast for a bronze horse, including Leonardo's drawing. Royal Library, Windsor 12351v (detail).

bivalve *cappe*, turbinated shells [*bovoli*] and similar, which are of innumerable species.'[25] When discussing his sculpture, he heads one of his notes, 'On removing the horse's *cappa*' ('*Dello spiccare la cappa* del cavallo').[26] He talks of ensuring that 'the *cappa* would not imprint brush strokes of ash'. The term to imprint ('*improntare*') is also shared. The horse's *cappa* consisted of valves or shells that can be opened, as in the case of a bivalve mollusc. The term 'mouth' ('*bocca*') also recurs. It is used to designate the opening through which the molten bronze is poured.[27] In the case of fossils, it describes the opening for the entrance for the 'viscous petrifying humour' or the 'gum' ('*gomma*').[28] Finally, writing of the process of sculptural casting, he explicitly uses the term 'viscous humidity'[29] corresponding to that used in relation to fossils: 'viscous humour'.[30]

Given these similarities, between natural processes (fossilization) and artistic procedures (casting a statue), it is not surprising that the note in the Codex Leicester, in which he remembers the fossils he received while working on the Sforza horse, is immediately followed by another passage in which he explains the tripartite process of fossilization, analogous to the indirect casting model planned for the horse:

> In the mountains of Verona the red stone is found all mixed up with shells, turned into this stone, which came through their openings, with the matter of the stone in the form of gum [*era gomata*], and, in some parts, they remained separated from the rest of the mass of the rock surrounding them, because of the covering of the shell that is interposed there, and they did not leave them joined up. And, in some other parts, this gum had petrified the aged and punctured covering.[31]

In the passage in Manuscript F he uses the term 'gum' ('*gomma*') instead of 'viscous humour', but the mechanism is the same. The petrifying 'gum' passes through the 'mouth' to fill the interior space, while the shell itself remains more or less completely immune to the petrification and is sandwiched between two petrified layers.

25 'Delli animali che han l'ossa di fori, come nichi, chiocciole, ostrighe, *cappe*, bovoli e simili, che sono di spezie innumerabili'

26 Madrid II, f. 144v (emphasis added).

27 Madrid II, f. 144r: 'Of the mouth for casting it . . .' ('*Della bocca per gittare . . .*'); ff. 146r, 149r, 152r.

28 Leicester, f. 9v: 'which came through their 'mouths' with the matter of the stone in the form of gum' ('delli quali [nichi] per la boca era gomata la materia d'essa pietra').

29 Madrid II, f. 141r: '*vischiose umidità*'.

30 MS F, f. 79r: '*umore vischioso*'.

31 Leicester, f. 9v.

When Leonardo was pondering the formation of fossils, another relevant sculptural technique was beginning to be explored, namely life casting from nature. The pioneer was the Paduan sculptor Andrea Riccio (1470–1532), virtuoso of large and small bronzes, including some representing amphibian animals which appear to be based on direct casts from the creatures' bodies.[32] One of Riccio's patrons was Isabella d'Este, who was Leonardo's host in Mantua in 1500. The sculptor also worked in Venice and Verona, and there may be a link with the notable Veronese physician and philosopher, Gerolamo Fracastoro (1478–1553) who advocated an organic interpretation of the fossils found in the Verona area in a way that is very similar to Leonardo.[33] Fracastoro was a close friend of the sons of the Veronese doctor Girolamo della Torre, and Andrea Riccio was the sculptor of the tomb of Girolamo and his son Marcantonio della Torre. Marcantonio was the young anatomist with whom Leonardo almost certainly worked in 1510–11.[34]

Life casting of plants and animals became a prominent and fashionable pursuit in sixteenth-century metalwork and ceramics.[35] In the case of Bernard Palissy, eccentric sculptor in ceramics, the process of life casting became a model for the formation of fossils. Working in the later sixteenth century for major patrons in France, most notably Marie de' Medici, Palissy created spectacularly naturalistic casts coloured with vivid glazes.[36] He particularly delighted in reptiles, which were used to adorn picturesque grottos in gardens and ceramic dishes. Like Leonardo, he claimed in his writings and lectures that fossils were cast in the 'womb' of the earth in essentially the same way that his ceramic creatures were baked in his roaring kilns.[37] Palissy did not, however, develop a historical view of fossils. He regarded the watery animals as arising from the action of 'salty freezing seeds' ('*semence salsitive et congelative*'), which were implanted in the earth by divine agency.[38]

By contrast, Leonardo generated a complex geological theory of a geodynamic type. He thought much more widely about what fossils can tell us, using their presence in strata to develop a geological theory that included both local changes, such as the lowering of the Mediterranean after the opening of Gibraltar, and global transformations through the uplifting of previously submerged lands in the northern hemisphere in order to compensate the subterranean collapse of lands southwards from the Northern hemisphere.[39] His development of the study of fossils in a theoretical and causal direction also differentiates Leonardo from Agostino Scilla, an artist-scientist in the seventeenth century, who was also interested in the study of fossils. Another significant difference between Leonardo and Scilla involves their use of visual representation. While the depiction of fossils and their visual comparison to living animals are the key aspects in Scilla's 'artistic' contribution to this field of study, Leonardo seems to eschew the drawing of fossilized creatures.

32 For Riccio, see Planiscig 1927, pp. 450–64; Shell 2004, p. 8 n. 17.

33 Fracastoro's ideas are mentioned in Benedetto Ceruti and Francesco Chiocco, *Musaeum Franc. Calceolari iun. Veronensis*, Verona, 1622, pp. 407ff. See Ligabue 1957, pp. 35ff; Rudwick 1985, pp. 40–1; Bordonali 2007, pp. 145ff.

34 Laurenza 2011.

35 See Kemp 1996.

36 See Rossi 2002, pp. 25ff; Kemp 1999; Smith 2004, pp. 100–6; Shell 2004.

37 *Oeuvres complètes de Bernard Palissy*, ed. P. A. Cap, Paris, 1844.

38 *Oeuvres complètes de Bernard Palissy*, p. 276. On Palissy's limited geological implications, see Rudwick 1985, pp. 41–2, and Gohau 1990, p. 34.

39 See Gould 1998; Laurenza 2015.

The Codex Leicester, extensively devoted to the study of geology and fossils, contains no drawing of fossils whatsoever. Rough sketches of fossils in Manuscript F (f. 80r) and some sketchy drawings in Manuscript I (ff. 24v–25r, Figure 35) represent his only surviving visual studies, whereas he devoted lengthy texts to the fossils, to the denial of the Deluge and to the changes in the history of the earth. Obviously, Leonardo would have been able to draw the fossils with the same high quality that he applied to the study of other aspects of nature, as evinced by the beautiful drawing of a gastropod shell in a sheet of the Codex Arundel (f. 33r, Figure 39). In the case of fossils, the key 'artistic' framework for Leonardo did not correspond to the observation-representation of reality in a drawing or a painting,

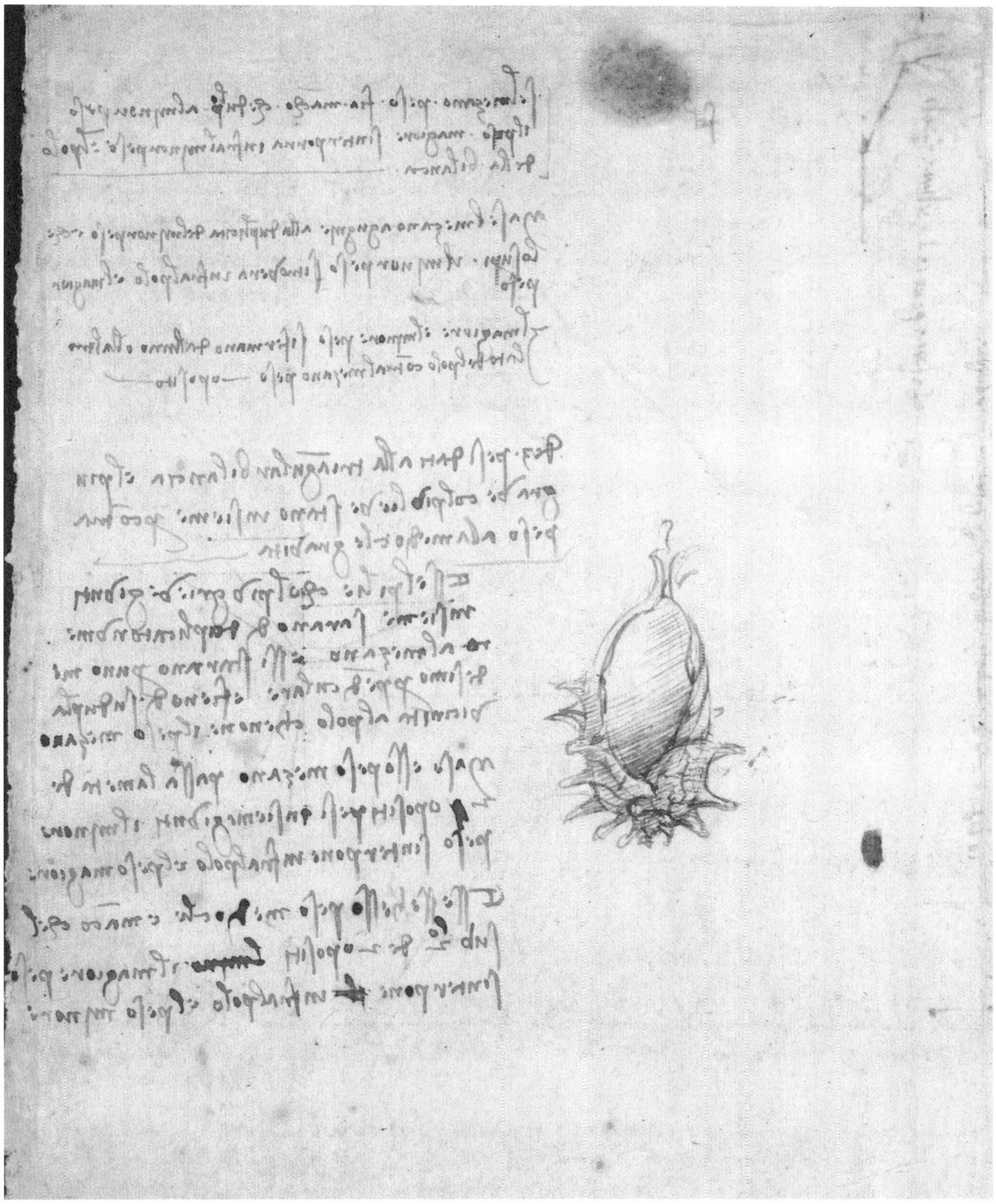

FIG. 39
Drawing of a gastropod shell. Codex Arundel, f. 33r.

but consisted of a conceptual model provided by casting techniques. It helped him to demonstrate the animal origin of fossils, which in turn was developed into a theory of the earth and its historical changes.

Geological Events

Formation in sedimentary strata

The first 'geological' event in the generation of marine fossils proposed by Leonardo is the formation of sediments, the ultimate result of the erosion of mountains and coasts by the waters. In both cases, the final products of erosion are sand and silt. The sand is partially thrown back by the sea on to the beaches, while all smaller and lighter silt passes on to the bed of the sea. These two granular materials form sedimentary strata, above all the silt. They are laid down especially after notable floods or marine tempests (ff. 35r, 10r). Leonardo highlights that the numbers of the strata in the mountains even allows us to count the number of years over which the strata have been deposited: 'one can count the winters of the years in which the sea multiplied the strata of sand and mud, carried there by the neighbouring rivers and which were discharging over its shores'.[40] The sediments deposited on the bed of the sea envelop the broken shells and kill those animals that are still alive.

Petrification

He also assigns a role to a more mechanical cause, namely the weight of the sediments. In Manuscript F (f. 79r) sediment is described as causing the death of animals: 'under the great weight of such mud it was necessary [for the animal] to die'. Afterwards petrification begins. This process is described in Manuscript F (ff. 78v–80v) and includes the following steps: the decay and consumption of fleshy parts ('*carnosità e intestine*'—'flesh and intestines', MS F f. 79v); the empty space or mould is then filled by silt or mud; the slow expulsion of 'aqueous humour' (MS F, f. 79v) is accompanied by the entrance of a 'viscous and petrifying humour' ('*umore viscoso e petrificativo*', MS F, f. 79r), which transforms the silt cast of the body of the animal into stone.

Leonardo explains the actual petrification through traditional 'petrifying virtue', which in the Codex Leicester is equated to a 'gum' (f. 9v). The terminology used by Leonardo—'*congelamento*' (freezing), '*seccarsi*': (drying up), '*qualità d'omore vischioso*' (quality of viscous humour)—alludes to the physics of qualities by which petrification is explained by Aristotle (*Meteorology*, 4) and in the following tradition by Avicenna and Albertus Magnus.

Re-emergence and uplift on the mountains

The third and more general level of Leonardo's theory of the origin of the fossils is wholly consistent with his general reconstruction of the history if the earth. He advocates a double cause: one more local, the other more global.

The local cause is the lowering of the Mediterranean and the Black Sea after the opening of the Strait of Gibraltar and the Bosphorus. Since part of their waters flowed respectively, into the Ocean and into the Mediterranean, the seas retreated, uncovering land that had been previously submerged. In addition to this 'passive' lowering of the sea level, there was, for static reasons, an 'active' uplift of land due to the

40 Leicester, f. 10r.

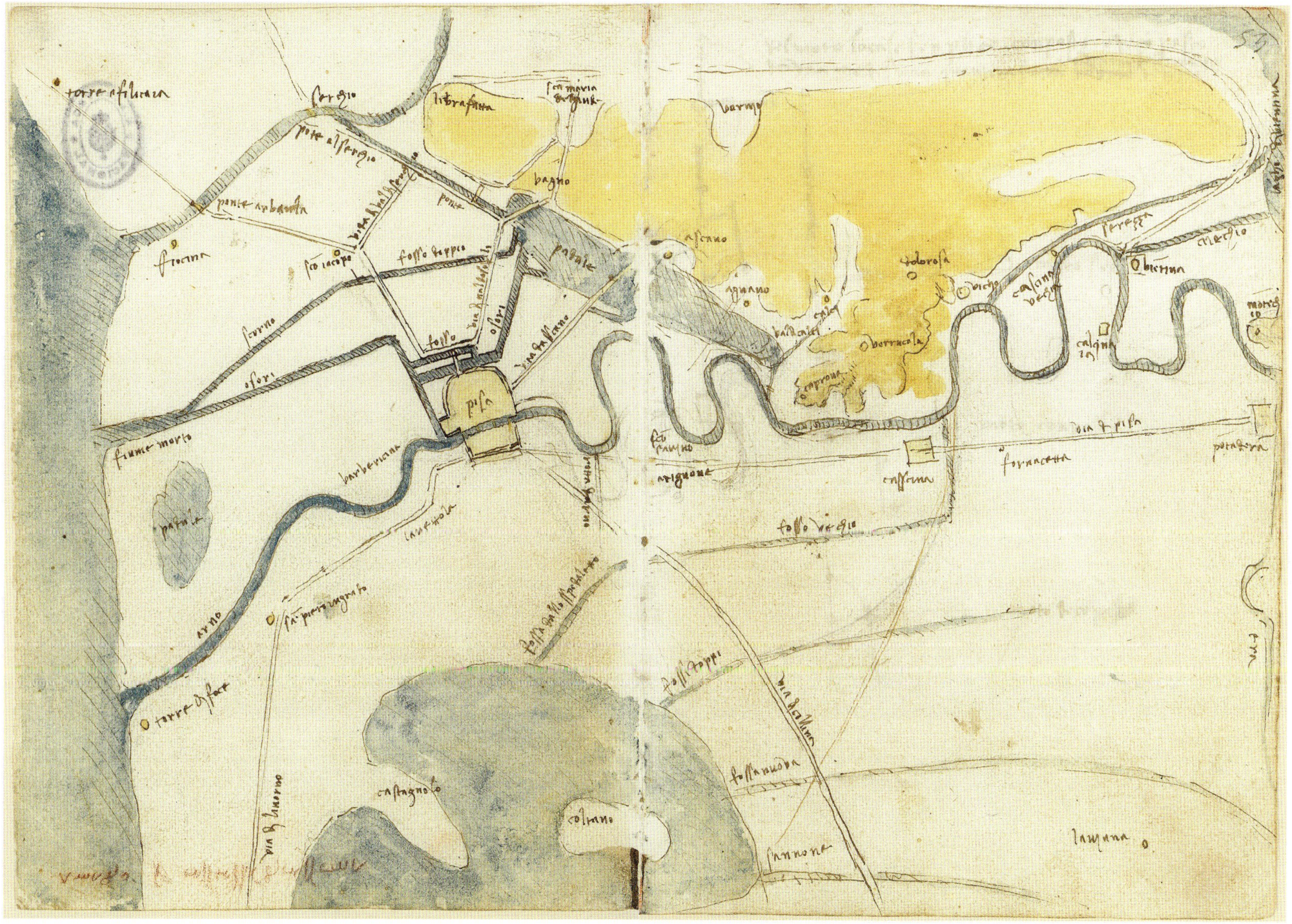

A. Map of the Arno showing Pisa and the proposed 'breaking' of the Arno. Codex Madrid II, ff. 52v–53r

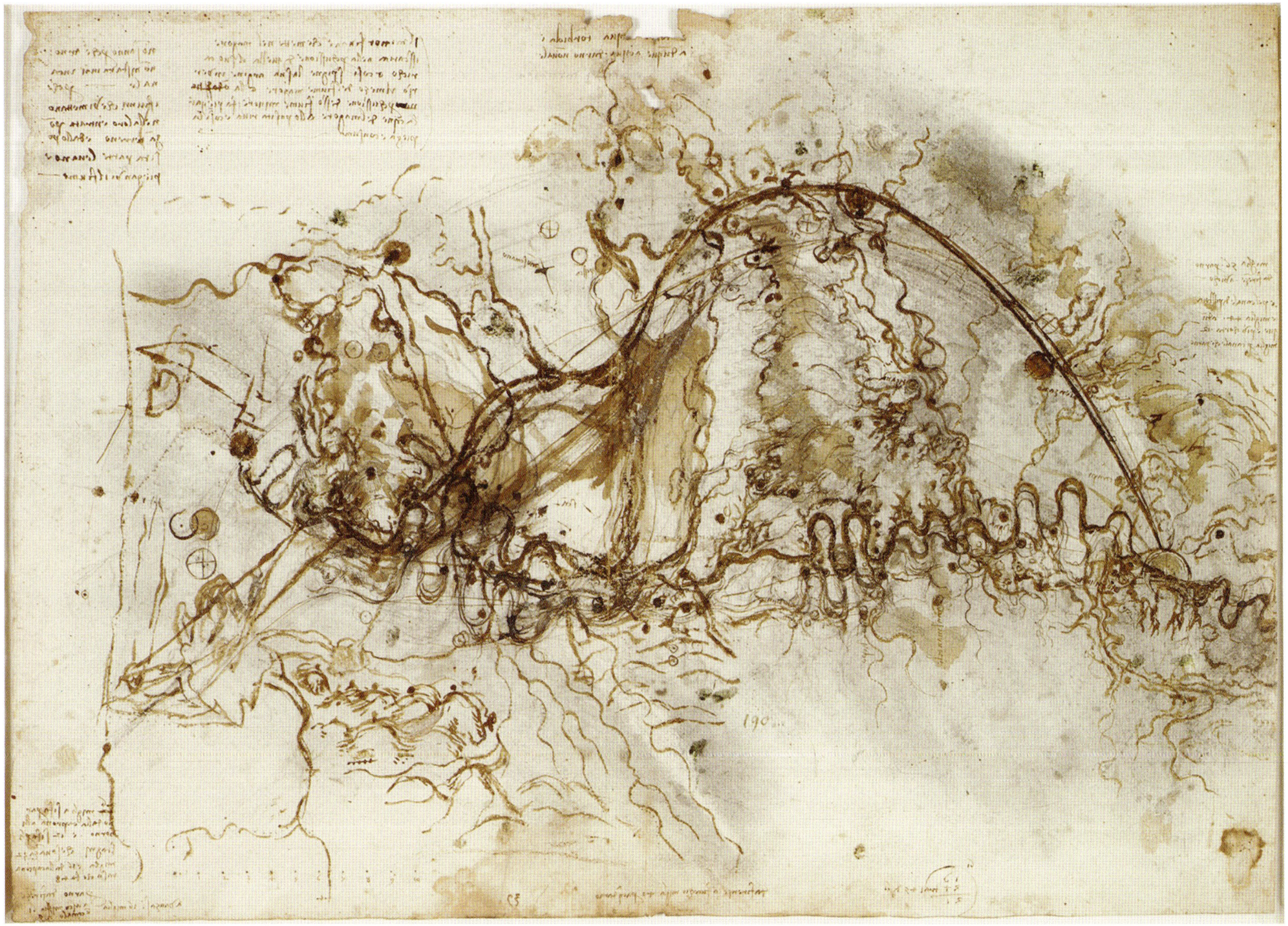

B. Map of the Arno between Florence and the sea with Leonard's proposed canal. Royal Library, Windsor 1227

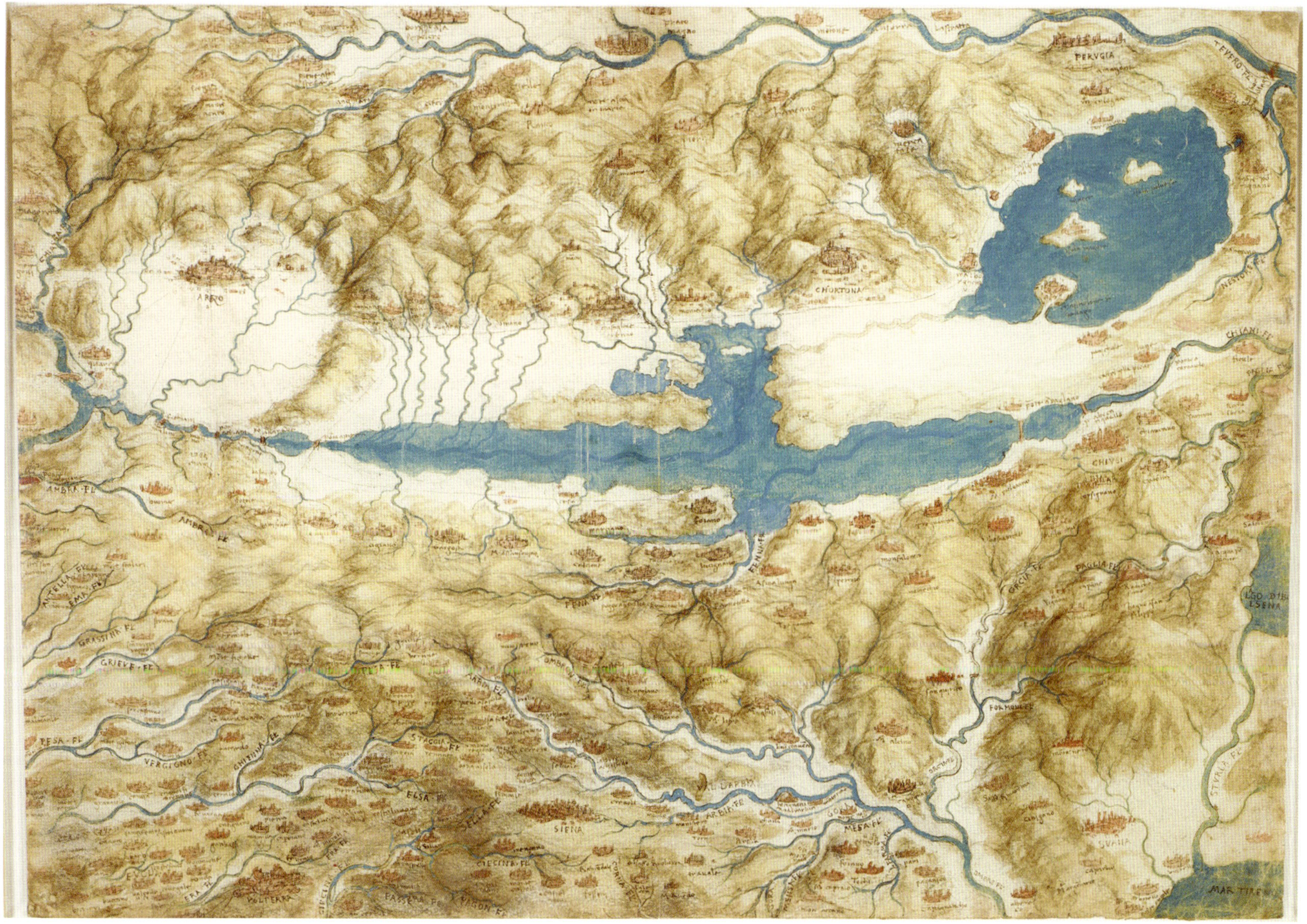

C. Map of the Val di Chiana with Arezzo and Perugia etc. and Lake Trasimeno. Royal Library, Windsor 12278

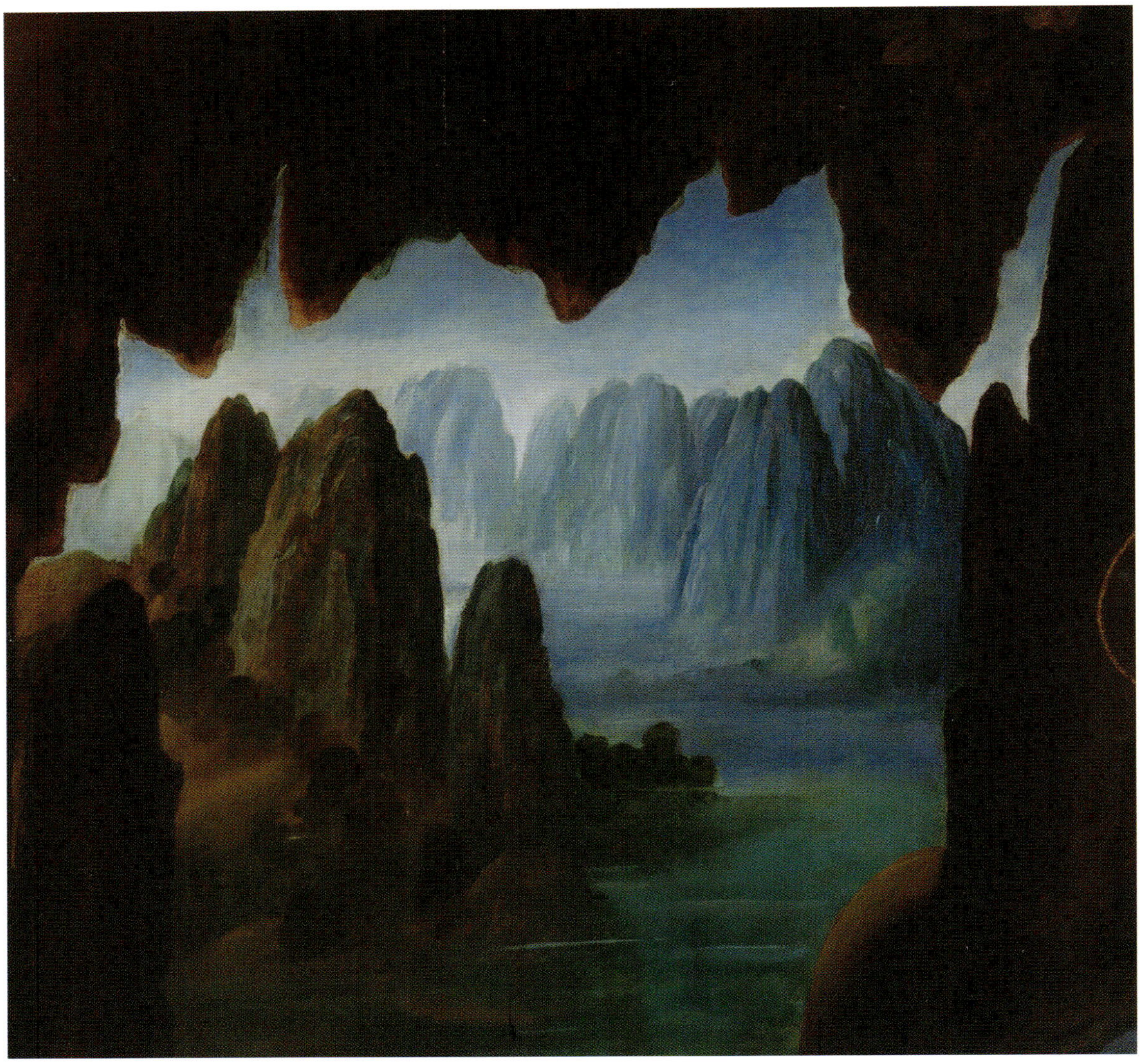

D. Virgin of the Rocks (detail showing blue of the sky and edge contrasts). London, National Gallery

decreasing weight of the Mediterranean area following the passage of waters into the Ocean.[41] Subsequent erosion by the rivers severs the mountains and exposes the strata with fossils on the walls of the valleys. Many notes dealing with the fossils in the 'inner set' of sheets in the Codex also allude to the opening of Gibraltar:

> And we find the oysters together, in very large families, among which you see a lot of those which still have their shells joined, which means that they were left there by the sea while they were still living, when the Straits of Gibraltar were cut.[42]

The opening of the Bosphorus and the consequent formation of fossils are considered on folio 1v.

The outermost sheet of the Codex Leicester (f. 36r) outlines Leonardo's global account of how dry land emerged in 'our hemisphere' as the result of a subterranean collapse, noting that 'this might yet be the cause of why the shells and the oysters that are visible on the high mountains'. Also in MS E, f. 4v, he describes the rising of the land rather than the retreat of the sea and its connexion with the fossils. Fossils, formed on the bottom of the sea, can also be supposed rise in this way:

> Of the sea which changes the weight of the earth
>
> The shells, oysters and other similar animals, which originate in sea-mud bear witness to the changes of the earth round the centre of our elements. This is proved thus: great rivers always run turbid, being coloured by the earth which is stirred by the friction of their waters at the bottom and on their shore.[43]

He continues by explaining that successive sediments discharged on the bottom of the sea form a series of strata supporting the animals on the surface of each stratum before it is covered by the following stratum. Afterwards,

> [T]hese strata were covered over again from time to time, with muds of various thickness . . . and thus these layers of mud became raised to such a height, that they came up from the bottom to the air. At the present time these beds are so high that they form hills or high mountains and so the uplifted side of the earth continually rises and the antipodes draw closer to the centre of the world and the ancient beds of the sea are made into the peaks of mountains.[44]

In a sense, fossils, among the natural forms, might have attracted Leonardo as exactly opposite to the waves of the water: forms created by nature over a very long time span and long lasting, as opposed to the instantaneous formation of fleeting waves. The study of fossils and that of the forms of the water are, from this point of view, the two opposite poles of Leonardo's reflection on the morphology of the natural world in the Codex Leicester, one dealing with present, the other with the past.

41 See Codex Atlanticus, f. 901r, discussed in Chapter 4 in this volume.

42 Leicester, f. 9v.

43 MS E f. 4v.

44 MS E f. 4v.

AFTER LEONARDO

PART II

The History of the Codex Leicester after Leonardo

(16th–early 19th centuries)

Towards a Reconstruction of Leonardo's Legacy as a Scientist

6

DOMENICO LAURENZA

THE HISTORY OF the Codex Leicester is different from most of Leonardo's manuscripts, which, after Leonardo's death in France in 1519, were inherited by Leonardo's pupil Francesco Melzi and then, between the late sixteenth and the seventeenth century, passed through various hands in Milan, Spain, England, and later Paris.[1] The first certain document regarding the Codex Leicester after Leonardo dates back to 1689–90 and informs us that the painter Giuseppe Ghezzi found the Codex among the papers belonging to the heirs of the sixteenth-century sculptor and hydraulic engineer Guglielmo della Porta. After its emergence, various manuscript copies of the Codex were produced. The research presented here, based on study of these copies and of their contexts of production, demonstrates a previously unsuspected circulation of the Codex Leicester, alongside other of Leonardo's scientific ideas, especially in the eighteenth century.[2]

Leonardo's legacy as an artist and an art-theorist has been widely studied, demonstrating the widespread production and circulation of manuscript copies of the *Treatise on Painting* prior to its first printed edition in 1651.[3] Much less studied has been his legacy as a scientist, a virtually taboo subject

1 For the history of the Codex Leicester, see Calvi 1909, pp. xxv–xxix, and Pedretti 1987a, pp. liii–lvi; for the other manuscripts, see Marani 2003; Bambach 2009.

2 See Appendix to this chapter for a summary of the copies.

3 Steinitz 1958 and, more recently, Massey 2003; Fiorani 2003; Bell 2009; Farago 2009; Sconza 2009; Barone 2011; Farago, Bell, and Vecce 2018.

over the past century of Leonardo scholarship, which has deemed it unlikely that Leonardo's scientific works had substantial circulation among scientists before Giambattista Venturi's first printed edition of excerpts in 1797.[4] This was also a reaction against a wave of studies that, during the nineteenth and twentieth centuries, argued without any real evidence in favour of Leonardo's influence on various fields of the Scientific Revolution[5] (not to mention the free speculations of 'week-end', non-professional scholars that still proliferate). Recent attempts to reconstruct the circulation of Leonardo's anatomical manuscripts have not gone beyond the interest in his anatomical drawings by later artists,[6] thus confirming by omission their non-circulation among scientists, while the seventeenth century manuscript anthologies copied on behalf of Cassiano del Pozzo, based on Leonardo's codices then in Milan, have long remained underestimated.[7]

The Codex Leicester, as far as we know, was the only manuscript repeatedly copied in its entirety, and its history, scarcely studied in comparison to Leonardo's other manuscripts and drawings, demonstrates that Leonardo's scientific ideas enjoyed some form of circulation well before the first printed edition of excerpts from Leonardo's scientific manuscripts in Giambattista Venturi's 1797 *Essai*.

Compiled in odd mirror handwriting, left in the disordered form typical of Leonardo's private notes and far from any planned publication, the understanding and circulation of Leonardo's scientific manuscripts could take place only through the heavy work of linguistic mediation: transcription in normal handwriting, reordering of the relationship between texts and images and so on. This was exactly what happened in the eighteenth century to the Codex Leicester: the production of manuscript copies which rendered the idiosyncratic mirror writing legible to all, enlarging his circle of readers, including scientists.

The content of the Codex Leicester roughly consists of three main scientific fields: science of water, geology, and astronomy. These three scientific fields each had a different historical development within the Scientific Revolution. Astronomy was a science that had been in full renovation since the mid-sixteenth century, after Copernicus had published his revolutionary work in 1543. The science of water enjoyed a real development only in the seventeenth century. Benedetto Castelli's fundamental *Della misura delle acque correnti* (*Mensuration of Running Water*) was first published in Rome in 1628. Not coincidentally, a few years later a manuscript collection of Leonardo's hydrological texts after various manuscripts was also realized in one of the seventeenth-century anthologies mentioned above.

Geology developed even later and more slowly. Doubtless, many theories attributed to later periods—from the changes of the earth over time to the organic origin of fossils—were already shared by the many authors who from antiquity down to the Renaissance studied the earth.[8] However, while it is nearly always possible to find precedents for a given vision or theory (the favourite sport of the *continuists*), what makes the difference is the overall place it occupies and, from this point of view, it was only between the eighteenth and nineteenth centuries that the historical vision of the earth generated

4 Venturi 1797.

5 e.g. Govi 1872.

6 Pedretti 2005; Clayton and Philo 2012, pp. 25–7; Laurenza 2004, 2006, 2009, 2012.

7 See Di Teodoro 2009 and 2018b. See 'Giovanni Ambrogio Mazenta, Cassiano dal Pozzo, and the Leonardesque *Del Moto et Misura dell'Acqua*' in the section on 'Guglielmo della Porta and Lombard-Ticinese hydraulic engineers', p. 135 of this chapter.

8 See Chapter 4 in this volume.

and, starting from a certain epoch, became part of a more general and new, historical, and dynamic vision of nature known in biology as *evolutionism*. Steno's main treatise was published in 1669 and the first 'theories of the earth' between the end of the seventeenth and the beginning of the eighteenth century. However, even in 1812, George Cuvier, a leading figure of the new geology and palaeontology, compared the historical development of geology in his time to the phase in which astronomers believed that the sky was made of stones and the moon was as big as the Peloponnese Peninsula.[9] When, at the end of the seventeenth century, the Codex Leicester emerged, and, copies began to circulate between the eighteenth and early nineteenth centuries, it would be as a document of merely historical interest for astronomy and the science of water. On the other hand for geology, a science still in an early phase, the Codex could also be the subject of comparison, if not of inspiration, for theories and problems still in the course of contested elaboration. We will also see how historical and scientific interests sometimes intertwined. For example, in late eighteenth century Paris, the *historical* curiosity towards the astronomical content of the Codex stimulated *scientific* attention for its geological content, in particular for Leonardo's theory of an active uplift of the earth's crust, a topical theory in the geological debates of the time. The reconstruction of the history of the Codex will therefore cover all three areas (astronomy, water, and geology), but, for the reasons above, will mainly concern geology.

The reconstruction of the contexts of production and circulation of the copies of the Codex Leicester reveals a chronological convergence, probably not accidental, between the re-emergence and circulation of the Codex and the development of modern geology. The Codex Leicester spells out an important series of scientific 'messages': the discovery that the Earth has been globally transformed over time; evidence that it has a long 'history' and its present aspect is not the one created by God; theories explaining how the earth changed over time and how the continents emerged from the water; and the empirical study of its structure and the processes behind these structures (strata, rocks, stones, subterranean veins of water, erosion, etc.). These themes were to be central in the development of geology between the seventeenth and the early nineteenth centuries, and it was precisely in this period that handwritten copies of the *Codex Leicester* enjoyed a wide circulation in key places for modern geology: England (London), Italy (Rome, Florence, Naples), France (Paris), and Germany (Weimar).

Alongside direct evidence, we will also present contextual reconstructions: significant historical junctures, scientists, and people in contact with the owners of the Codex Leicester and its copies as potential readers and so on. These more general reconstructions aim, on the one hand, to promote future research about the impact of the Codex, and, on the other, to highlight scientific historical events little known because of their eccentric and interdisciplinary nature and therefore deserving attention in their own right, beyond Leonardo studies.

Guglielmo della Porta and Lombard-Ticinese hydraulic engineers in sixteenth- and seventeenth-century Rome

Thanks to documents dating back to the period in which, at the end of the seventeenth century, Giuseppe Ghezzi acquired the Codex Leicester, we know that the Codex had previously belonged to Guglielmo della Porta (*c.*1515–77), who lived in Rome since around 1537–46.[10]

9 Cuvier 1812 (Cuvier 1992, p. 47).

10 Brentano 1989a.

Known above all as a sculptor, Guglielmo della Porta was also an architect and hydraulic engineer. In sixteenth century Rome, marred by repeated floods of the River Tiber and burdened by the problem of the distribution of drinking water, many of the engineers and architects who dealt with the control of the waters were Lombard or Ticinese. Their shared names do not necessarily imply kinship, but may denote common geographical origin. For example, we have various Della Portas and various Fontanas, and there are at least two Castellis, the founder of modern hydraulic science, Benedetto Castelli (1578–1643), author of *Della misura delle acque correnti* (*Mensuration of running waters*, Rome, 1628), and the architect Francesco Castelli, better known as Borromini (1599–1667), whose father was an hydraulic technician.

Guglielmo della Porta, native of Ticino and active as a young artist in Milan, belongs to this community, and it is very likely that he brought the Codex Leicester to Rome. A resident of Rome in the first half of the seventeenth century, the Milanese hydraulic engineer Giovanni Ambrogio Mazenta wrote *Memorie*, a famous text dealing with Leonardo and his manuscripts, especially those on water. Moreover, in the seventeenth century, *Del Moto et Misura dell'Acqua*, an anthology after various manuscripts in Milan copied on behalf of Cassiano del Pozzo, arrived at the Roman library of cardinal Francesco Barberini. A tradition of study and interest in Leonardo's hydrological studies was therefore formed in Rome in the sixteenth and seventeenth centuries. As we will see, later on, at the beginning of the eighteenth century, the title page of the Codex Leicester will emphasize Leonardo's water studies and continue this tradition. In this period the science of the water was in the midst of its own renaissance and, not by chance, it was this aspect of Leonardo that attracted the most interest in this period.

One of the Roman locations linked to Leonardo is in Via Giulia: the house where, after 1556, Guglielmo della Porta lived and where at least until 1609 the Codex Leicester was housed. The name of its ancient owner is still visible in the inscription on the windows: '*GUILELMUS D P ME [DIOLANENSIS] S [CULPTOR] CI [VIS] RO [MANUS]*', that is 'Guglielmo della Porta Milanese sculptor and a Roman citizen'.[11]

Born in Porlezza, near Como in northern Italy, Guglielmo della Porta spent his youth between Milan and Genoa and, from 1537–46, he lived in Rome. His father (not his uncle as Vasari wrongly claims) Gian Giacomo della Porta, was a sculptor and between 1524 and 1530 was engaged as 'engineer and sculptor' in the Milanese cathedral. It is reasonable to set the acquisition of the Codex Leicester in this Milanese period of the two Della Portas. Vasari writes that Guglielmo 'in Milan devoted much study to Leonardo da Vinci's works, around the year 1530, which was of very great benefit to him'.[12] However, we know that Gian Giacomo left a grandson as his heir rather than his son Guglielmo.[13] Therefore the Codex did not come into his possession as part of his father's inheritance None of the three inventories of Guglielmo's belongings in Via Giulia's house, compiled after his death, include a Leonardo manuscript,[14] but at the beginning of the 1609 inventory, Teodoro della Porta (1567–1638), Guglielmo's son and heir,[15] emphasizes the fact that Guglielmo left 'many drawings of those professions, made on

11 Now 163 and 167 via Giulia. See Salerno 1973.

12 Vasari 1568, vol. 3, part II, p. 843.

13 Brentano 1989b; Kruft and Roth 1973; Coppel Aréizaga 2012.

14 Masetti Zannini 1972; Bertolotti 1881, vol. 1, pp. 142–3 [1578] and vol. 2, pp. 120–61 [1609].

15 Brentano 1989c.

paper by his hand, and by the best various Painters and Sculptors, which at that time and before flourished in Rome and elsewhere'.[16] The Codex Leicester would belong among these drawings and letters.

Another item in one of the inventories (1577) reads: 'A model of a mill invented by the former gentleman, n. 1' ('Uno modello de uno molino de inventione del qm. messer, n. 1'). This clearly relates to Guglielmo's work as an engineer and can be tentatively connected with a note, contained in Guglielmo's albums in Düsseldorf, dealing with his invention of a mill that 'grinds and separates flour without water'.[17] Although this particular mill specifically did not use water, it is natural that Guglielmo had to be familiar with the set of watermills that succeeded each other along the banks of the Tiber. Perhaps his activity as an inventor of mills had something to do with the later misidentification of the Codex Leicester as a manuscript containing studies of water and mills, supposedly realized by Leonardo at the time he took charge of the Milanese canals.[18]

Guglielmo's activity as a hydraulic engineer is well documented in 1571, in relation to his involvement in the renovation works of the ancient Roman Aqueduct of the Virgin water ('acqua Vergine'), which entered the city from the north, piercing the Pincio and then reaching the Trevi fountain. Guglielmo was in charge of the extension of water supplies to public and private fountains located at the foot of the Pincio, near the Campo Marzio. Underground pipelines were planned, and Guglielmo proposed to make them in travertine, an idea that would prove faulty because of the low water resistance of this material. The pipeline was later replaced by terracotta by other engineers. We know that Guglielmo deposited his drawings at the institutions who took care of the work, one of which was located in the Capitol.[19] In the Capitoline Archive there are drawings which, for various reasons, we may connect with Guglielmo's project (Figure 40). The drawings represent round conduits in cross-section, and include the section of earth to be removed as well as the base and the roof.[20] It is hard to say if these documents also include Guglielmo's autograph drawings or contemporary copies. They would be the only surviving visual evidence of Guglielmo's work as a hydraulic engineer,[21] an activity that suggests that his interest in Leonardo and the Codex extended beyond than that of an artist.

16 Damaged and therefore incomplete inventory. State Archive, Rome, Notai A. C. Curtc, 2277 (not 2227 as in Masetti Zannini 1972): 'molti disegni di cette Professioni, fatti in carta di sua mano, e da piu e diversi valentuomini Pittori e Scultori, che in quel tempo e prima fiorivano in Roma et fuori'.

17 This note is included in the smaller (ff. 255–6) of the two sketchbooks by Della Porta now in Düsseldorf, Museum Kunstpalast, Graphische Sammlung, Sammlung der Kunstakademie; Gramberg 1964, p. 216.

18 See the section 'Thomas Coke's Acquisition of the Codex Leicester', p. 166 of this chapter.
in this Chapter.

19 State Archive, Rome, *Notai di Acque e di Strade*, vol. 6, not. Gracchi, 1570–1, protocollo II (or part of it with different numbering), ff. 18ff.: '. . . et come appare in disegno fatto da frate Guglielmo della Porta del Piombo, et detti disegni saranno sottoscritti dall'Ill.mi Deputati accio se habbino da osservare et lassati in Campidoglio, et la copia a me infrascritto della Congregatione. . . .'

20 L 1087 (Figure 40) 'Le pietre di travertino insieme con la sua [vitatura ?]. . . .' Many measurements correspond to those in the State Archive document. Another document (L 1094), connected to the drawings by handwriting and other considerations, is dated 1571 the period of Guglielmo's involvement. D'Onofrio 1986, p. 38, considers the drawing L 1097 predates 1565, and therefore Guglielmo's intervention. Indeed, there are also documents from 1563 (e.g. L 1088–9), but, on the basis of the evidence collected, both L 1097 and 1087 are datable to *c.*1571.

21 Wentworth Rinne 2010, pp. 68 ff briefly mentions, without further indication, artistic drawings by Guglielmo representing the Tiber and its mills. I have been unable to verify their existence. On Della Porta as hydraulic engineer see also Long 2018, pp. 86–9.

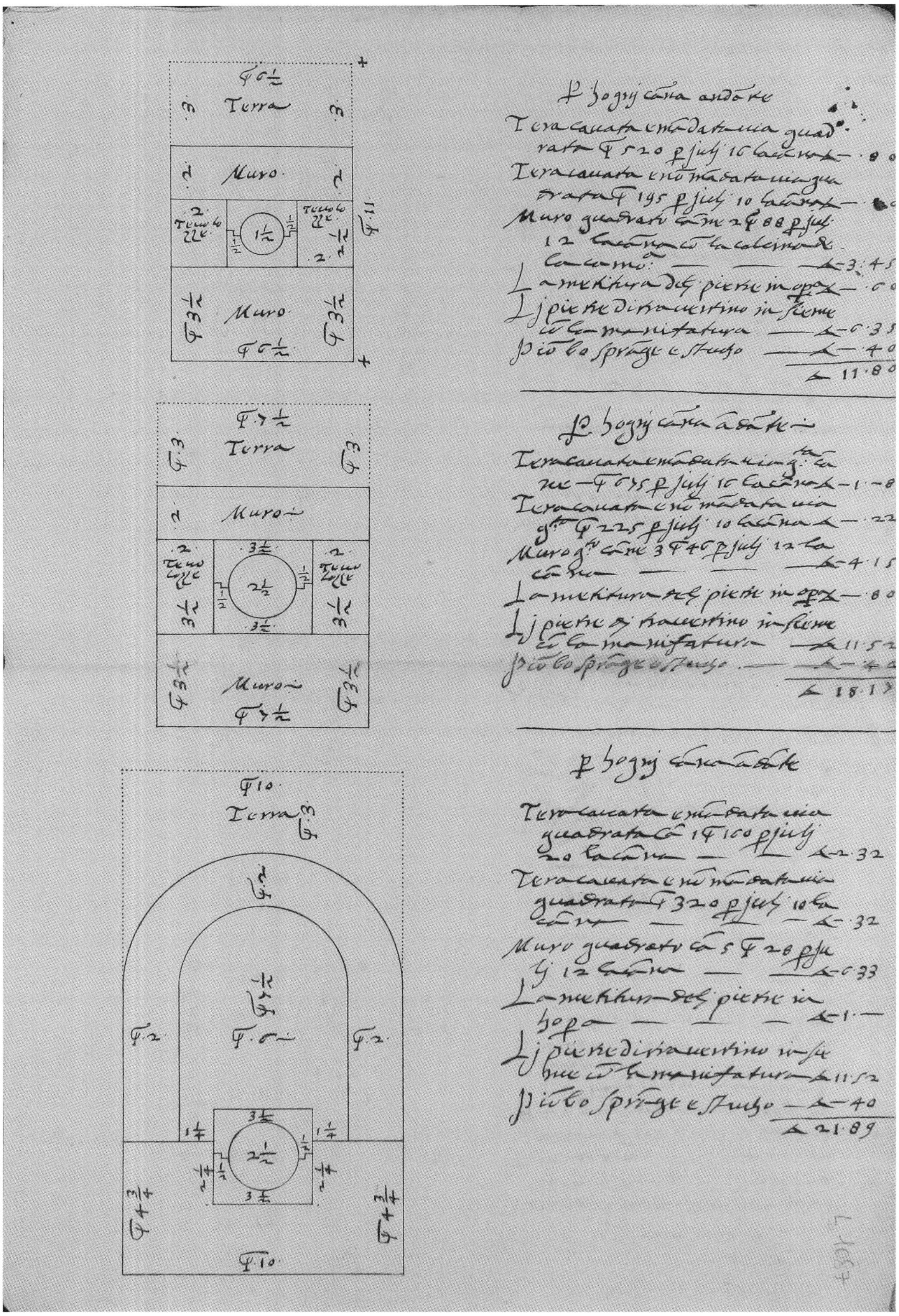

FIG. 40.
Study for an aqueduct, pen and ink drawing. Capitoline Archive, Rome, Carte Boccapaduli, Arm. II, Mazzo IV Busta 104 B, fasc. 66, L 1087.

Guglielmo della Porta's contacts in Rome: Ulisse Aldrovandi, Annibal Caro, and the hydraulic engineer Camillo Agrippa

In 1556 the Bolognese Ulisse Aldrovandi visited Guglielmo's house and workshop in Rome. The purpose of the visit was a sculpture of an ancient mask owned by Guglielmo, of which, Aldrovandi writes, 'It is the moon'.[22] Aldrovandi was a collector of a wide range of natural and artificial things, a naturalist, and the author of an important treatise on fossils.[23] He was one of Guglielmo's contacts potentially interested in the content of the Codex.

An important representative of Roman culture who was in touch with Guglielmo, and who, like him, was interested in Leonardo, was the writer and poet Annibal Caro. Caro gravitated around the Farnese family, like Guglielmo. The restoration of ancient statues in the Farnese Palace was amongst Guglielmo's early tasks in Rome. His most famous work was the monument to the Farnese pope Paul III in St Peter's. Caro was directly involved in the conception of the monument, and therefore in contact with Guglielmo.[24] Vasari mentions a scene conceived by Caro for the Paul III monument which Guglielmo was to make in bronze relief: it was an allegorical scene of the Farnese estates dominated by 'hydrological' representations of a river and a lake.[25] In a letter addressed to the medallist Alessandro Casati in Rome, in 1540, Caro suggested that a medal in honour of Bishop Giovanni Guidiccioni, should include '. . . a Neptune with his chariot and his sea horses and his trident in the way that you know and in the act of commanding the winds and set up the storm. . . . And as for Neptune, you can use, if you like, Leonardo da Vinci's drawing'[26]—almost certainly the lost presentation drawing for Antonio Segni, then belonging to Monsignor Giovanni Gaddi in Rome.[27] A sketch at Windsor, on which Leonardo reminds himself to 'lower the horses', is probably preparatory for the Segni *Neptune*.[28] A drawing of an *all'antica cuirass*, recently on the art market with a convincing attribution to Guglielmo della Porta, includes a detail that may well be derived from Leonardo's Neptune drawing.[29]

In 1547, Caro was asked to read a draft of Vasari's *Lives*. The two had been in touch since Vasari's Roman journey. This is obviously the first edition of the *Lives*, published in Florence in 1550. Now, in the Life of Leonardo, Vasari writes:

> And so many were his whims, that he, philosophizing about the natural things, expected to understand the properties of the herbs, and continued by observing the motion of the sky, the course of the moon and the paths of the sun. He therefore made in his soul such a heretical concept that it did not approach any religion and judged of higher value to be a philosopher than a Christian.[30]

22 'Delle statue antiche che per tutta Roma in diversi luoghi e case si veggono', in L. Mauro, *Le antichità de la città di Roma*, Venice, 1562, pp. 231–2.

23 *Musaeum Metallicum*, Bologna, 1648 (posthumous).

24 Caro mentions Guglielmo in 1551 and 1560. See Caro 1957, vol. 2, pp. 100–2 n. 368 and pp. 104–7 n. 372; vol. 3, pp. 41–3 n. 604.

25 Vasari 1568, vol. 3, part II, p. 844:

> . . . Fecevi similmente una storia pur di metallo, e con ordine del detto Caro, che aveva a essere messa in opera, con due fiumi, l'uno fatto per un lago, e l'altro per un fiume, che è nello stato de' Farnesi. Et oltre a tutte queste cose vi andava un monte pieno di gigli con l'arco vergine, ma il tutto non fu poi mai messo in opera. . . .

26 Caro 1957, vol. 1, n. 134.

27 Pedretti 1991, p. 219.

28 Windsor 12570.

29 Christie's London, *Old Master And 19th Century Drawings*, London, Tuesday, 8 July 2003, no. 22.

30 Vasari 1991 [1550], p. 547.

There might be a connection between the increasing control of the Counter Reformation Church and the cutting of this passage from the second, 1568, edition of the *Lives*; however, more likely this came from corrections suggested by Melzi who had a big impact on the second edition.[31]

What interests us is the inclusion in Leonardo's 'heretical' science of his studies of astronomy mentioned by Vasari.[32] The Codex Leicester, as well as the bold denial of the Biblical Deluge, attributed an 'earthly' nature to the moon, which according to the dominant theological vision at the time, was part of the 'immaculate' celestial world. Did Vasari, through his relationship with Annibal Caro, become aware of the Codex Leicester and its potentially heretical content? It seems plausible.

The astronomical and cosmological content of the Codex could have attracted another Lombard hydraulic engineer active in Rome in these years and who was acquainted with Caro, namely the Milanese Camillo Agrippa (*c.*1516–95).[33] Agrippa, known to Leonardo scholars because his *Trattato di Scientia d'Arme* . . . (*Treatise on Fencing*, Rome, 1553), seems to have been influenced by Leonardo's lost studies on the kinematics of the human body, as echoed in the Codex Huygens, attributed to the Cremonese painter Carlo Urbino (1525/30–1585). Urbino was also the author of the plates in Agrippa's treatise on fencing. Much less known and studied are Agrippa's other scientific treatises and his work as a hydraulic engineer in Rome. Agrippa moved to Rome when quite young, perhaps as early as 1535, and it was mainly in Rome that he carried out his work. One of his treatises, published in Rome in 1553 together with the treatise on fencing, is a dialogue between himself and Annibale Caro. The full title of the book is *Trattato Di Scientia d'Arme, con un Dialogo di Filosofia* (*A Treatise on Fencing, with a Dialogue of Philosophy*). In the *Dialogue* Agrippa emphasizes that it lay outside the conventional scholarly world, which had become something of a topos and specifically recalls the well-known statement by Leonardo that he was 'a man without letters' ('*uomo sanza lettere*'). Agrippa writes: 'I do not belong to the learned people, but among those who wish to reason of everything'.[34] He includes a plate representing a nightmare in which he is assailed by pompous philosophers who 'judged me presumptuous for thinking about such issues when I had not studied'.[35]

Another plate represents Agrippa arguing with Caro (Figure 41). The globe under Agrippa's foot and the armillary sphere in his hands allude to the cosmological content of the *Dialogue*. In general, Agrippa's cosmological and 'geological' ideas, presented in this and later works, recall some aspects of the Codex Leicester. The *Dialogue* begins with a series of statements, which, Agrippa writes, would have scandalized his learned enemies. Even the planets and the moon have 'heaviness' ('gravezza') and are attracted to the centre of the world, but, unable to descend, actually revolve. Likewise, connecting mechanics and cosmology, and adding the framework of Aristotelian meteorological vision, he claims that the centre of the world and the earth were not the same.

31 See Vasari 2019, pp. 33–5.

32 On science and Counter-Reformation, see Clericuzio 2005. For possible connexions with Leonardo's anatomical dissections in Rome, see Laurenza 2004, pp. 19–20.

33 For the dates, see Barni 1960; Nenci 1992, p. 78.

34 Camillo Agrippa, *Trattato di Scientia d'Arme, con un Dialogo di Filosofia*, Rome, 1553, p. lxxxiii recto: '& l'esser mio non si connumera fra li dotti, ma volunterosi di ragionare d'ogni cosa'.

35 Camillo Agrippa, *Trattato di Scientia d'Arme*, p. lxiii verso: 'reputandomi Presuntuoso il voler ragionare di materie simili, non havendo io studiato . . .'.

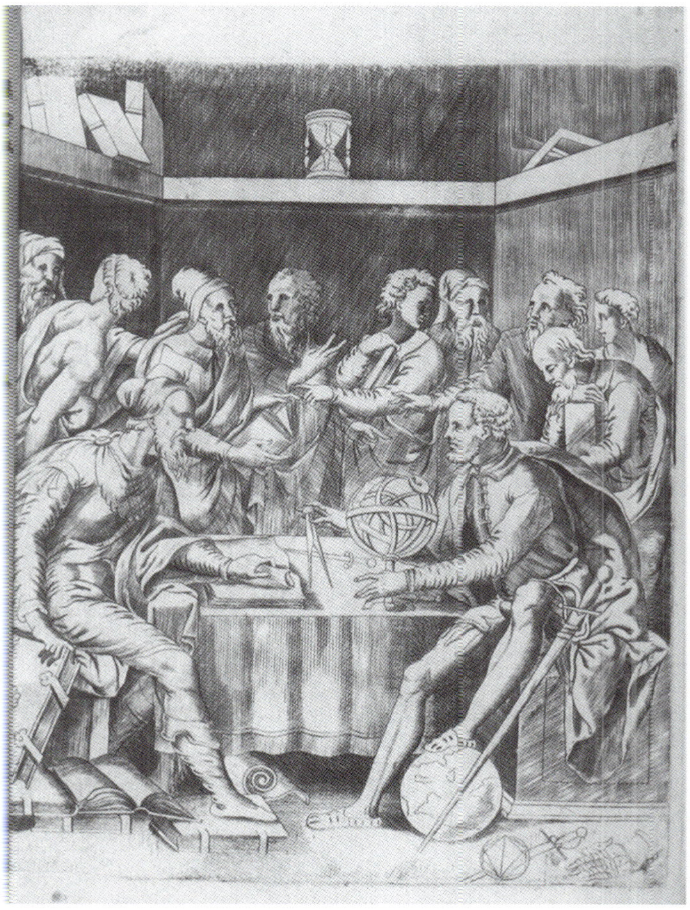

FIG. 41
Camillo Agrippa arguing with Annibal Caro. From Camillo Agrippa, *Trattato di scientia d'arme, con un Dialogo di Filosofia*, Rome, 1553.

The same theory was taken up a few years later by Agrippa in a pamphlet, in which he summarized his proposal on how to move the Vatican obelisk,[36] a problem finally solved in 1586 by another Ticinese active in Rome, Domenico Fontana.[37] Agrippa addresses a mechanical-cosmological conception when discussing the practical problem of how to lift and transport the obelisk, and illustrates this with a diagram in the lower left corner of the large plate in the book (Figure 42a and b).

The obelisk, once moved thanks to the machine, alters, albeit minimally, the static-dynamic equilibrium of the whole globe! The small area of the earth on which the obelisk is set in motion increases its heaviness and thus tends to move downward with respect to the centre of the world (indicated by *A* in the diagram in Figure 42b). Then, in order to counter-balance this downward movement, the opposite area of the earth becomes lighter and moves in the opposite direction, upwards. In the diagram the continuous circumference represents the starting situation with the centre in *A*; the dotted circle *C* indicates the displacement induced by the movement of the obelisk; the dotted circumference *D* represents the move of the earth in the opposite direction. The small circle *F*, around the centre of the world *A*, represents these little movements of the earth. The diagram, although over-simplifying the theory, is effective in representing with dotted lines the reversible movements of the earth, compared

36 Camillo Agrippa, *Trattato . . . di trasportar la guglia in su la piazza di San Pietro [Treatise . . . on How to Carry the Obelisk up to the Square of Saint Peter]*, Rome, 1583.

37 On the involvement of Francesco Masini, master of the painter and Leonardo scholar Matteo Zaccolini, see Laurenza 2016a.

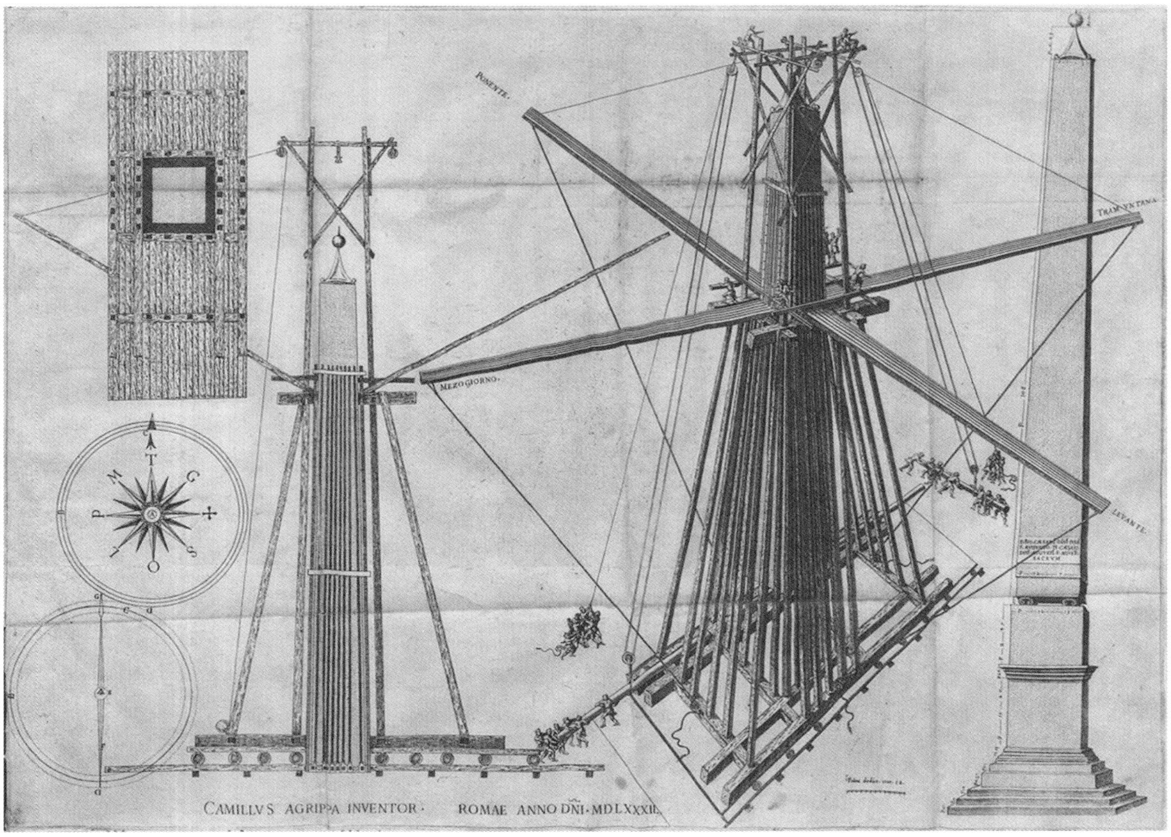

FIG. 42A
Proposal on how to move the Vatican obelisk. From Camillo Agrippa, *Trattato di Camillo Agrippa milanese di trasportar la guglia in su la piazza di San Pietro*, Rome, 1583.

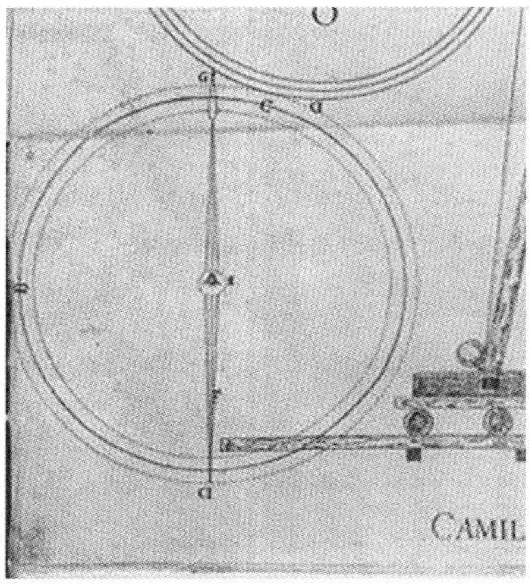

FIG. 42B
Mechanical-cosmological diagram showing how the Vatican obelisk can be moved (detail).

to the baseline and constant state indicated by the continuous circumference. The machine invented by Agrippa, as illustrated in Figure 42a, consisted of lever arms that would move the obelisk thanks to the laws of statics.

Agrippa's plate generically recalls Leonardo's drawings of digging and military machines, both due to the giant size of the machine and the detailed representation of the proportional relationship between the machine and its operators.[38] Furthermore, Agrippa's mechanical-cosmological conception is reminiscent of Leonardo. Leonardo's static/mechanical theory formed the most important part of

38 e.g. Windsor 12647 and CA, ff. 3r, 4r, 149br. On the latter, see Landrus 2010.

the cosmological and geological vision contained in the Codex Leicester, not least through its diagrams. One diagram on folio 36r was meant to explain, according to statics and the dynamic movements of the earth and with respect to the centre of the world, the origin of dry land.[39] Similarly, Agrippa's diagram served to illustrate, on the basis of comparable movements, the displacement of the obelisk and its effects on the static-dynamic equilibrium of the earth.

Neither the conception of the moon and the planets as 'heavy' ('gravi') bodies like the earth nor the association between mechanics and cosmology originated specifically with Leonardo. In particular, the tradition inaugurated by John Buridan had broadly developed the Aristotelian conception in the direction of physical theory.[40] Nevertheless, the frequency with which Agrippa uses, in relation to practical problems, the theory of small movements of the earth and the distinction between centre of gravity of the earth and centre of the world is very close to Leonardo.

Other aspects recalling the Codex Leicester are Agrippa's geological theories. Agrippa seems to be an almost unknown author in the history of geology, and it is worth outlining his ideas as presented in the *Dialogue . . . on the generation of Winds . . .*[41] and in *The virtue: Dialogue . . . on the Statement of the Cause of Motion*.[42] In the *Dialogue* Agrippa describes how, starting from a smooth ball-shaped earth, 'rainwater' excavated the valleys and formed the mountains and the rivers, by discharging the eroded sediments into the sea, increasingly silting it up.

This, together with the withdrawal of some of the waters into underground 'caves' ('caverne sotterranee') might lead to the complete drying up of the earth. The frame is the usual Aristotelian 'meteorology', as confirmed by the title of the book. What is original, and recalls Leonardo, are the details with which Agrippa supports his theory: a compound of geology and geography that goes back to Strabo and at the same time develops it. This is sketched in the *Dialogue* of 1584 and expanded in the later *Virtue*.

In the latter, the description of the lowering of the Mediterranean after the opening of the Strait of Gibraltar, with uncovering of lands previously submerged, all witnessed by the fossils,[43] and the silting up of the sea by the rivers,[44] is based on Strabo. But while Strabo examined zones of the Nile and the Black Sea,[45] Agrippa above all studies Italy and other areas,[46] just as Leonardo had done in the Codex Leicester. And, like Leonardo, Agrippa develops Strabo in a predictive way, foreseeing for the Adriatic Sea and for the Po a destiny similar to that in Egypt for the Nile.

39 See Chapter 4 in this volume.

40 See Chapter 4 in this volume.

41 Camillo Agrippa, *Dialogo . . . sopra la generazione de' Venti, Baleni, Tuoni, Fulgori, Fiumi, Laghi, Valli, et Montagne* [*Dialogue . . . on the generation of Winds, Lightning, Thunders, Thunderbolts, Rivers, Lakes, Valleys, and Mountains*], Rome, 1584.

42 Camillo Agrippa, *La virtù: Dialogo . . . sopra la dichiarazione de la causa de' Moti, tolti da le parole scritte nel Dialogo de' Venti* [*The Virtue: Dialogue . . . on the Statement of the Cause of Motion, developped from the words written in the Dialogue on the Winds*], Rome, 1598.

43 Agrippa, *La Virtù*, p. 15.

44 Agrippa, *Dialogo . . . sopra la generazione de' Venti*, p. 38: '. . . riempiendo il mare per la materia che portano . . .'.

45 See Chapter 5 in this volume.

46 Agrippa, *Dialogo . . . sopra la generazione de' Venti*, pp. 37–8; Agrippa, *La Virtù*, p. 16.

Agrippa writes:

> In this way, by the time it will be possible to say that the Po will go to the tip of Italy, *and that the Po will receive all the rivers that are to the right & left, making a plain of land where the Adriatic Sea is now.*[47]
>
> In Lombardy [are] these: Po, Brenta, Adige, Piave, Tagliamento, Torre & Ars. In Schiavonia [are] these: Poie, Zelina [. . .] and all of them enter the Adriatic. These rivers will make the Adriatic Sea into countryside, like those of Padua and in time towards Venice . . . [and] In Africa the Nile.[48]

And here is Leonardo:

> How the river Po in short time will cause the Adriatic Sea to dry up, in the same way it dried up a great part of Lombardy.[49]
>
> Only the lowest part of the Mediterranean will remain as the bed and course of the Nile, the greatest river that pours into that sea. And it will be joined by all its nearby [rivers], which previously used to pour their waters into that sea, as it is seen that the Po with its nearby [rivers] does, which once poured into the sea that was enclosed between the Apennines and the German Alps and was united to the Adriatic Sea.[50]

The legacy of Strabo and the sixteenth-century earth sciences is still understudied. However, this development of Strabonian issues in a predictive sense and with such a degree of geographic detail in both Agrippa and Leonardo seems rather unusual. Cardano only cites some passages from Strabo; Gororopius mentions Strabo in order to reject his ideas and sustain the '*in situ*' generation of marine fossils;[51] while Fracastoro merely refers to the most obvious case of Ravenna as a well-known example of silting up.[52] Another aspect of Agrippa that comes close to Leonardo is the 'eco-friendly' approach to hydraulic engineering. One of the pages of the Codex Leicester in which Leonardo describes the geological future of the Po and the Adriatic (f. 27v) also contains passages on how to modify a river course by limiting artificial intervention (walls, banks, etc.) to a minimum and by exploiting the course and force of the river itself. This is a strategy similar to that with which Agrippa, avoiding 'walls', proposes to govern the Po: '. . . I say that with the course of the Po itself one can make a channel in the middle . . .'[53]

These correspondences between Agrippa's scientific works and the Codex Leicester might be seen as independent reactions to a common tradition in geology that is not well understood at present. Two factors, however, lead us at least to consider the hypothesis of some influence exercised by the Leonardo's

47 Agrippa, *Dialogo . . . sopra la generazione de' Venti*, p. 43 (emphasis added): 'A questo modo mediante il tempo si potrà dire, ch'l Pò andrà infin alla ponta d'Italia, è ch'l Pò riceverà tutti i fiumi che sono a destra & sinistra, facendo la pianura di terra dove adesso è il mare Adriatico.'

48 Agrippa, *La virtù*, p. 16:

> Nella Lombardia questi, il Pò, la Brenta, l'Adice, la Piaue, il Tagliamento, Torre & Ars. Nella Schiavonia questi, Poie, Zelina [. . .] & tutti questi sboccano nell'Adricatico [*sic*]. Li quali fiumi faranno il mare Adriatico campagne, come quelle di Padoua con il tempo verso Venezia . . . Nell'Africa il Nilo.

49 Leicester, f. 27v.

50 Leicester, f. 10r.

51 Ellenberger 1988, pp. 178ff.

52 See Benedetto Ceruti and Andrea Chiocco, *Musaeum Franc: Calcolari*, Verona, 1622, pp. 407–10.

53 Agrippa, *La Virtù*, p. 20:

> . . . io dico che si può con il proprio corso del Pò far fare un canale nel mezzo, di doi canne e mezza fondo, e più, per fare che si abbassi, & inanischi, acciò non cavalchi l'argini, & le macchine per far questo fondo saranno in modo fatte, qual zapparanno in fondo una quantità di zappe, si come mostrerà il modello. . . .

Codex on Agrippa. Firstly, Agrippa and Della Porta, at the time the owner of the Codex Leicester, attended the same professional and cultural circles in Rome. Secondly, as Leonardo's scholars have long recognized, Agrippa's treatise on fencing has strong links to Leonardo's work.[54]

Both Agrippa and Della Porta belonged to the community of Lombard-Ticinese hydraulic engineers in Rome, and both of them were involved, during the same period, in the construction of the Aqueduct of the Virgin. Agrippa, in the early 1570s, built a large hydraulic wheel and a pump to draw water from the aqueduct over the Pincian Hill to feed the gardens of the Roman villa of Cardinal Giovanni Ricci from Montepulciano.[55] The Cardinal, as papal superintendent for aqueducts and fountains,[56] contributed to Guglielmo della Porta's appointment, while, as we have seen, both Agrippa and Guglielmo were familiar with Annibale Caro and the Farnese circle.[57] It is therefore likely that Agrippa had access to Leonardo's Codex in the house of Guglielmo della Porta.

Carlo Urbino in Rome and the astronomical-optical section of the Codex Huygens

In 1552 the painter Carlo Urbino was in Rome.[58] He is the author of the plates in Agrippa's treatise on fencing,[59] published in Rome the following year. Urbino is almost certainly the author of the Codex Huygens, a manuscript containing various studies, from the proportions and motions of the human body to perspective—some copied from Leonardo, others influenced by him.[60]

The section on perspective begins with studies that straddle astronomy and optics. The topics include: how the shadow of a body varies according to the successive positions of the sun;[61] the rising speed with which the sun seems to descend from the zenith to the horizon;[62] the variable size that the sun seems to assume when approaching the horizon;[63] how shadow varies in the various positions occupied by the moon;[64] and shadows cast by the fire of a candle (Figure 43).

These studies, with their optical-astronomical conceptions applied to the sun–earth–moon system, recall the Codex Leicester. For example, the main drawing in the Codex Leicester, f. 1r, studies how, in some parts of the earth, the sun appears partially occluded when it rises and when it sets, a problem tackled in the Codex Huygens, f. 91r (in particular in the drawing to the lower left; Figure 43). However,

54 Farago 1997; Pedretti 1997; Landrus 2018; Farago 2018.

55 He realized hydraulic works for the same villa *c.*1578, after the Medici had acquired it. See Lombardi 2008; Lincoln 2014, pp. 61ff.

56 Lombardi 2008, p. [illegible]. Agrippa's and Guglielmo's engineering work receives praise in C. Cantù, *Milano storia del popolo pel popolo*, Milano, 1871, p. 200.

57 Both Della Porta and Agrippa were members of the Artistic Academy of the Virtuosi in the Pantheon. See *Statuto della insigne artistica Congregazione de' virtuosi al Pantheon*, Roma, 1839, p. 45.

58 Cirillo 2015, pp. 18–19.

59 Lomazzo 1587, Bk IV, p. 230.

60 Panofsky 1940; Marinelli 1981; Zöllner 1986; Farago 1997; Cremante 1998–9, 1998; Barone 2002, 2012; Farago 2018; Landrus 2018.

61 Huygens, f. 88r (*Del sole. Per l'ombra prima dimostrazione* [*On the sun. First demonstration of the shadow*]).

62 Huygens f. 88v.

63 Huygens, f. 91r.

64 Huygens, f. 89r (*Della Luna. Per l'ombra seconda dimostrazione* [*On the moon. Second demonstration of the shadow*]).

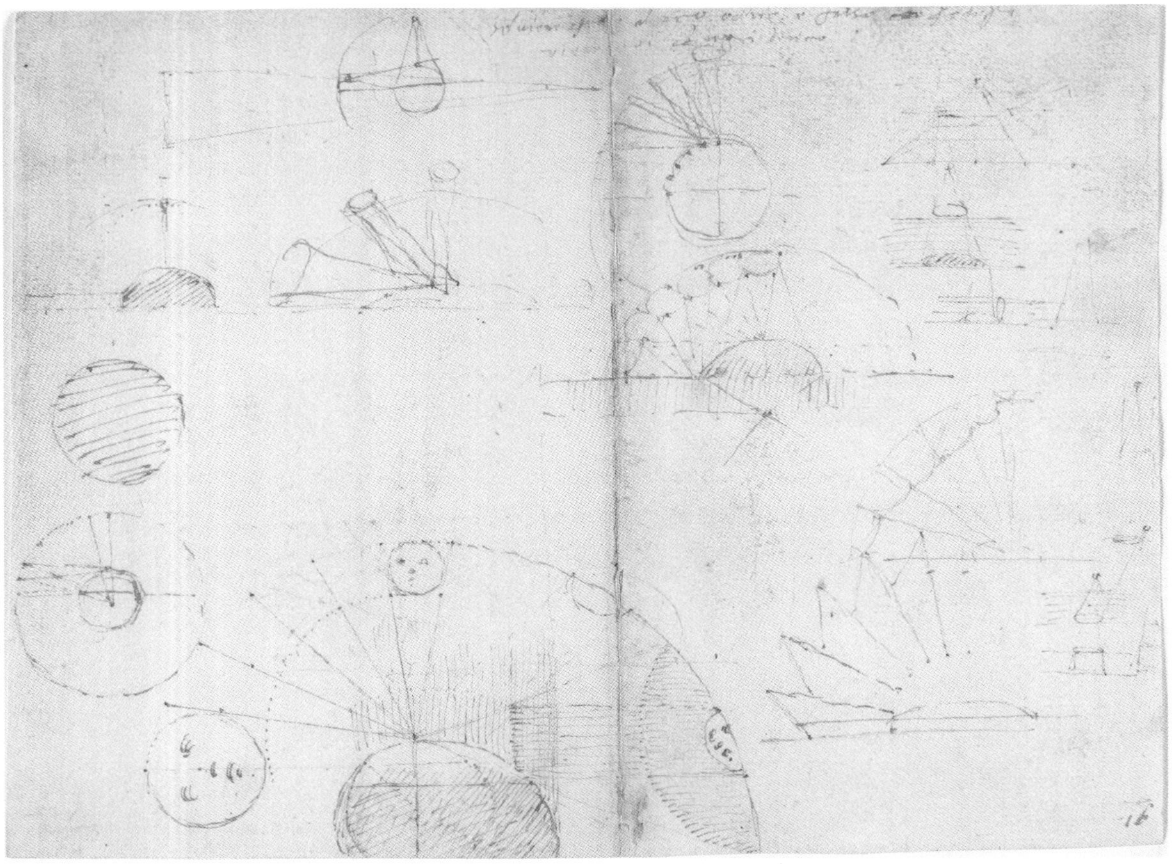

FIG. 43
From Carlo Urbino, *Studies of Perspective and Astronomy*. Morgan Library, New York, Codex Huygens, f. 91r.

the use of dotted lines, arcs, and numbers to indicate various positions, times, speeds, and dimensions of the sun and moon relative to the earth are visual tools typical of the Codex Huygens, and are not present in Leonardo's astronomical studies in the Codex Leicester.

First-hand examination of the Codex Leicester has revealed diagrams in metalpoint or faint black chalk in folio 7r (Figure 44). These were later additions—not by Leonardo—to the main astronomical drawing. They consist of rays starting from the circle representing the Earth (upper left corner) and of several marks and numbers along the main arc representing the orbit of the Sun '1 [?] . . . 2 . . . 3 . . . 4[?] . . .'. In one of the Codex Huygens's astronomical studies (f. 88r), similar numbers are visible along the arc of the Sun's orbit marking the successive hours of daylight and/or the various positions of the sun (Figure 45).

This opens the possibility that Carlo Urbino himself added the diagrams to the Codex. As previously mentioned, Carlo Urbino spent some time in Rome in 1552, where he prepared the drawings for the plates to Camillo Agrippa's treatise on fencing, printed in Rome the following year. Could the late additions (numbers, etc.) on folio 7r of the Codex Leicester date back to this mid-sixteenth Roman period and connexion?

I also suspect that Urbino could have brought the Codex Huygens to Rome at the time when he designed the human kinematic schemes for Agrippa's book on fencing, so close to those in the section of Codex Huygens devoted to the kinematics of the human body.[65]

65 Farago 1997.

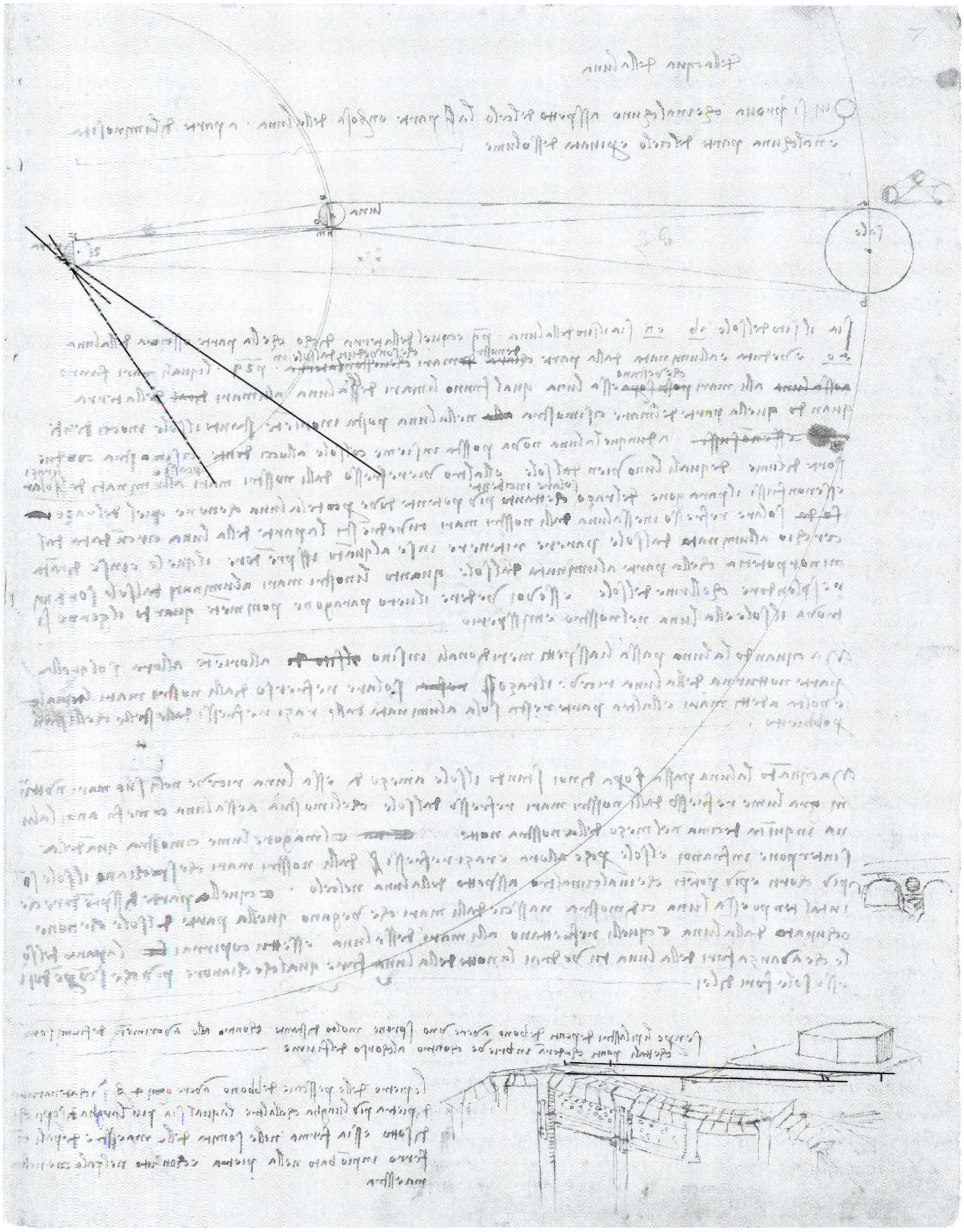

FIG. 44 Later additions to Leonardo's astronomical diagram. Drawings in faint black pencil are highlighted (dark lines); close to the marks added along the main arc of the sun's orbit (slightly visible in the figure) is a series of numbers (1 . . . 2 . . . . 3 . . .). Collection of Bill Gates, Seattle, Codex Leicester, f. 7r. (Digital reconstruction by Curtis Wong and Domenico Laurenza).

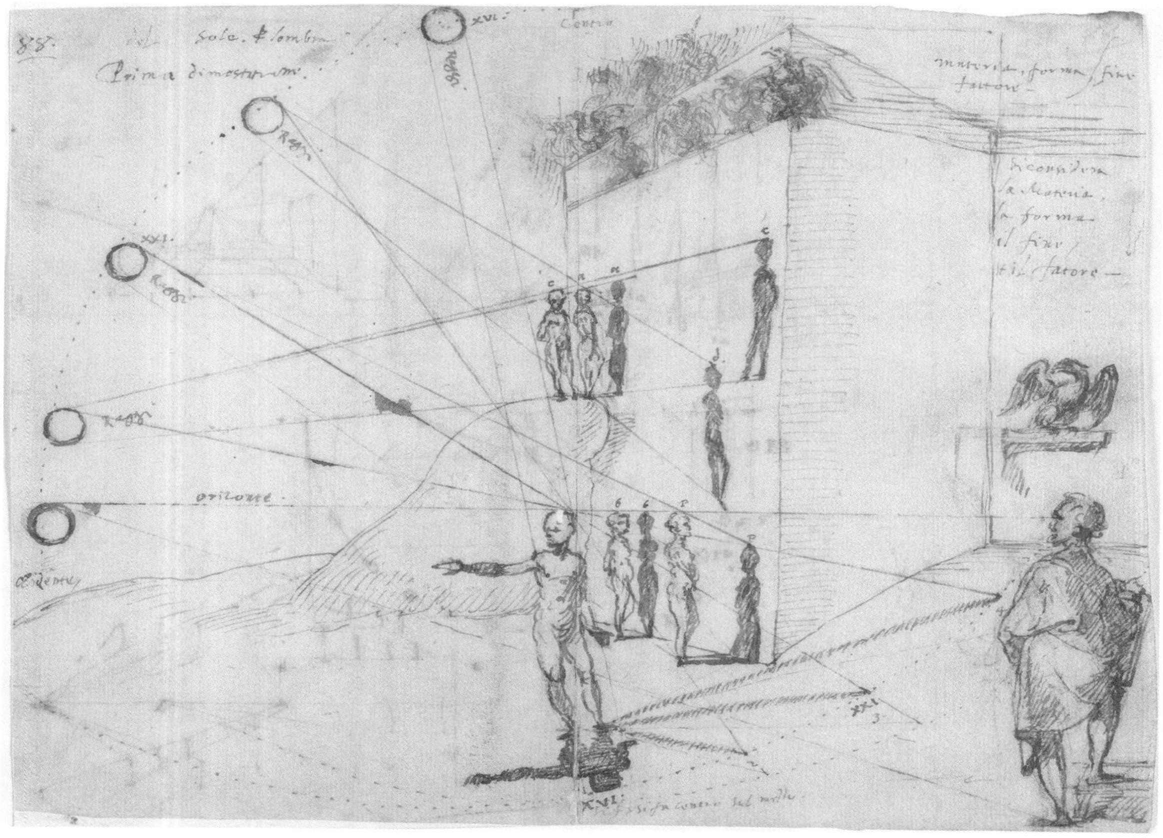

FIG. 45
From Carlo Urbino, *Studies of Perspective and Astronomy*. Morgan Library, New York, Codex Huygens, f. 88r.

Giovanni Ambrogio Mazenta, Cassiano dal Pozzo, and the Leonardesque *Del Moto et Misura dell'Acqua*.

An important episode in Leonardo's legacy as a scientist took place in seventeenth-century Rome, and, not by chance, its most pronounced outcome involved Leonardo's studies on water.

Cassiano dal Pozzo (1588–1657), the learned librarian of cardinal Francesco Barberini, was very interested in Leonardo. His manuscript copy of Leonardo's *Treatise on Painting* was the source of the first printed edition (Paris, 1651). Cassiano was also interested in Leonardo's scientific manuscripts, and promoted a campaign of study and copying of the manuscripts in Milan. Unlike the Codex Leicester, these manuscripts were anthologies from various sources. Among the copyists was Luigi Maria Arconati, son of Galeazzo Arconati. The latter, after having acquired these manuscripts, donated them to the Ambrosiana Library in Milan in 1637.

The correspondence exchanged by Cassiano and Galeazzo Arconati still exists,[66] as well as letters between Cassiano and Giovanni Ambrogio Mazenta, which, among other subjects, also deals with Leonardo's manuscripts in Milan.[67] Mazenta was a Milanese Barnabite friar whose family, according to

66 Carusi 1929–30; Sparti 2003; Barone 2011; Farago, Bell and Vecce 2018.

67 Accademia dei Lincei Library, Rome, Ms. Dal Pozzo VI bis, f. 322r. See Niccolò 1991. In 1635, Mazenta writes about having asked the Canonico Alfieri in Milan to send Cassiano 'as much as he could find of Leonardo' ('quel tanto che poteva ritrovar di Leonardo').

his *Memorie* came for a time into possession of thirteen of Leonardo's manuscripts, some of which were subsequently to pass to Pompeo Leoni and Galeazzo Arconati.[68] Mazenta spent a period in Rome from 1620 to 1626 and then again in 1635, the year he died. In addition to being an architect, he was a hydraulic engineer. His brother Guido, who came into possession of six of the thirteen Leonardo manuscripts, wrote a *Discorso intorno al far navigabile il fiume Adda*.[69] Almost certainly on behalf of Cassiano, Mazenta wrote the *Memorie dei fatti di Leonardo da Vinci a Milano e dei suoi libri*, in which he discussed Leonardo and in particular the history of the thirteen manuscripts that at some point came into possession of his family.[70]

Two aspects of this Roman episode should be emphasized. Firstly, much of Mazenta's *Memorie*, written while in Rome, concerned Leonardo as a hydraulic engineer. Secondly, the most thematically consistent among the copies made in Milan on behalf of Cassiano was an anthology of Leonardo's texts on the science of water, which, with the title *Del Moto et Misura dell'Acqua*, was realized by Luigi Maria Arconati in 1643 or 1634.[71]

This is not surprising. The study of water was in fact, in these years, a science undergoing its own vigorous renaissance. This is the period in Rome in which Benedetto Castelli published treatises in 1628 and 1639, founding the modern science of water in a newly mathematical manner, following a theoretical and Galilean approach.[72] It was in Rome that Mazenta and Cassiano developed the myth of Leonardo as a great hydraulic engineer, which, though less substantiated, had already been circulating in the previous century.

Mazenta does not mention the Codex Leicester in his *Memorie*, perhaps since he is mainly concerned with the history of the thirteen Leonardo's manuscripts formerly owned by his family. It is surprising to think that Cassiano, who struggled hard to obtain copies of Leonardo's manuscripts in Milan, should not know of the existence of a Leonardo manuscript in Rome under his nose!

These two Roman episodes, one concerning Leonardo's manuscripts in Milan and the other the Codex Leicester, intersected at some point. Mazenta, in his *Memorie*, after describing Leonardo's work on the Martesana canal and other Milanese hydraulic works, quotes Leonardo's writings 'about nature, weight, motion and revolutions of the water' ('circa la natura, peso, moto e giri dell'acque').[73] This information, supplemented by Giovanni Paolo Lomazzo's mention of Leonardo's drawings of mills moved by water and without water,[74] was used by Du Fresne in his life of Leonardo published in the first edition of the *Treatise on Painting* in 1651: 'the enterprise of the Martesana canal gave him the opportunity to write a

68 Gramatica 2008 [1919], p. 73 n. 1 and p. 54 n. 34.

69 Guido Mazenta, *Discorso intorno al far navigabile il fiume Adda* [*Discourse on How to Render the River Adda Navigable*], Milan, 1599. See Gramatica 2008 [1919], pp. 33–4.

70 Gramatica 2008 [1919]; Barone 2013.

71 Vatican Library, Ms. Vat. Barberiniano Lat. 4332. See Carusi 1929–30, p. 506; Sparti 2003, pp. 145ff. for the date 1643; Gramatica 2008 [1919], p. 82 for the date 1634; Buccaro 2011 and 2016; Di Teodoro 2009.

72 Arredi 1933; Maffioli 2008 and 2010

73 Gramatica 2008 [1919], p. 61.

74 Lomazzo, 1585, p. 652: '. . . de i quali Leonardo ne disegnò trenta carte di chiaro e scuro che sono pervenuti nelle mani d'Ambrogio Figino, dove si veggono alcuni molini che macinano con acqua e altri senza, tutti fra se diversi'.

book on the nature, weight, and movement of the water, full of many drawings of various wheels and mill machines in order to regulate the course of the waters and lift them up'.[75]

As we will see in the next section, it was on this basis that Sebastiano Resta and Giuseppe Ghezzi, around 1690, devised the title page of the Codex Leicester, which thus extended this tradition of interest in Leonardo's water science that had been established in Rome between the sixteenth and seventeenth centuries. It should also be noted that the letters exchanged by Mazenta and Cassiano dal Pozzo touched on geological issues. A letter by Mazenta to Cassiano dated 1632 deals with the philosophical and scientific works by cardinal Carlo Borromeo, in press in Milan.[76] It is followed by two printed pages with the index of Book 3 of a work including many geological subjects:

> Chapters of Book 3 [. . .] Which are the most important theories on the origin of the world. Chapter 2 [. . .] On the changes of the earth. Chapter 8. On the end of the world [. . .] On Leone Ebreo and Galeotto Marzio theories [. . .] On gems [. . .] On the rivers and the surface of the Earth.[77]

This printed work has not been identified and could be one of the few copies of Carlo Borromeo's scientific works published only as proofs.

It also is worth remembering that in 1635 Athanasius Kircher arrived in Rome, where he was to write *Mundus Subterraneus*,[78] a work whose importance for the history of geology has been increasingly recognized by historians. It is not known if this 'geological' side of Cassiano-Mazenta relationship involved Leonardo's manuscripts as well. Many other aspects of this seventeenth-century Roman episode in Leonardo's legacy as a scientist are awaiting clarification.

Cassiano was also an admirer of Matteo Zaccolini's optical and perspective writings. Zaccolini (1574–1630) lived for a long time in Rome and was so much influenced by Leonardo that he imitated his mirror writing, as reported by early sources. This may be confirmed by a seventeenth century manuscript written in reverse, in the manner of Leonardo, recently attributed to him.[79] Was Zaccolini aware of the Codex Leicester in Rome? A study of his extensive and largely unpublished manuscripts might give an answer.

Rome 1689: Sebastiano Resta and the Re-emergence of the Codex

Although the Codex Leicester had almost certainly been in Rome since the first half of the sixteenth century, the fully documented history of the Codex begins only at the end of the seventeenth century, when Giuseppe Ghezzi acquired the Codex, in 1689. Shortly thereafter, a copy was realized. The Codex remained in Rome until *c.*1713–17. All of this happened at the time of important cultural and scientific

75 'Vita di Lionardo da Vinci descritta da Rafaelle Du Fresne', in *Trattato della Pittura di Lionardo da Vinci*, Paris, 1651: 'L'impresa del naviglio di Mortesana gli diede occasione di scrivere un libro della natura, peso e moto delle acque, pieno di gran numero di disegni di varie rote e machine per molini, e regolar il corso dell'aque, e levarle in alto.'

76 Accademia dei Lincei Library, Rome, Ms. Dal Pozzo VI bis, f. 301. The letter also mentions the 'canonico Alfieri', later quoted in relation to Leonardo's manuscripts in a letter of 1635.

77 Accademia dei Lincei Library, Rome, Ms. Dal Pozzo VI bis, ff. 302–3. 'Libri Tertii Capita [. . .] Quaenam sint praecipuae opiniones de Mundi exortu Cap. II [. . .] De mutationibus terrae caput VIII. De fine mundi [. . .] De opinione Leonis Hebrai & Galeoti Martij [. . .] De Gemmis [. . .] De fluminibus & superficie Terrae [. . .].' On other letters dealing with Borromeo's works see Lumbroso.

78 Athanasius Kirchner, *Mundus Subterraneus* [*The Subterranean World*], Amsterdam, 1664–65.

79 Laurenza 2016a. On Zaccolini, see Bell 1985.

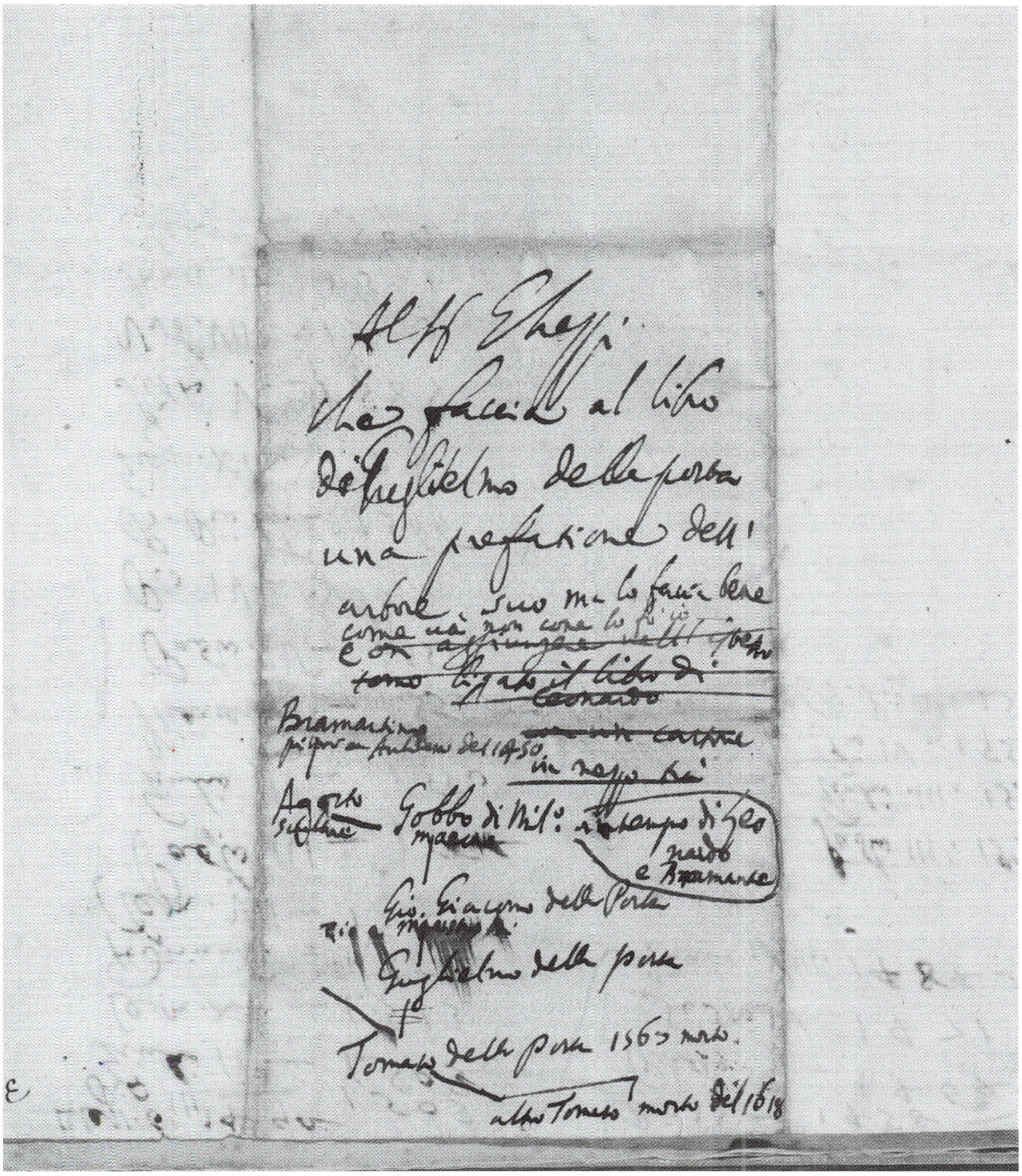

FIG. 46 Attributed to Sebastiano Resta, Letter to Giuseppe Ghezzi concerning the Codex Leicester, opening lines. Museum Kunstpalast, Graphische Sammlung, Sammlung der Kunstakademie, Düsseldorf, Guglielmo della Porta's sketchbook, f. 3.

intersections in Rome. On the one hand two protagonists of the nascent geology, Gottfried Wilhelm Leibniz and Agostino Scilla, happened both to be in Rome in this period; on the other, Rome saw an increasing cultural and editorial interest towards empirical and scientific knowledge hidden in old and unpublished manuscripts. Interest in a manuscript by Leonardo, as a scientist and hands-on professional of the Renaissance, was not an isolated case.

Two documents, one in Milan and the other in Düsseldorf, inform us that Giuseppe Ghezzi rediscovered the Codex in Rome in 1689, according to the latter, or in 1690 according to the former (Figure 46).

The document in Milan reads:

> In the life of Leonardo written by the French Raphael Du Fresne it is said that the occasion of the Martesana canal was the reason why [Leonardo] wrote about the nature, weight and motion of the water with various drawings of wheels and mills and machines to raise the waters and run their course. This book, written backwards and left-handed as Leonardo used to, has been found in the current year 1690 by Mr. Giuseppe Ghezzi in a chest with manuscripts and drawings by Guglielmo della Porta, sculptor of the tomb of Paul III . . .[80]

The document in Düsseldorf reads:

> To Mr. Ghezzi. . . . In a chest remained in the inheritance of Fra Guglielmo della Porta's heirs, in the current year 1689, together with books and drawings by Guglielmo, a Leonardo manuscript book on the motions of the Sun and on the motion of the waters was found and it is highly esteemed by Mr. Giuseppe Ghezzi, as it is worthy.[81]

According to these documents, the Codex was found among documents that had belonged to Guglielmo della Porta, a sculptor and a hydraulic engineer living in Rome, who died in 1577. Both these documents also include general information on Guglielmo della Porta and Leonardo's life and work, including a mention of the theft of the thirteen Leonardo manuscripts (not including the Codex Leicester) in Milan at the end of the sixteenth century by Lelio Gavardi.[82]

The documents in Milan and Düsseldorf were published as anonymous by Bonelli and Gramberg and they have been considered as such in later studies of the Codex Leicester. An unknown document (Figure 47) allows us now to give a name to the author of the other two documents and link them to the present title page of the Codex Leicester, which, as is known, dates back to the Ghezzi's period (Figure 48).

The document is a short note addressed to Giuseppe Ghezzi by Sebastiano Resta (1653–1714), collector, great connoisseur of drawings, and a key figure in the late seventeenth-century Rome art market. In his note, Resta describes a manuscript compiled by Leonardo at the time of his hydraulic engineering work in Milan and quotes, as source, Raphael Du Fresne's life of Leonardo in the first print edition of the *Treatise on Painting* (Paris, 1651).

80 State Archive, Milan, Miscellanea Lombarda, vol. III, n. 91:

> Nella Vita di Leonardo descritta da Raffaele Du Fresne francese si dice che l'occasione del Naviglio della Martesana li fu motivo di scrivere della natura, del peso e moto delle acque con varij disegni di Ruote e Molini e machine per aque di alzar l'aque, e regolar il corso loro. Questo libro manuscritto all'indietro alla mancina al solito di Leonardo l'ha di presente quest'anno 1690 il Signor Giuseppe Ghezzi ritrovato in un coffano di manuscritti e disegni di Gugliemo della Porta, scultore del sepolcro di Paolo III . . .
>
> See Bonelli 1906 and Calvi 1909, pp. xxv–xxvii.

81 Graphische Sammlung, Sammlung der Kunstakademie, Museum Kunstpalast, Düsseldorf, Guglielmo della Porta's sketchbook, Binding, ff. 1–3, in particular f. 2:

> Al Signor Ghezzi . . . in un coffino rimasto nell'eredità de' descendari di fra Guglielmo della Porta si sono trovato quest'anno 1689 con disegni e libri di Guglielmo un libro manuscritto de moti del Sole e del moto dell'acque di mano di Leonardo, che li tiene, come porta il dovere, in grande stima il Signor Giuseppe Ghezzi.
>
> See Gramberg 1964, vol. 1, pp. 144–5 n. 244.

82 Milan document, full transcription in Bonelli 1906: 'Didactic texts for the Milanese academicians remained at Vaprio, among which thirteen volumes were stolen by Lelio Gavardo' ('A Vavero restarono scritti d'istruttione agli accademici di Milano, tredici volumi de' quali . . . Lelio Gavardo . . . li rubò . . .').

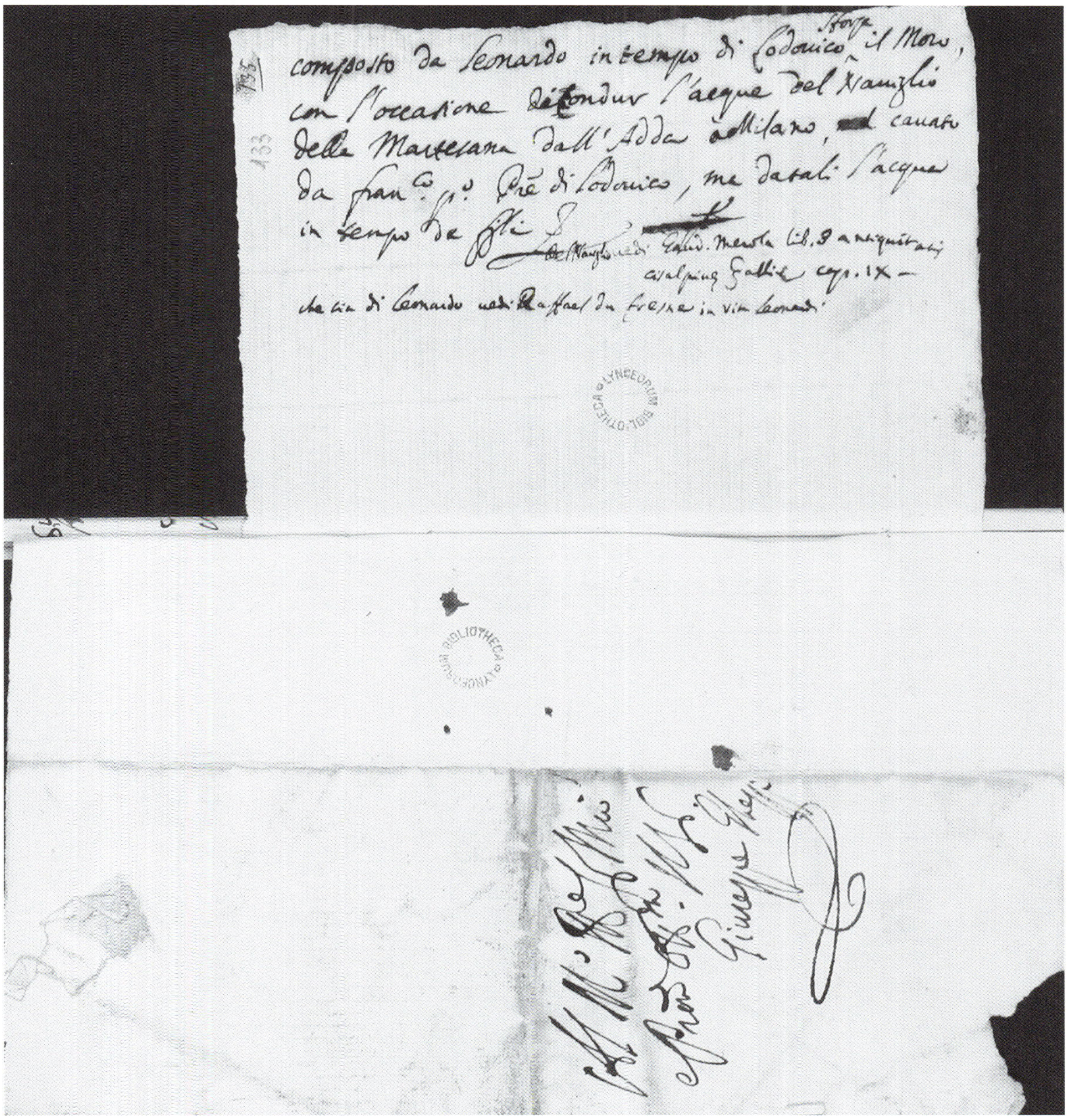

FIG. 47
Sebastiano Resta, Note concerning the Codex Leicester, Corsiniana Library, Rome, MS 31 B 9, f. 133r.

The document reads:

> *Composed by Leonardo at the time of Lodovico ^ Sforza ^ il Moro, on the occasion of the conducting of the waters of the Martesana canal from Adda to Milan* [. . .] excavated by Francesco [Sforza] the First, father of Lodovico, but the water was provided at the time of his sons. On the Naviglio see Equid. [?] Merola [?] Book 3 antiquitatis cisalpinae Gallicae chapter IX. *On the fact that [this codex] is by Leonardo see Raphael Du Fresne's Leonardo's life.*[83]

The unmentioned subject of Resta's note is, no doubt, the Codex Leicester and the text highlighted in italics corresponds to the title page of the Codex Leicester (Figure 48), which reads:

83 Corsiniana Library, Rome, Ms 31 B 9 (former Cors. 1403), f. 133r (modern numbering in pencil) (Emphasis added): Composto da Leonardo in tempo di Lodovico ^Sforza^ il Moro, con l'occasione di condur l'acque del Naviglio della Martesana dall'Adda a Milano [. . . ?] cavato da Fran co p.° P.re di Lodovico, ma datali l'acqua in tempo de fili. [. . . ?]. De Naviglio vedi Equid. [?] Merola [?] lib. 3 antiquitatis cisalpinae Gallicae cap. IX. Che sia di Leonardo vedi Raffael Du Fresne in Vita Leonardi.

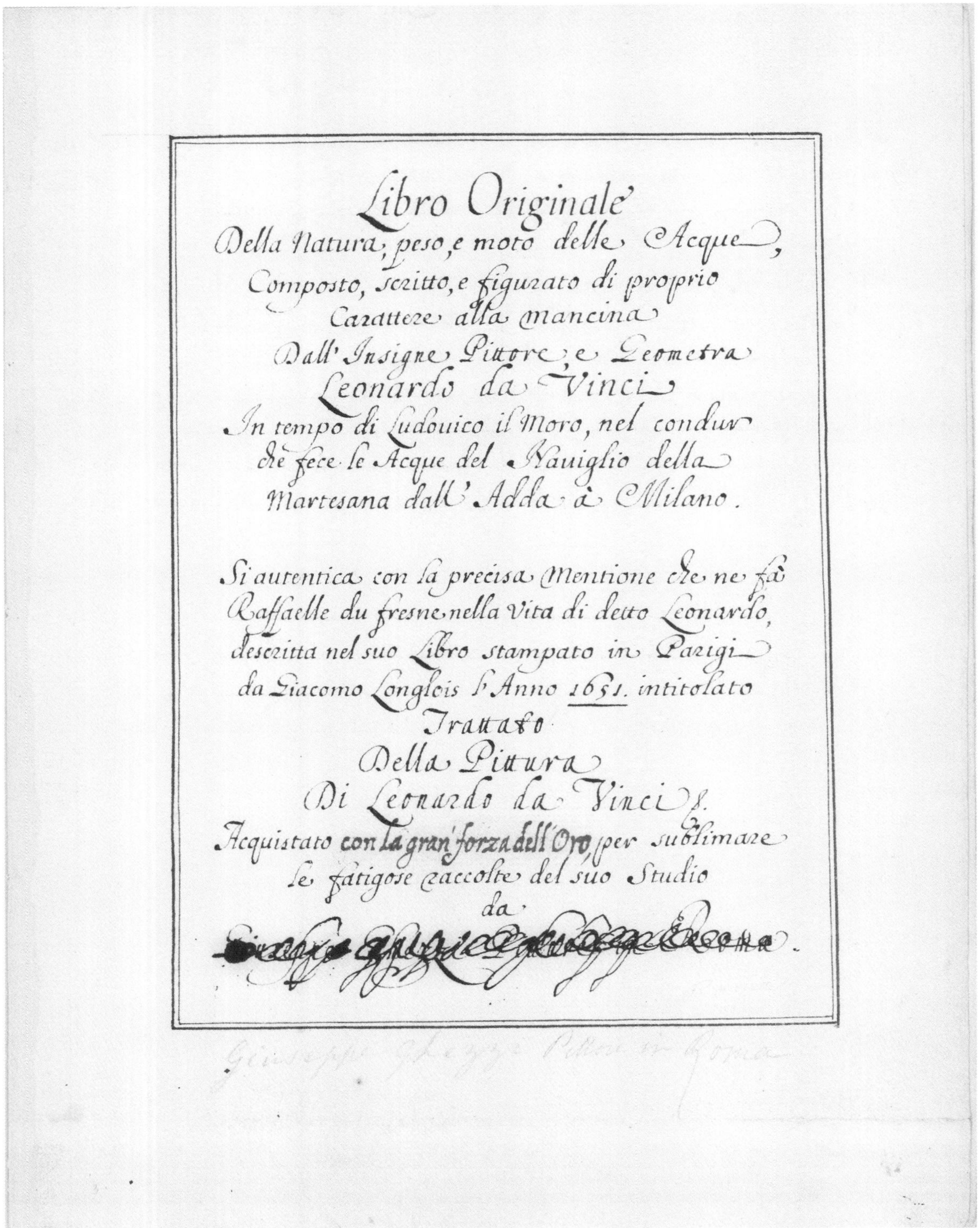
Libro Originale
Della Natura, peso, e moto delle Acque,
Composto, scritto, e figurato di proprio
Carattere alla mancina
Dall'Insigne Pittore, e Geometra
Leonardo da Vinci
In tempo di Ludouico il Moro, nel condur
che fece le Acque del Nauiglio della
Martesana dall'Adda à Milano.

Si autentica con la precisa Mentione che ne fà
Raffaelle du fresne, nella Vita di detto Leonardo,
descritta nel suo Libro stampato in Parigi
da Giacomo Langlois l'Anno 1651. intitolato
Trattato
Della Pittura
Di Leonardo da Vinci &.
Acquistato con la gran forza dell'Oro, per sublimare
le fatigose raccolte del suo Studio
da
[redacted].

Giuseppe Ghezzi Pittore in Roma

FIG. 48
Codex Leicester, Title page, late seventeenth–early eighteenth century. Collection of Bill Gates, Seattle.

> [. . .] Composed [. . .] by the distinguished [. . .] Leonardo da Vinci in the time of Ludovico il Moro, on the occasion of the conducting of the waters of the Martesana canal from Adda to Milan. This is certified in the precise mention by Raphael Du Fresne in the Life of the aforementioned Leonardo.[84]

Thanks to this new document it is now possible, for reasons of content, context, and handwriting, to attribute to Sebastiano Resta the two known documents, including similar notes on Leonardo, the

84 Leicester, title page. For a full transcription, see the Introduction to this volume.

Codex Leicester, and its previous ownership by Guglielmo della Porta.[85] Thus, a new figure emerges in the history of the Codex Leicester, namely Padre Sebastiano Resta. This is important, since both Resta and Ghezzi were at the centre of an extensive web of cultural contacts in Rome. Thanks to these connections, the Codex Leicester re-emerged and began to circulate.

The new document concerning the Codex Leicester is included among other papers and letters exchanged by Resta and Ghezzi, some of which related to various aspects of Leonardo's work and life.[86] Resta was a close friend of Giuseppe Ghezzi, who also collected drawings and manuscripts. In a caricature by Pier Leone Ghezzi, Giuseppe's son, they are portrayed examining drawings and manuscripts. Resta, on the left in the drawing, regularly sent Ghezzi scholarly notes, such as the three documents concerning the Codex Leicester.[87]

The documents demonstrate what happened shortly after Giuseppe Ghezzi acquired the Codex Leicester. Just as he did with other works in his collection, Ghezzi lent the Codex Leicester to Resta for a while, as suggested by a passage in the Düsseldorf document, in which Resta writes that he plans to send Ghezzi a family tree of Guglielmo della Porta together with various other papers including 'the book of Leonardo with a cartoon in the middle'. Resta, after examining the Codex, drew up the three notes in question, which formed a collection on which the final title page of the Codex Leicester was based.[88]

The verso of the Codex Leicester's title page provides other possible evidence of this exchange of information between Resta and Ghezzi (Figure 49).[89] It contains a note that summarizes the content of the Milan document: the left-handed handwriting, the fact that the Codex was compiled at the time of hydraulic engineering works in Milan and 'those thirteen Volumes that I have already mentioned' ('quei tredici Volumi de quali de già abiamo fatto menzione'), the thirteen stolen manuscripts described in Resta's note in Milan.[90]

The note on the verso of title page of the Codex Leicester does not seem to be in Resta's handwriting. It can be attributed to someone (perhaps Ghezzi) who, on the title page, summarized Resta's notes as received by Ghezzi. The title page, given the corrections at the bottom, should perhaps be considered a provisional version, never replaced by a final one. It is difficult to say if it was written by Ghezzi or Resta, but certainly the latter supplied the content.[91]

The year of acquisition of the Codex was more likely 1689, and not 1690 as so far assumed. The note in Düsseldorf, bearing the 1689 as date of acquisition, given its fragmentary and less specific content about both Leonardo and Della Porta, was probably written first, when Ghezzi, shortly after its acquisition,

85 Further evidence is the fact that in the Milan document Resta quotes himself in the third person as he often did: 'Il P. R. [i.e. Padre Resta] havrebbe gola di questa gloria [. . .]'. See also Prosperi Valentini 1999, p. 114 n. 9, for a problematic attribution of the Düsseldorf document to Resta.

86 On some of these letters, see Bottari 1822–5, vol. 3, pp. 487, 489, 512, 515–18.

87 Cod. Ott. Lat. 3114, f. 16r. See reproduction in Prosperi Valentini 1999, p. 113.

88 Graphische Sammlung, Sammlung der Kunstakademie, Museum Kunstpalast, Düsseldorf, Guglielmo della Porta's sketchbook, Binding: 'il libro di Leonardo con un cartone in mezzo'. See Gramberg 1964, vol. 1, p. 144, no. 244.

89 For full transcription see the Introduction to this volume.

90 For full transcription see the Introduction to this volume, n. 6.

91 A further possible document on Ghezzi and Resta about Leonardo is in Rome, Gabinetto Comunale delle Stampe (Ms. 2452, f. 47v, new numbering in pencil). Among Ghezzi's lists of paintings for exhibitions, the following note seems by Resta: 'cose del med[esimo] Leonardo' followed by 'P Sebastiano Resta Dev [or Rev.] del Cav [?] Giuseppe [. . . ?]' .

Soleua il Vinci scriuere Alla Mancina, secondo l'uso degl' Ebrei, nella qual Maniera erano scritti quei tredici Volumi de quali gia abiamo fatto Mentione, et essendo il cara ttere buono, si leggeua assai facilmente. Mediante un spe chio grande e' probabile ch' egli facessi questo, accio tu =tti non legessero cosi facilmente i suoi Scritti. L'impresa dell' nauiglio di Morteseana gli diede ocasione di scriuere un libro della Natura, pero e Moto delle Aque, pieno di gran Numero di disegni di Varie cose, Machine e Molini, e Regolar il Corso dell' aque, e leuarle in Alto =

FIG. 49
Codex Leicester, title page verso, late seventeenth–early eighteenth century. Collection of Bill Gates, Seattle.

sent Leonardo's Codex and Della Porta's drawings to Resta so that he could study them. Then, in 1690 Resta elaborated the more complete note now in Milan, specifically focused on Leonardo and the Codex Leicester, saying mistakenly that the acquisition of the Codex took place in 1690. The title given by Resta to the Codex in 1689 ('a manuscript book on the movement of the sun and on the movement of the waters')[92] is more general; that used in 1690 and consequently in the title page of the Codex ('on

92 Graphische Sammlung, Sammlung der Kunstakademie, Museum Kunstpalast, Düsseldorf, Guglielmo della Porta's sketchbook, Binding, ff. 1–3 (Gramberg 1964, vol. 1, pp. 144–5 no. 244): 'un libro manuscritto de moti del sole, e del moto dell'acque'.

the nature of the weight and movement of the waters')[93] is more precise. It borrows information contained in Du Fresne's life of Leonardo and erroneously connects it with the Codex, emphasizing its hydrological and 'gravitational' content. Resta did the same in another later hand written note regarding the Codex Leicester. He recorded in a copy of Pellegrino Orlandi's *Abecedario Pittorico* (Bologna, 1704) that 'Mr. Ghezzi owns Leonardo's writings on weight and on water'.[94]

A favourable conjunction: Scilla and Leibniz in Rome, new geology, and editorial interest in unpublished manuscripts

None of the notes exchanged by Resta and Ghezzi mentions the geological content of the Codex, and only Resta's later note, which lists 'weight' as a separate topic, seems to go beyond water and astronomy. This is understandable, since at the time geology existed neither as a term nor as an autonomous science. It was in a very embryonic phase of its development, compared to the more developed science of water. By connecting the Codex with the latter, Resta and Ghezzi emphasized the role of Leonardo as a hydraulic engineer, following a tradition that, as we have seen, originated in the sixteenth century. The study of the earth was part of a broader cosmological science, to which Resta's title of 1689 seems allude.

Rome in 1689 was the site of an important conjunction both in the history of geology and in the relationship between practical and theoretical knowledge. This conjunction deserves attention, because the protagonists were directly or indirectly in contact with Ghezzi and Resta. At the time of the re-emergence of the Codex Leicester in the late seventeenth century, there was still a strong permeability between different branches of knowledge. For example, the study of the history of the earth was deeply influenced, among other things, by the most advanced research methods in civil history, while Agostino Scilla, by profession a painter who was living in Rome during these years, was the author of a paleontological treatise that would influence later geological studies.[95]

The painter Giuseppe Ghezzi, the owner of the Codex Leicester, was a learned man, who possessed a degree in law and acted as a notary for a time. Then he became a painter, and joined several important cultural institutions, including the Arcadia, one of the most important literary academies of the time. He was also Secretary to the powerful artists' Academy of St Luke in Rome. The painter-geologist Agostino Scilla was among the members of the Academy of St Luke at the time, and therefore certainly in contact with Giuseppe Ghezzi. The records of the Academy's activity in 1689–90 testify to Scilla's and Ghezzi's regular attendance at the Academy's sessions.[96]

Then, in May 1689, the great philosopher, mathematician, and 'geologist', Gottfried Wilhelm Leibniz (1646–1716) arrived in Rome,[97] and, in the six months he spent there, got to know various representatives of the scientific and cultural communities, including Agostino Scilla.[98] Two aspects of

93 State Archive, Milan, Miscellanea Lombarda, vol. III, n. 91 (Bonelli 1906 and Calvi 1909, pp. xxv–xxvii): 'della natura del peso e moto delle acque'.

94 'Il Sig. Giuseppe Ghezzi tiene scritti di Leonardo de peso e dell'Acqua' (quoted in Nicodemi 1956, p. 312). Whereabouts unknown.

95 *La vana speculazione disingannata dal senso*, Naples, 1670. See Chapter 5 in this volume. On history and geology, see Rappaport 1997a.

96 Archive of St Luke Academy, Rome, vol. 45.

97 Robinet 1988.

98 Leibniz's answer to a letter of M. Thévenot dated August 10, 1691 ('Vous aurez entretenu aussi à Rome Mr. Scylla [. . .]') reads: 'Je suis entièrement du sentiment de M. Stenonis, du peintre Scilla et d'autres qui tiennent les coquillages des montagnes et les

Leibniz's work, both focused in these years, are relevant for our reconstruction: the geological treatise *Protogaea*, written shortly after his return from Italy, in 1691–3,[99] and the *Discours sur la méthode* composed in 1689–90 and concerning the relationship between hands-on professionals and theoretical knowledge.[100]

The *Protogaea* project originated as a part of a work devoted to the history of the house of Hanover and the Duke of Brunswick, therefore within civil history, which Leibniz decided to extend back to remote times. In *Protogaea*, Leibniz presents a theory of the earth emphasizing the huge changes undergone by our planet over time. He starts from the study of marine fossils in the mountains, denying theories that explain them as 'jokes of nature', and devotes long discussions and numerous illustrations to the demonstration of the animal origin of fossils, using them as evidence that the sea once covered what subsequently became mountains. And, within the section on the marine fossils, Leibniz cites with admiration the work of 'a learned painter', namely Scilla.[101]

Indeed, Scilla's treatise *La vana speculazione disingannata dal senso*[102] contained a systematic demonstration of the natural origin of marine fossils. Scilla is proud to present himself as a painter 'naked of learning' ('nudo di buone lettere'), and juxtaposes his visual acumen to the vain mental speculations of the philosophers. All of this might resemble Leonardo's stance, but Scilla's specific approach to the study of fossils is methodologically different from Leonardo's. He avoids overarching theories about the past of the earth and limits himself to demonstrating the animal origin of fossils through visual comparisons between fossils and marine animals.[103] The *Vain Speculation* was published in 1670, before the documented re-emergence of the Codex Leicester in 1689. Therefore, any suggestion of a possible influence on Scilla remains hypothetical, dependant on the earlier circulation of the Codex Leicester when it still belonged to Della Porta's heirs, or on the reconstruction of Scilla's late scientific studies, after the *Vain Speculation*.[104]

Leibniz's *Protogaea* recalls some aspects of the geological content of the Codex Leicester, to a greater extent than Scilla's treatise. We have already noted his demonstration of the animal origin of marine fossils in the mountains. Another common aspect is the hypothesis of many diluvial events, in addition to the Biblical Flood, for which Leibniz provides a scientific explanation. Finally, Leibniz considers, but does not favour, the possibility of a change in the earth's centre of gravity as a possible explanation

glossopètres de Malte et autres pour des dépouilles d'animaux marins' (24 August/3 September 1691). *Akademie-Ausgabe: Leibniz. Sämtliche Schriften und Briefe.* Reihe 1: *Allgemeiner politischer und historischer Briefwechsel*, Band 7: *1691–1692*, ed. G. Scheel, K. Müller, and G. Gerber, Berlin, 1992, pp. 330 and 353; Robinet 1988, pp. 79–80.

99 Gottfried Wilhelm Leibniz, *Discours touchant la méthode de la certitude et de l'art d'inventer pour finir le disputes et pour faire un peu de temps des grands progrès* [*Discourse Touching the Method of Certitude and the Art of Discovery in Order to End the Disputes and Make Progress Quickly*], integral text printed posthumously in 1749. Cf. Leibniz 2008 [1749], pp. xxxvii–xxxviii; Rappaport 1997b; Laurenza 2019b.

100 *God. Guil. Leibnitii, Opera Philosophica quae Exstant Latina Gallica Germanica Omnia*, ed. J. E. Erdmann, Bern, 1840, pp. 172–6.

101 Leibniz 2008 [1749], pp. 75 and 79. See Findlen 2013, p. 124.

102 Scilla, *La vana speculazione disingannata dal senso* [*The Vain Speculation Undeceived by Senses*], Naples, 1670.

103 Findlen 2013; Laurenza 2018b and forthcoming b.

104 Alternatively, the publication date of 1670 for Scilla's book may be incorrect, since, as recently proved, the treatise was published in Messina (Sicily), and not in Naples as indicated in the printed edition.

of the variations between mountains and sea levels that occurred in the past, a hypothesis on which Leonardo based his theory of the earth at the end of the Codex Leicester (f. 36r).[105]

Again, the state of our knowledge does not allow us confidently to claim that, while in Rome, Leibniz gained direct or indirect knowledge of the contents of the Codex Leicester. Leonardo is never mentioned. But, it should be recalled, Leibniz published an image directly based on a plate in Scilla's book without citing his source;[106] this is an example of the otherwise well-known phenomenon of plagiarism in nascent geology. The coincidence of place (Rome), date (1689), and people (Scilla–Ghezzi) remains suggestive, potentially connecting Leibniz to the re-emergence of the Codex Leicester. We should also take into account Leibniz's interest in hands-on professionals' knowledge that remained buried in manuscripts, of which Leonardo's unpublished codex would have been a perfect example.

Leibniz developed the latter position in a text written at the time of his trip to Italy in 1689–90, namely his *Discours*. He shares a Baconian conception of the useful connection between practical and theoretical knowledge, and in the *Discours*, by developing a programme envisioned previously, he presents a eulogy and a history of technology, and advocates knowledge of the *arts*.[107] He also mentions wisdom, little known or completely ignored, that was contained in manuscripts of difficult access ('hors de leur place') and promotes a search programme for these manuscripts with the aim of finally making their content public.[108]

Now, this is just what took place in Rome in these very same years. The re-emergence of the Codex Leicester and its later history can be linked to this cultural horizon. In the very same years in which Leonardo's Codex re-emerged, the manuscripts of two Renaissance scientists were printed in Rome for the first time: the anatomical treatise by Bartolomeus Eustachius (1500/10–1574),[109] and a treatise on mineralogy and fossils by Michele Mercati (1541–93),[110] both edited by the Roman scientist and anatomist Giovanni Maria Lancisi. The title page of the former showed a scene of dissection after a drawing by Pier Leone Ghezzi, Giuseppe's son. In 1667, Steno had already published a plate illustrating fossils after the unpublished manuscript by Michele Mercati (a plate in turn reused in Leibniz's *Protogaea*), praising the rediscovery of a manuscript that remained unpublished.[111] As we will see, the future owner of the Codex Leicester, Thomas Coke, promoted a similar operation when he acquired and arranged the printing of an unpublished manuscript on Etruscology written at the beginning of the seventeenth century.

A similar interest in Leonardo among editors in Rome during this period is suggested by an unknown document regarding the history of the Codex Leicester. Giuseppe Ghezzi, as we have seen, was also a member of the literary Arcadia Academy and it is to one of the founders of the Arcadia, Giovanni

105 Leibniz 2008 [1749], p. 15.

106 Leibniz 2008 [1749], plate VI, fig. *a*, based on Scilla's *Vain Speculation*, plate VI. Laurenza 2018b and forthcoming b.

107 *God. Guil. Leibnitii, Opera Philosophica*, pp. 173–4: 'Il y a une infinité de belles pensées et observations utiles, qui se trouvent dans les auteurs, mais il y en a encore bien plus qui se trouvent dispersées parmi les hommes dans la pratique de chaque profession.'

108 *God. Guil. Leibnitii, Opera Philosophica*, p. 174.

109 Bartolomeus Eustachius, *Tabulae anatomicae*, Rome, 1714.

110 Michele Mercati, *Methalloteca Vaticana*, Rome, 1717.

111 Mercati's manuscript 'aeterna nocte latuisset sepultum, nisi literatissimi Dati solertia ab Orco reductum fuisset'. Steno, 1667, p. 70.

Mario Crescimbeni (1663–1728), that we owe the following note: 'and it is unusual that he [Leonardo] wrote backwards, as the writers mention and as I realised in an original by him, which is preserved by Giuseppe Ghezzi [. . .]'.[112] Crescimbeni is obviously referring to the Codex Leicester. Ghezzi evidently showed the Codex to his learned friends. Crescimbeni adds: '[Leonardo] left behind various extremely noble Treatises, which are in press'.[113]

We are at the beginning of the eighteenth century and Crescimbeni mentions a contemporary project to print Leonardo's scientific manuscripts. This is, as far as I know, the earliest definite piece of news about such an enterprise, after Cassiano dal Pozzo's failed attempt in the seventeenth century to publish a collection of texts from Leonardo's manuscripts then in Milan (Ambrosiana Library) and now in Paris. It is unclear whether this previously unnoticed—and aborted—project to publish Leonardo's manuscripts was conceived in Rome at the time of the re-emergence of the Codex Leicester and whether the latter was part of it. But certainly Crescimbeni's simultaneous mention of the Codex Leicester in Rome and of an editorial project to print the unpublished manuscripts by a Renaissance scientist and artist such as Leonardo, connects the re-emergence of the Codex to contemporaneous editorial practices concerning other Renaissance scientists such as Eustachius and Mercati, in keeping with the ideas expressed by Leibniz in his contemporary *Discours*.

Potential and actual readers: Francesco Bianchini, Lodovico Antonio David, and Pier Leone Ghezzi

Another leading figure of the scientific and cultural Roman world, with whom Leibniz would remain in epistolary relationship, was Francesco Bianchini (1662–1729). Astronomer and author of a fundamental *Istoria universale* (Rome, 1698), which was innovative in its use of empirical data in archaeology and geology, Bianchini was, like Scilla, in contact with Ghezzi and Resta.

In a letter dated 1711, part of the Resta–Ghezzi correspondence, which, as we have seen, also mentioned Leonardo and the Codex Leicester, Resta informs Ghezzi of his intention to send Bianchini a 'quinterno' (five folded sheets) on Bramante.[114] Bianchini therefore participated in Resta and Ghezzi's exchange of information on artists and engineers of the past. Furthermore, Ghezzi's son, the painter Pier Leone, was engaged by Bianchini as illustrator in his archaeological research on the Palatine in Rome, and Bianchini will be portrayed while holding a telescope in one of Pier Leone's famous drawn caricatures.[115]

Bianchini could have more than one reason to be interested in the Codex Leicester, if he happened to look at it in Ghezzi's house. The Deluge, which he did not question and which he used to explain marine fossils, was part of his research across history and geology.[116] At the same time, as an astronomer,

112 Giovanni Mario Crescimbeni, *Comentarj . . . intorno alla sua istoria della Volgar Poesia* [*Commentaries . . . on his History of the Vernacular Poetry*], Rome, 1710, p. 200: 'ed è cosa singolare, che egli scrivesse con caratteri mancini, come riferiscono gli Scrittori, e noi abbiamo riscontrato in un suo originale, che si conserva da Giuseppe Ghezzi Segretario dell'Accademia del Disegno di Roma, e pittore insigne del nostro secolo'.

113 Crescimbeni, *Comentarj . . .*, p. 200: 'e lasciò dopo di se varj nobilissimi Trattati, i quali vanno in istampa'.

114 Corsiniana Library, Rome, Ms. 31 B 9, f. 47r.

115 Vatican Library, Cod. Ottobon. Lat. 3113, f. 106.

116 For example *Istoria universale*, pp. 246–8.

he published images of Venus, showing a developed sensitivity for light-and-shadow contrasts. A thorough search in Bianchini's manuscript documents, still largely unexplored and kept mostly at Vallicelliana Library in Rome, will perhaps provide further evidence.

An obstacle to existing or potential readers of the Codex was obviously the odd handwriting. A transcription of the Codex was a critical requirement for better comprehension and circulation. We know that at least one transcription was underway in these years. Ludovico Antonio David (1648–1720/8) was a Ticinese painter who, from 1686, spent a part of his life in Rome. He also had scientific interests.[117] In a well-known letter of 15 March 1704, addressed from Rome to the learned Ludovico Antonio Muratori in Modena, David writes that he was copying a 'Leonardo manuscript which was in Rome' ('un Ms. di Lionardo che è in Roma'), which both Calvi and Bassoli identify with the Codex Leicester.[118] This identification is confirmed by David's unpublished *L'Amore dell'arte*, a text of art theory in which, during a long discourse against the Academy of St Luke's coat of arms designed in 1704 by Giuseppe Ghezzi, he adds:

> [. . .] but if he [i.e. Giuseppe Ghezzi] would deign to give me some hints in the past winter, in which several times I came to his house to read some sheets from a small manuscript book by Lionardo Vinci at the time and even earlier than he was inventing the coat of arms, I would have told him my thoughts about that, as I did about other things regarding the Academy.[119]

David published an astronomical pamphlet,[120] which is consistent with his interest in the Leonardo Codex. Leonardo was for David above all a prestigious model, qualifying and legitimizing his status of artist-scientist. Indeed, at the beginning of his anti-Vasarian *L'Amore dell'arte*, Leonardo is first mentioned as an example of an artist with strong scientific interests unfairly undervalued by Vasari.[121] David then came into conflict with Ghezzi, as well as with other people associated with him, including Resta and Bianchini. As regards the Bianchini, David's criticism of the technically challenging attempts to lift the ancient Column of Antoninus Pius, successfully accomplished in 1705, may also refer to Bianchini, who was directly involved in the work. In contrast to works that required 'excessive expenses and the arms of more than 500 men', David counterpoises an invention by Leonardo.[122]

117 Bassoli 1954; Engass 1987; Spiriti and Cappelli 2004.

118 Campori 1866, pp. 534–5; Calvi 1909, p. liv n. 9; Bassoli 1954; Pedretti 1987a, p. liv; Marani 2004.

119 Archive of St Luke Academy, Rome, Ms. n. 35, f. 13:

> [. . .] e però se egli [i.e. Giuseppe Ghezzi] si fosse degnato di darmene qualche cenno nel passato inverno, nel quale più volte, era portato a casa sua per leggere alcune carte d'un Libretto M. S. di Lionardo Vinci nel qual tempo e molto prima si sa, che egli andava ricercando l'Impresa; sopra questa gli averei detto liberamente il mio parere: come aveva fatto d'altre cose attinenti alla stess'Accademia.

See Bassoli 1954 on the Modena version of the same manuscript.

120 Lodovico Antonio David, *Ad Inclytam Rempublicam Literariam . . .*, Frankfurt, 1716.

121 Archive of St Luke Academy, Rome, Ms. n. 35, f. 3.

122 For Resta, see David's letter to Muratori (28 April 1703). For Bianchini, see David's letter to Muratori (10 October 1705): 'Il proietto della colonna [. . .] Nella seconda parte [of L'Amore dell'arte] darò la spiegazione di ciò che è creduto ideale ed enigmatico perchè solo da pochi dotti inteso, e dimostrerò essere invenzione di Archimede spiegata da Lionardo Vinci, ed approvata da tutti i più insigni matematici, e praticata ancora ove ben s'intendono le mecaniche'. See Campori 1866, p. 542; Marani 2004, p. 92.

David's copy of the Codex Leicester is now lost or its whereabouts unknown.[123] Perhaps a copy, even partial, was also produced by Ghezzi or Resta, who were so interested in the Codex and Leonardo. However, a 1762 inventory of Ghezzi's library includes no copy of the Codex, but it does include a *Trattato della pittura di Leonardo da Vinci* (i.e. an apparently unnoticed manuscript copy of the Treatise), as well as a *Summary of geometry, of the quality of fruits and herbs, the movements of the rivers and other*.[124] Did the latter manuscript include passages from the Codex Leicester? The inventory is incomplete,[125] but the hypothesis remains plausible. As we will see, there is more evidence of interest in the Ghezzi house towards Leonardo and scientific topics related to the Codex Leicester.

Pier Leone Ghezzi, Giuseppe's son, besides being famous for his caricatures, of which we have mentioned an example, was interested in the sciences. The list of manuscripts in his library, between Leonardo's *Treatise on Painting* and the scientific *Summary*, includes a now-lost manuscript by Pier Leone: *Mattematica, Sfera, Pesi, e numeri del Cav. Ghezzi* [*Mathematics, the Sphere, Weights and Numbers by Cav. Ghezzi*].[126] He was interested in archaeology and was one of the first authors of stratigraphic drawings, including those made on behalf of Bianchini.[127] The birth of archaeology from antiquarianism took place in the same period as the birth of geology and fruitful exchanges occurred between these two new-born sciences. Pier Leone was also interested in geological phenomena. In his diary he collected news of the 1732 earthquake in South Italy and realized paintings in which he reconstructed with a certain realism the effects of other notable earthquakes, such as that of 1688 in Benevento. In 1696 he achieved first place in a competition at the Academy of St Luke with a painting representing the effects of the Deluge.[128]

In any case, it is worth considering that the Codex Leicester was in Ghezzi's home during the years of Pier Leone's training and education. Pier Leone represented himself and his books in a self-portrait when he was fourteen or fifteen years old: alongside treatises by Vitruvius and Euclid we also find one by Leonardo da Vinci ('L. VINC', on the spine) (Figure 50a and b).[129]

This could be a generic reference to Leonardo, or it might be a reference to the *Treatise on Painting*, which, as we have seen, was listed in the 1762 list of manuscripts in Ghezzi's library. Alternatively it could be a reference to the Codex Leicester. Indeed, it is worth noting what Pier Leone represented in a later drawing of the artist's studio, dated 1712 (Figure 51a and b).[130]

123 On the belongings of David's son, Antonio, who lived in Rome, see Cola 2005 and *Inventarium bonorum hereditorum bone memorie Antonii David factum ad instatiam . . .*, State Archive, Rome, Notai Tribunali A.C. Vol. 7219, ff. 332–52.

124 *Compendio di Geometria, qualità de' frutti, et Erbe, Moto de' Fiumi et altro.*

125 Dorati da Empoli 2008, pp. 455ff.

126 Dorati da Empoli 2008, p. 456.

127 Fusconi 1994; Lo Bianco 2010; Pigozzi 2010.

128 The Diary is in Angelica Library, Rome, Ms. 2136, f. 24. Geological texts were in Pier Leone's library, including Scilla, and also books linked to the later history of the Codex Leicester (see Sections 'Thomas Coke's Acquisition of the Codex Leicester' and 'Scientists Versus Artists', pp. 166 and 192 of this chapter): *Thomae Dempsteri De Etruria Regali*, Florence, 1726, and Filippo Maria Bonini, *Il Tevere incatenato*, Rome, 1663.

129 On Pier Leone's age at the time of the selfportrait, see Lo Bianco 2010.

130 Fusconi 1994.

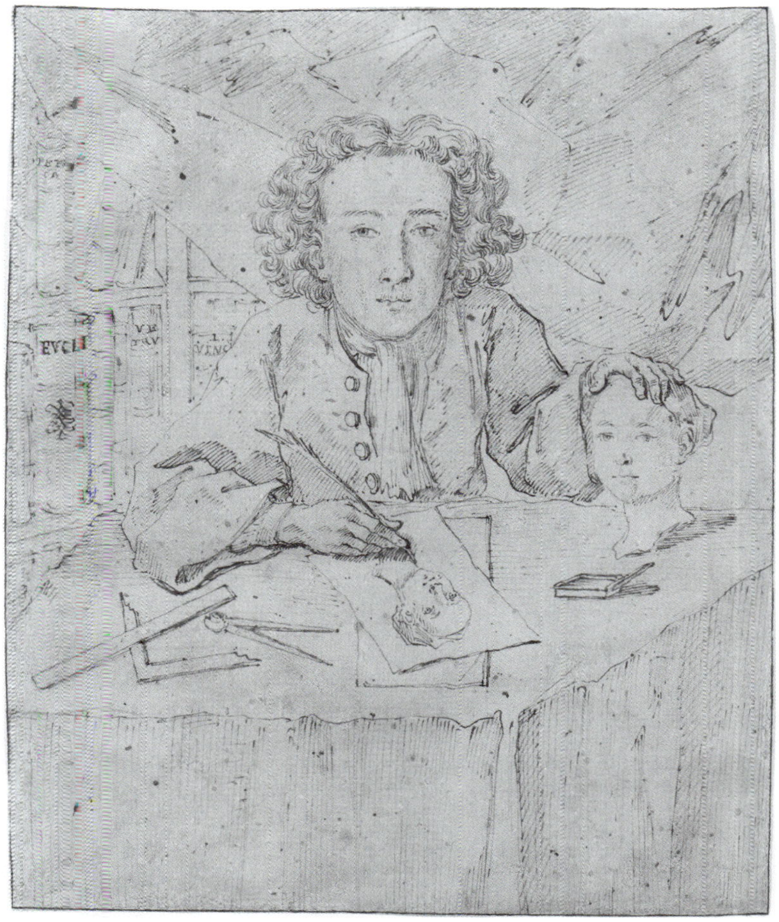

FIG. 50A
Pier Leone Ghezzi,
Selfportrait, pen and
ink drawing. Private
collection, London.

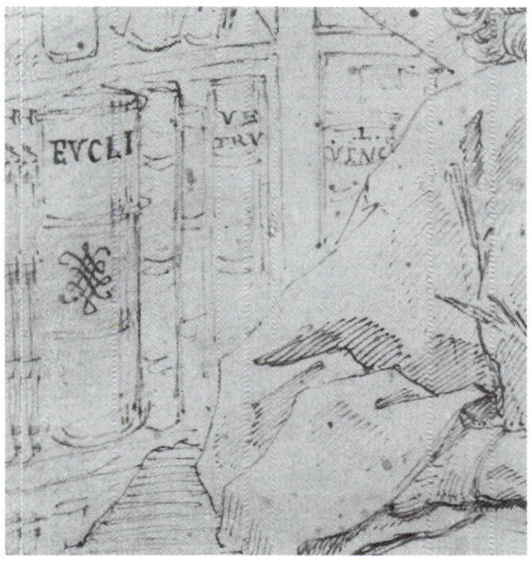

FIG. 50B
Pier Leone Ghezzi,
Selfportrait (detail).

The drawing does not include any volume by Leonardo, but, surprisingly, there is a volume attributed to one Apianus ('Appiano') written backwards as if in Leonardo's handwriting ('ONAIPPA'). This might refer to the ancient Roman historian Appian or, more likely, to the Renaissance geographer and astronomer Petrus Apianus, whose scientific images might have reminded Pier Leone of those in the Codex Leicester. The drawing might suggest that Leonardo's Codex had already left Ghezzi's house by 1712. The eccentric Pier Leone, so accustomed to making caricatures, may well have made a witty allusion to the departure from his library of the precious manuscript by Leonardo.

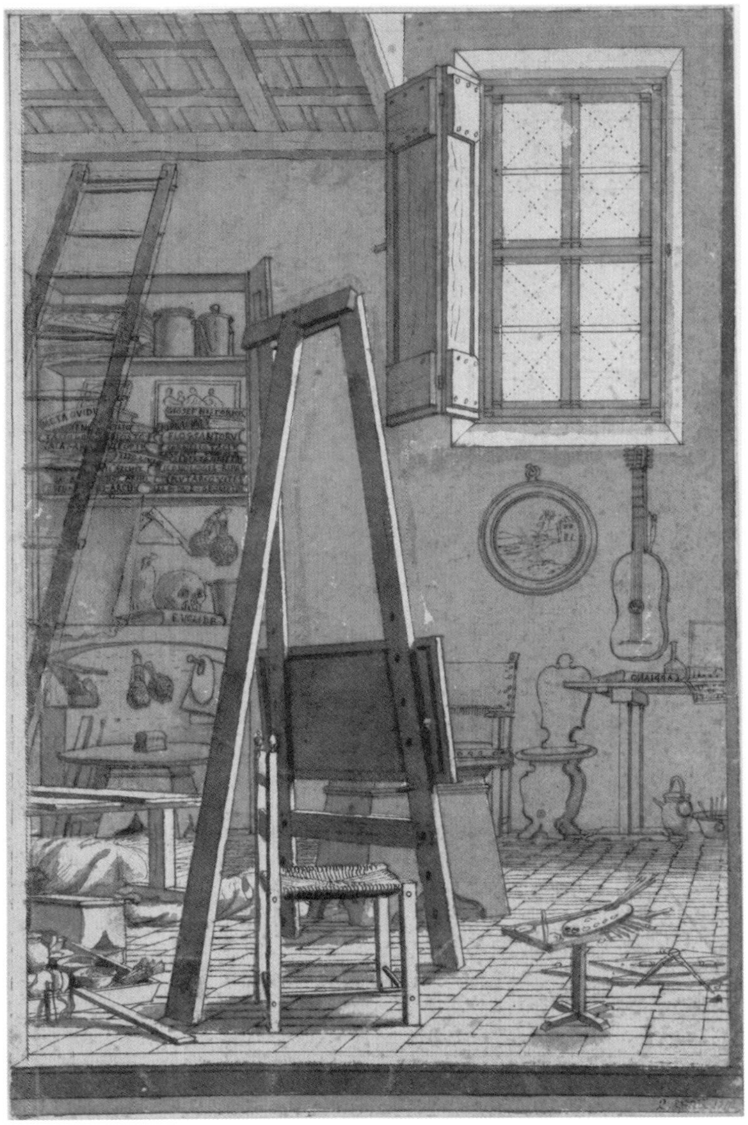

FIG. 51A
Pier Leone Ghezzi, *The artist's studio*. Graphische Sammlung Albertina, Vienna.

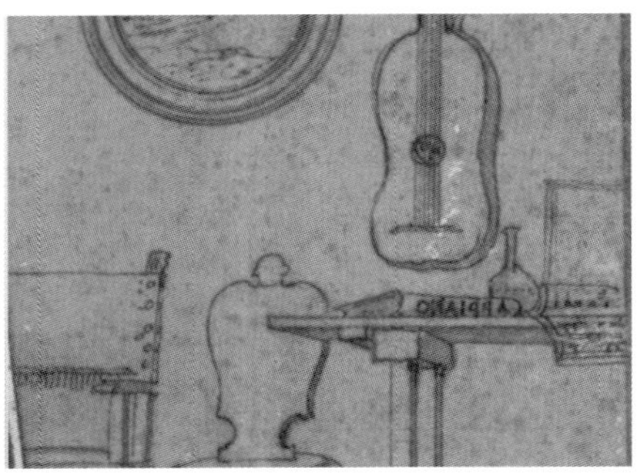

FIG. 51B
Pier Leone Ghezzi, *The artist's studio* (detail).

Another drawing by Pier Leone represents the sun as it appeared in the sky of Rome on 11 August 1732, seemingly surrounded by a strange halo in the shape of the iris of the eye (Figure 52).[131] Pier Leone's contact with Bianchini may explain his interest in astronomy. The phenomenon of the sun's halo, explained as a meteorological event resulting from the so-called terrestrial vapours (an Aristotelian

131 Angelica Library, Ms. 2136, f. 171. On the same event, see Thomas Dereham, *Philosophical Transactions*, London, 1735, p. 118.

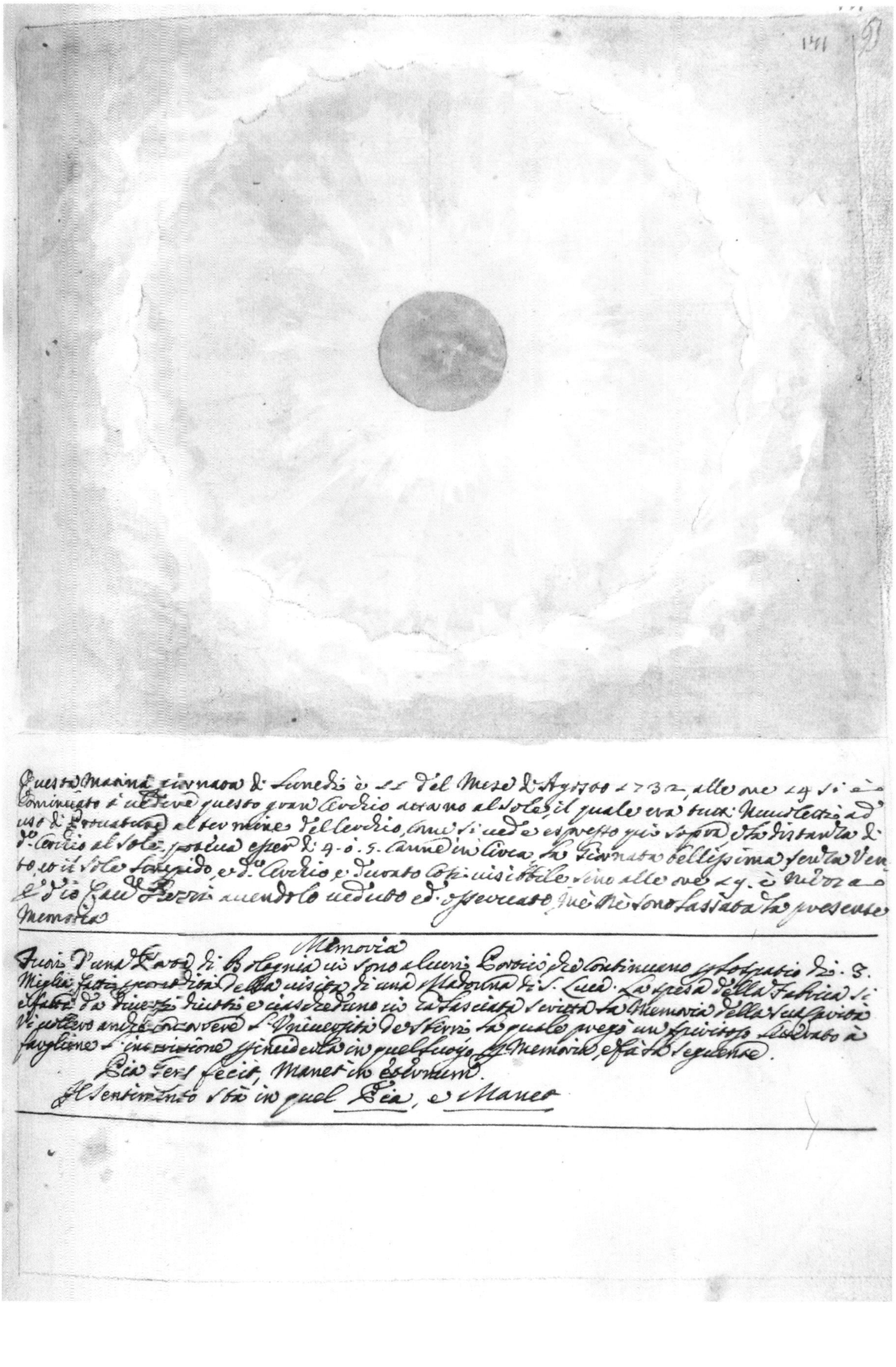

FIG. 52
Pier Leone Ghezzi, *The sun surrounded by a halo*, watercolor, 1732. Biblioteca Angelica, Rome, MS 2136, f. 171.

'topic'), is represented using heightened contrasts: the dark brown of the Sun in the centre, surrounded by a lighter brown of the inner ring of the halo which darkens gradually toward the edge, where a luminous fringe separates the halo and the sepia wash of the sky.

This is clearly not copied after the Codex Leicester. Nevertheless, in the Codex (f. 2r, lower right corner) the '*lumen cinereum*', or the pale light visible in the darkened portion of the moon during its crescent phase, is represented and also partially explained by Leonardo through enhanced tonal contrasts: the bright crescent makes the adjoining portion of moon darker, which then appears a little brighter where it is surrounded by the dark sky.

Pier Leone applies a similar perceptual approach to the study of the sun surrounded by a halo. Both Ghezzi in his study of the sun and the author of the drawings in the copy of the Codex Leicester, now in Holkham Hall, used a watercolour technique (Figure 54).

Thomas Coke's Acquisition of the Codex Leicester: Rome and Florence, *c.*1714–18

At an unknown date, but certainly by 1717, Thomas Coke, later Earl of Leicester, purchased the *Codex Leicester* during his grand tour in Italy and commissioned a copy that made Leonardo's odd handwriting understandable (Figures 54–8, 60, 65).

Coke's is the finest copy of the Codex, enriched by the drawings here attributed to the Florentine artist Tommaso Redi. The commission of the copy demonstrates that the young Thomas Coke, who was studying Italian, was interested in the content of Leonardo's manuscript. Moreover, during his long stay in the peninsula, Coke acquired, *inter alia*, the seventeenth century manuscript of Etruscology by Thomas Dempster, which he had copied and enriched with drawings for publication in print. The simultaneous interest in a cosmological and geological manuscript and in antiquarian manuscripts confirms the general connection between historical and antiquarian studies and emerging geology, already encountered in Ghezzi's Roman context.

A document concerning the Codex Leicester, after its rediscovery in Rome in 1689–90 by Giuseppe Ghezzi, dates back to 1717 and informs us that in that year the Codex was in Florence in order to be copied on behalf of a 'Cavaliere Inglese'.[132] The 'Cavaliere Inglese' was Thomas Coke (1697–1759), who was in Italy on his grand tour—with some intervals—from 1713 to 1717.[133] We do not know when and how Coke came into possession of the Codex.[134] We can only be certain that by 1717, the Codex was already in his hands and that in 1710 the Codex was still owned by Giuseppe Ghezzi, as proved by

132 Favaro 1917.

133 Ingamells 1997; Hiskey and Reynolds 2014; Moore 2014; Gialluca 2014.

134 Calvi 1909, p. xxv, thinks that the Codex passed directly from Ghezzi to Coke and, on the base of information by the librarians Alexander W. Napier and Léon Dorez, mentions, as possible intermediary, 'the sculptor and restorer Cavaceppi', who, however, if he corresponds to Bartolomeo Cavaceppi, cannot be considered since he was born in Rome in 1715–17. Pedretti 1987b, pp. liii–liv, thinks that Coke's acquisition dates back to 1717, on the base of the document above, in which, however, only the production of a copy is mentioned.

Crescimbeni's *Comentarj* printed in Rome that year.[135] We have also seen that two drawings by Pier Leone Ghezzi might suggest that the Codex left Ghezzi's hands by 1712 (Figures 50a–b, 51a–b). At this time Coke was not yet in Italy and this would imply an intermediate passage between Ghezzi and Coke.

Alternatively, there are interesting clues pointing to a direct purchase through the contacts in Rome between Sebastiano Resta and William Kent (1685–1748), the painter who was to become a great architect and garden designer. The contacts between Resta and Kent date back to 1713, and are known to us through the Resta–Ghezzi correspondence, which, as we have seen, includes information on the Codex Leicester. Resta informs Ghezzi of Kent's desire to copy a work of art owned by Ghezzi[136] and that he had recommended him for an artistic competition at the Academy of St Luke, of which, as we know, Ghezzi was the secretary.[137] In the following year, Kent entered the service of Thomas Coke, who had just arrived in Rome, working for a long time for the noble Englishman with various tasks, from teacher to consultant and intermediary in the purchase of drawings and works of art.[138] In 1714, the death of Resta, a great collector of drawings, marks, according to Warwick, the start of extensive acquisition campaigns by northern Europeans, including Coke, who bought drawings from Resta's acquaintances, by Ghezzi, Maratta, and others.[139] Kent may have introduced Coke to Resta's circle and in particular to the Ghezzi family.

This would favour a direct passing of the Codex from Ghezzi to Resta. But when? At Holkham Hall in Norfolk, the magnificent country mansion designed by Kent in the Palladian style for Thomas Coke, are account books of the grand tour. One of them was compiled by the valet, William Jarret;[140] the other by Thomas Hobart, Coke's tutor who accompanied him on the grand tour.[141] These are the most useful documents for reconstructing Coke's movements and contacts. From these notebooks we know that Coke was in Rome in 1714 (7 February–8 March, 24 March–6 June), 1716 (2 June–12 September), and 1717 (8 January–5 April). However, these are rather uninformative documents with respect to the description of the authors and the content of manuscripts, books, or works of art purchased by Coke. Browsing Hobart's notebook, at most we gain a clear impression of Coke's expanding expenditure 'for books' in Rome in 1716.[142]

Hobart's notebook contains also clues in favour of an indirect purchase. Among the names frequently mentioned are Williams and Smith, sometimes in relation to Coke's stays in Venice. Joseph Smith, a

135 See section on 'Rome 1689', p. 150 of this chapter. Further evidence of the Codex in Ghezzi's house are David's letter of 1704 and Resta's note in a printed copy of Orlandi's *Abecedario* (1704), on which see Nicodemi 1956, p. 312.

136 Ms. Cors. 1403 (now 31 B 9), f. 31: 'Ieri diedi copia della lettera magnavacca all'Inglese Kent, che venne col compagno suo miniatore, perspicace, virtuoso. Vorrebbe il Kent copiare speditamente in carta le Nozze [or "Notte" by Correggio, mentioned in letters of 1713] di VS. Vedendo, lo riceva, che credo con la lettera la manderà a Londra dove tratta il mio Cristo all'Orto in 750 scudi e con riverirla.'

137 Ms. Cors. 1403 (now 31 B 9), f. 27r. Prosperi Valentini 1983–4, p. 239 nn. 25–6 and 2007, p. 16.

138 Gialluca 2014, p. 108.

139 Warwick 2000, p. 16.

140 Holkham Hall Archive, FT/C4.

141 FT/C6. The third one (FT/C5) is a copy after FT/C4. I owe this information to the Holkham Hall archivist Christine Hiskey. For FT/C4 see now the transcription in Gialluca 2014, pp. 112–21.

142 A summary of acquisitions from 1716 and 1717 is in FT/C6, ff. 14v–15r.

British consul in Venice, was a famous collector of books and works of art who, at some point in his Italian stay, was associated with Thomas Williams in the import-export trade and a merchant-banking house.[143] Smith is known to Leonardo scholars because of a note in a letter to the Count of Caylus, attributed by Bottari to the art collector Jean-Pierre Mariette and published by Bottari himself in 1757.[144] In the note, Smith is indicated as the possible owner of a Leonardo manuscript containing drawings of mills, once belonging to the Lombard painter Giovan Ambrogio Figino. Smith's ownership is then mentioned by various authors.[145]

The hypothesis reported by Bottari was probably based on a passage in Lomazzo's *Treatise* (1585) or on Du Fresne's life of Leonardo in the printed edition of the *Treatise on Painting* (1651), in which a Leonardo manuscript on mills with and without water is mentioned as belonging to the painter Figino. Perhaps the information that Smith had acquired Figino's manuscript[146] induced Mariette to associate Smith with Leonardo's manuscript on the mills? Or did Smith come into possession of the Codex Leicester before Coke? Although the Codex contains no drawings of mills, it did deal with water, and in the cover-page added by Ghezzi and Resta it was connected with the manuscript on the mills quoted by Du Fresne. If so, did Coke acquire the Codex during his stay in Venice (1–29 January and July 1714) or even later through intermediaries? But it is also possible to imagine a different role for Smith. In an account book of 1719–23, Smith and Williams are again mentioned in relation to the dispatch to England of various items, including 'books'.[147] Perhaps Smith simply played an intermediary role in sending the Codex Leicester to London?

We do know for sure that by 1717 the Codex Leicester had left Ghezzi's house in Rome and was in Florence in order to be copied on behalf of Thomas Coke. In letters of 29 June and 1 October 1717 addressed to Guido Grandi, the Florentine Tommaso Buonaventuri, says that he had seen the Codex in Florence and adds that 'an English cavalier left it here in order to have it copied according to modern usage'.[148] Grandi was a Tuscan mathematician, hydraulic engineer, and one of Galileo's pupils, while Buonaventuri was a learned man and director of the Grand Ducal printing house. During his journey to Italy, Coke studied Italian and, in 1715, he decided to spend one more winter in Italy 'to confirm myself in the language and the virtuosoship of that Country', as he wrote in a letter to his grandfather.[149] Nevertheless, Leonardo's idiosyncratic writing was an obvious obstacle. Thus, Coke left the original

143 For Coke's acquisitions of manuscripts after the grand tour, in 1721, through Smith, see De Ricci 1932, p. x.

144 'Ambrogio Figini pittore famoso n'ebbe un altro, che forse passò in potere del Signor Giuseppe Smit, che comprò tutti i disegni dello stesso Figini'. Bottari 1754–73, vol. 2 (1757), p. 171.

145 Amoretti 1804, pp. 128–9 ('Quest'opera de' mulini è forse quello stesso libro d'Idrostatica che dicesi venduto al Sig. Smith Inglese'); Govi 1872, p. 6; Uzielli 1884, pp. 366–7; Gramatica 2008 [1919], pp. 94–5 n. 29.

146 *Bibliotheca Smithiana*, Venice, 1755, p. 174: 'FIGINUS Jo Ambrosius, Medionalensis Pictor, Schedae & experimenta, ms. fol. m. lig. Corio Turcico rubro deaur.' The library included also a collection of praise poems dedicated to Figino ('Raccolta di Poesie di varj Autori in sua lode, mss, fol. leg. MNaroc rosso dor.').

147 A/5, f. 78: 'Paid to Mr. Williams the merchant for postage of letters etc. from abrod.' (1719); f. 180 (1722): 'Williams & Smith for books Brought over Sea'.

148 'Il Cav.re Inglese l'ha lasciato qua per farlo copiare all'uso moderno' (29 June 1717). University Library, Pisa, Ms. Grandi 86, ff. 94 and 123 (esp. f. 94r). See Favaro 1917.

149 Warburton James 1929, p. 191.

Codex in Italy, in Florence, for some time and only when the copy was realized did the Codex Leicester and its copy reach London in 1719, as attested by a document now at Holkham Hall.[150]

Coke's cultural and editorial interests in Italy: From the acquisition of the Codex Leicester to the edition of a manuscript on Etruscology

The acquisition of the Codex Leicester by Thomas Coke was not only the act of a wealthy collector who wanted to acquire a precious 'object' by Leonardo da Vinci (assuming that Leonardo was really a desideratum for an English collector of the early eighteenth century).[151] His acquisition was also motivated by a serious cultural interest in the manuscript's content. As we have seen, Coke commissioned the transcription of the Codex in clear handwriting. Therefore he considered it not just as an object to be contemplated for its drawings and the strange mirror handwriting, but a text whose content was to be understood. The cultural context of the Codex's acquisition and the making of its copy seem to confirm this. It is also necessary to consider that Coke was less than twenty years old when he travelled to Italy, accompanied by Thomas Hobart, a bibliophile and a man of science, a medical doctor educated in Cambridge, who served as Coke's mentor and who exercised an important advisory role in his acquisition of books in Italy and elsewhere.[152]

While in Italy, Coke was planning an edition of Livy based on various manuscripts, some of which were purchased during the grand tour. Although the project for the illustrated edition of Livy collapsed—perhaps because of its excessive ambition—another initiative promoted by Coke in these years was accomplished, namely the edition of an unpublished manuscript of Etruscology compiled at the beginning of seventeenth century by Thomas Dempster, a Scot who had lived in Tuscany. Coke promoted the printed edition of this manuscript and asked the Tuscan antiquarian Filippo Buonarroti to add archaeological and visual evidence to Dempster's text, which was mainly based on literary sources. Coke purchased the manuscript in Florence in 1719 through Hobart, having in the meantime come back to England. The purchase was followed by his edition. The manuscript was copied in fine script by Biscioni and Giovanni Bottari, while two Tuscan artists, Tommaso Redi and Giovanni Domenico Campiglia, were asked to realize drawings of various Etruscan antiquities for Dempster's treatise, which was devoid of illustrations. The whole thing was overseen by Filippo Buonarroti and, as far as possible, by Coke himself, who possibly played a prominent role also in the choice of artists.[153] Finally, the work was printed in Florence in 1726 as *Thomae Dempsteri De Etruria Regali*.

Buonaventuri, the learned director of the Grand Ducal printing house, who in 1717 showed interest in the Codex Leicester for its copying and possible publication, was at the same time interested in publishing Dempster's manuscript. Coke however, in the meantime, had acquired the latter, and it was he whom Buonaventuri was obliged to contact. Buonaventuri's request helped stimulate the positive

150 Holkham Hall Archive, A/5, p. 67.

151 Woodfield 2009, p. 481 claims that a work by Leonardo would not have been coveted, but the collecting of Leonardo in Britain was strong from the seventeenth century onwards. See Barone and Avery-Quash 2019.

152 Conner Bulman 2003; Reynolds 2014a.

153 Gialluca and Reynolds 2014, p. 321.

attitude of Coke, who would promote and finance an editorial project with far more ambitious illustrations than originally planned by Buonarroti.[154]

We have seen how in Rome manuscripts by Renaissance scientists, from the anatomist Eustachius to the antiquarian scholar Michele Mercati on 'fossils', were 'rediscovered' and printed for the first time. Coke seems to share this interest in rediscovering knowledge hidden in earlier manuscripts as explicitly advocated by Leibniz shortly after his trip to Italy and his stay in Rome in the very same year that Leonardo's Codex Leicester was rediscovered. In the specific case of Coke, there are also shared cultural aspects between the three editorial initiatives he promoted in Italy, namely the production of a fine manuscript copy of Leonardo's Codex Leicester, the collocation of Livy's manuscripts enriched by images, and the illustrated printed edition of Dempster's *De Etruria Regali*.

This was the period of publication in London of the geological treatises by Thomas Burnet (1680–89, 1691) and John Woodward (1695, 1714, 1726). Woodward, in addition to being a geologist, was also an antiquarian (in the older sense of scholar of ancient Greek and Roman culture and history). One of the most famous concepts in Burnet's treatise was the comparison of the present state of the earth to the ruins of a long-distant past. Perhaps thanks to the advice of his tutor, Thomas Hobart, Thomas Coke, on his 'grand tour', seems to embody the links between the history of the Earth and the antiquarian history of man, one of the key feature of the nascent geology.[155] On the one hand, he acquired a manuscript dealing also with geology, the Codex Leicester by Leonardo, ordering a copy of it, and on the other, he rediscovered and published an early-seventeenth-century manuscript on Etruscology.

It is to the most famous Tuscan 'geologist' of the eighteenth century, Giovanni Targioni Tozzetti (1712–83), that we owe a record regarding Hobart and Coke in Italy. Targioni, in the section on cosmography and geography in his monumental *Selva di notizie spettanti all'origine dei progressi e miglioramenti delle scienze fisiche in Toscana* [*Collection of News Concerning Origin and Development and Improvements of Physical Sciences in Tuscany*],[156] writes that a manuscript with the cosmological treatise by Goro Dati (1362–1435), in his time at the Magliabechiana library, had been donated to its former owner, Anton Francesco Marini, 'by Tommaso Hobart Englishman while he was in Florence with Tommaso Cock on 16 December 1716'.[157] Hobart was therefore interested also in cosmological manuscripts, and his presence at Coke's side in Italy might have played a role in the acquisition of the Codex Leicester, alongside another manuscript containing various writings of geography in Italian, including those by Carlo Giangolini from Fano (1591–1652).[158] The link between antiquarian studies and emerging geology reverberates further in other aspects of Thomas Coke's purchases. It was recently noticed that Thomas Coke was the first English collector of copies of ancient Roman paintings, a

154 Gialluca and Reynolds 2014, p. 319.

155 Rappaport 1997b.

156 Giovanni Targioni, *Selva di notizie spettanti all'origine dei progressi e miglioramenti delle scienze fisiche in Toscana*, composed over a long period of time, National Library, Florence, Ms. Targioni 189.

157 National Library, Florence, Ms. Targioni 189, volume on Cosmography and Geography, f. 622:
Il Cav. Anton Francesco Marini, per Legato del quale esso codice venne nella Magliab. vi ha notato d'averlo avuto in dono da Tommaso Hobart Inglese, mentre stava in Firenze con Tommaso Cock, à 16 xbre 1716. Ha notato inoltre che di Leonardo Dati ha veduto appresso il suddetto Hobart: Carmen ad Pontif. Max. Nicolaum V, in Mahomet regem Turcarum, che principia 'Maxime Pontifocuem. . . .

158 Holkham Hall Archive, MS 546 (on the spine: *Cosmografi*): 'Introduttione di Geografia del Sig.r Carlo Giangolino da Fano'.

pursuit that will include famous names such as Lord Burlington and Richard Topham.[159] Holkham Hall has two albums of drawings and watercolours of ancient paintings by Pier Santi Bartoli (1635–1700) and by his son Francesco. Although Bartoli's name is already included in the account books of the grand tour,[160] the two albums were probably purchased by Coke later on, between 1747 and 1754,[161] shortly before Coke in London, bought on the Roman market other works that recalled his grand tour period: two paintings by Agostino Scilla with allegorical-naturalistic representations of summer and winter, still at Holkham Hall (Figure 53).[162]

Scilla was the Sicilian painter we met above in reconstructing the context of the re-emergence of the Codex Leicester in Rome in 1689–90, and who is best known for his scientific treatise on fossils, with plates engraved, after Scilla's drawings, by Pier Santi Bartoli.[163] The graphic techniques developed by Bartoli to 'reintegrate' damaged ancient wall paintings are paralleled in Scilla's images by his

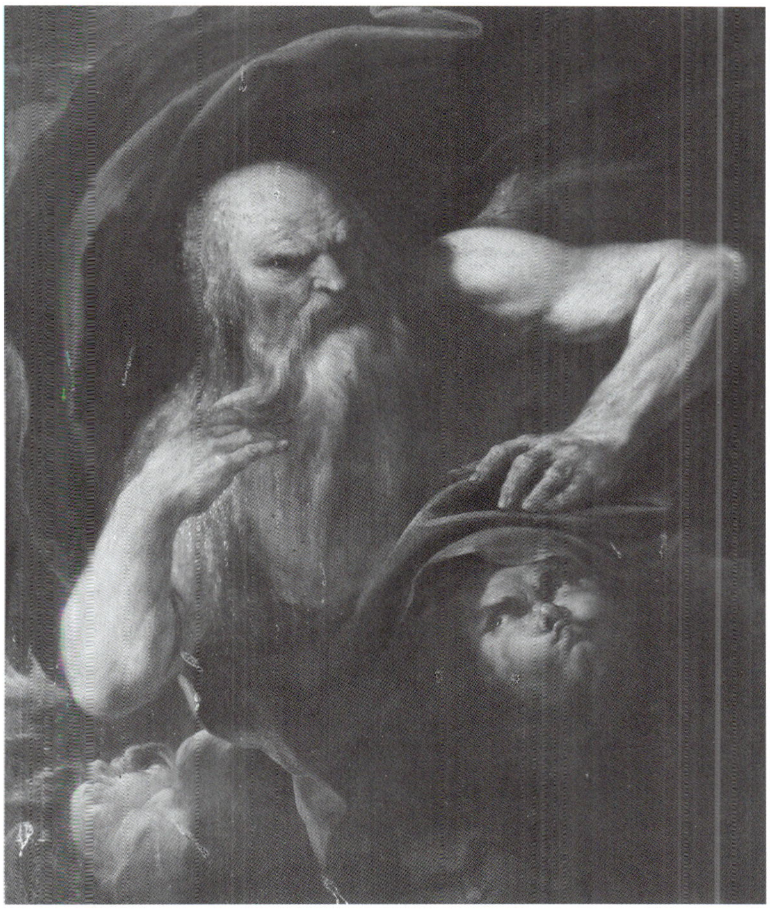

FIG. 53
Agostino Scilla, *Winter*, painting. Holkham Hall, London.

159 Modolo 2014, p. 1[illegible]1.

160 If he is the person mentioned in FT/C5, f. 79r as 'Signor Bartoli' (Rome, 30 June 1716).

161 Modolo 2014, p. 1[illegible]9.

162 The paintings arrived at Holkham in 1760, after Coke's death and on behalf of his widow (Murdoch 2006, p. 236), but Coke acquired them in 1759, before he died. Holkham Hall Archive, H/Inv 4: 'Account of Pictures sent from Italy to Holkham in november 1755 and in may 1760 [. . .] S of Winter, S of Summer Both Sylla'. See Hyerace 2007, pp. 163–4 and 168 n. 71, for a different document.

163 Carpita 2006.

'reintegration' of the damaged parts of fossils.[164] Pier Santi also worked for the scientist and archaeologist Francesco Bianchini who led excavations in Rome, in the area of the Palatine Imperial palaces, which also involved, as we have seen, Pier Leone Ghezzi, son of the previous owner of the Codex Leicester, whose drawings span archaeology and geological stratigraphy.

Francesco Ducci's copy of the Codex Leicester and a possible attribution of its drawings to Tommaso Redi

Coke commissioned a copy of the Codex Leicester from Francesco Maria Ducci (?–1718), chaplain in the basilica of San Lorenzo in Florence, librarian in the Laurentian Library, and an expert in Hebrew, which, like Leonardo's texts, is written from right to left. This copy, now at Holkham Hall, has not previously been studied (Figures 54–8, 60, 65).[165]

The attribution to Ducci of the copy commissioned by Thomas Coke, relies first of all on a note inserted before 1752 into another eighteenth-century copy of the Codex, the Weimar copy, which reads that the copy 'sent to London' was made by Ducci.[166] The attribution to Ducci is also confirmed by an unpublished document, which also allows us to set it within other important editorial enterprises and acquisitions promoted by Coke in Florence. One of the account books of the grand tour includes the following note of expenses in Florence on 13 April 1717: 'Paid to father Danney [or Danndy] for paper to copy out a manuscript 05-4-0'.[167]

It is very likely that 'Danny' or 'Danndy' is a mistake for Ducci, as in many other cases of Italian names written very inaccurately by William Jarret in his Account Books.[168] The year coincides with the letters of Buonaventuri to Grandi which concern the copy (1717). As recorded in the same account books, in 1719 the Tuscan literary man, Antonio Maria Biscioni was likewise paid by Coke to buy paper. Biscioni was in charge of making a copy of Dempster's Etruscological manuscript.[169] Other adjacent notes of expenses are also interesting. On 13 April Biscioni is also paid for his work collating various manuscripts and editions of Titus Livy.

The Laurentian Library in Florence holds many of Ducci's transcriptions from Hebrew works. Reading the right-to-left texts by Leonardo was therefore not hugely problematic for him.

Ducci died in 1718,[170] a *terminus ante quem* for the date of this copy, which was therefore made in 1717–18, given Buonaventuri's letter and the document almost certainly indicating that Ducci bought the paper to copy out the manuscript.[171] Ducci's death may account not only for the invoice of the

164 Laurenza, forthcoming b.

165 On Ducci, see Nolasco Cianfogni 1804, p. 125 n. 2; Bottari 1754–73, vol. 2 (1757), p. 180; Laurenza 2019a. On the copy, see also Appendix to this chapter. It is briefly mentioned in Jordan 1873, p. 104; Uzielli 1884, pp. 324–5; Calvi 1909, p. xxvii; Pedretti 1987b, p. liii.

166 See Appendix to this chapter on the Ciurini and Weimar copies.

167 FT/C4, f. 199 and FT/C5, f. 126r.

168 e.g. the artist Campiglia is mentioned as 'Comlia' (FT/C4, f. 197); Luigi Garzi as 'Guzzi' or Gazzi' (FT/C5, f. 78 and FT/C4, f. 151). See also Gialluca 2014, p. 112.

169 Gialluca and Reynolds 2014, esp. p. 320.

170 Nolasco Cianfogni 1804, p. 125 n. 2.

171 See earlier in this Section.

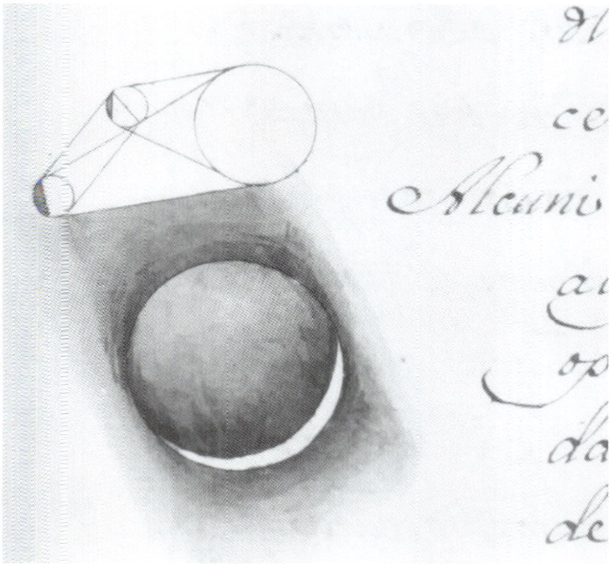

FIG. 54
Francesco Ducci (text) and Tommaso Redi (drawings). Copy of Codex Leicester, f. 2r. Holkham Hall, MS 70c, f. 7r.

original to London in the following year, but also for the unfinished state of the copy, since almost all the drawings lack all of their letters and in many instances their captions, as in the case of the astronomical figure copied from f. 1r of the Codex (Figure 54). However, this was not fatal to the reading of the text. The letters accompanying some drawings are not referenced in Leonardo's own text. And the copy could be read by Coke side-by-side the original with its labelled drawings.

Ducci first transcribed the main text, leaving a blank space on the left margin to accommodate the figures, which would have been added not by him but by an artist following his instructions. The study of this copy and of the system adopted by Ducci and by the artist charged with the copy of the figures has provided an explanation for the faint numbers visible throughout the pages of the Codex Leicester. They are close to Leonardo's drawings, but are clearly not in Leonardo's hand. It is clear now that the numbers were inserted by Ducci as indications for the artist: indeed, they correspond to the numbers that can be seen close to the drawings in the copy at Holkham Hall. For example, in the case of the representation of the 'ashen light' of the moon a '6' is visible on the original drawing by Leonardo and, in the copy, close to the place where the illustration was copied. Ducci was showing the artist where to insert each drawing in his carefully planned transcription (Figure 54).

The choice to have the figure in the left margin, rather than on the right as in the original, was consistent with the transcription of Leonardo's text, reversed from the original right-to-left script to the normal left-to-right. In most of the sheets of the Codex, Leonardo inserted the figures in the blank space in the right margin of the main text; Ducci moved them to the left. Consistent with this decision to reverse not only the text but also the whole content of the page, Ducci asked the artist to copy each drawing in reversed form, but this was only followed for the first nine sheets of the Codex. Indeed, the choice turned out to be problematic. In f. 10r, Leonardo inserted a drawing that, usually considered by scholars a generic sketch of the earth, is actually a precise geographical representation, in which, taking into account that Leonardo orients the globe with the East up, we can recognize the Euro-Asiatic and African lands surrounding the Mediterranean Sea, with the Strait of Gibraltar to the west (therefore towards the bottom of the drawing), Britain to the north (left in the drawing), all lands associated with geological changes described in the text. A reversed copy of this figure would have altered the topographical relationships, and the artist copied it in its original orientation (f. 43v).

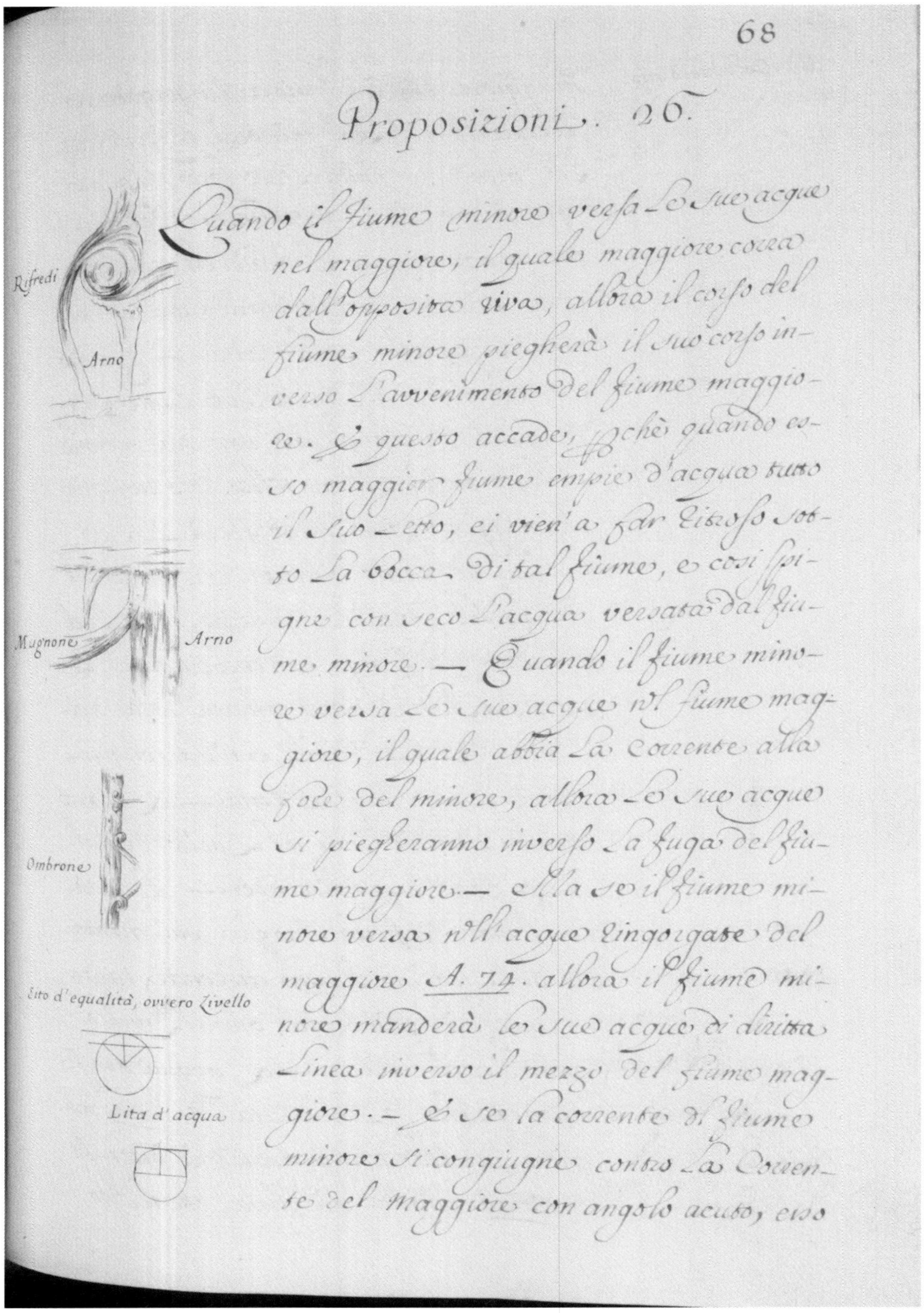

68

Proposizioni. 26.

Quando il fiume minore versa le sue acque
nel maggiore, il quale maggiore corra
dall'opposita riva, allora il corso del
fiume minore piegherà il suo corso in-
verso l'avvenimento del fiume maggio-
re. E questo accade, perchè quando es-
so maggior fiume empie d'acqua tutto
il suo letto, ei vien a far ritroso sot-
to la bocca di tal fiume, e così spi-
gne con seco l'acqua versata dal fiu-
me minore. — Quando il fiume mino-
re versa le sue acque nel fiume mag-
giore, il quale abbia la corrente alla
foce del minore, allora le sue acque
si piegheranno inverso la fuga del fiu-
me maggiore — Ma se il fiume mi-
nore versa nell'acque ringorgate del
maggiore A. 74. allora il fiume mi-
nore manderà le sue acque di diritta
linea inverso il mezzo del fiume mag-
giore. — E se la corrente di fiume
minore si congiugne contro la corren-
te del maggiore con angolo acuto, esso

FIG. 55
Francesco Ducci (text) and Tommaso Redi (drawings), Water studies. Copied after Codex Leicester, f. 15r, Holkham Hall, MS 700, f. 68r.

This occasioned a re-think the choice, because, starting with f. 13r, no figure was reversed in the copy, with a few exceptions in those rare cases where Ducci had already transcribed and reversed the words connected with the figures. In these instances, not reversing the figure would have resulted in a mistake, with the words referring to the wrong side of the figure. Such an error occurred in at least in one case in the copy of a figure with the geographical names 'Arno' and 'Rifredi' from f. 15r, names that were previously copied and reversed by Ducci with the intention of a reversing the copy of the drawing as well (Figure 55).[172]

The fact that Ducci and the artist carefully considered the geographic content of Leonardo's little sketch in f. 10r is revealing of the importance assigned to the images, which received a care and attention equal to the transcription of the text. It is possible that this was done at the explicit request of the commissioner Thomas Coke. The printed edition of the Thomas Dempster manuscript, promoted by Coke in the very same years, was enriched by new images, and the same visual richness was planned for the aborted project for an edition of Livy. As we will see, another copy of the Codex made outside Coke's commission, reveals a lesser concern with the images.

Just as Ducci's transcription aimed to render Leonardo's difficult writing in legible form, the copy of the images aimed to clarify the content of drawings that are often just schematic sketches. The copyists' correct understanding of the geographic content of Leonardo's quick sketch on f. 10r, which has tended to escape modern scholars, indicates the care taken. The copy of the drawings was not a passive act, but implied a careful consideration of their content. And at least in one case (f. 7v in the copy) this attempt to clarify the meaning of the drawings involved a certain degree of freedom. One example is the experiment on the behaviour of a fluid in a container subjected to heat in f. 3v of the Codex Leicester, where Leonardo drew a piece of coal and wrote the word 'fuoco' (fire), while the copyist added the flames. This addition involved careful reading of the main text, since Ducci had not yet transcribed the word 'fuoco'. Similarly, Ducci in order to clarify Leonardo's text, chanced on more or less the original word, by adding letters where Leonardo used an abbreviated word, for example transcribing 'ovvero' instead of the abridged 'ov'. The copy of the drawings was a kind of visual philology, providing what in modern time is defined as critical or interpretative transcription.

As in the case of transcription of the text, the original drawings were all slightly enlarged, but the artist was very careful in respecting the original proportions between them. In general, in order to clarify the original sketches the artist limited himself to translating Leonardo's often-schematic sketches into a more pictorial and naturalistic style. The copy of the drawing in which Leonardo studies the ashen light of the moon brilliantly translates the light contrasts expressed by Leonardo with linear hatching into the pictorial language of watercolour; the copy respects the original and once more demonstrates a deep comprehension of the content, in which Leonardo explains the origin of the ashen light and the resulting tonal contrasts between areas that seem brighter if placed on a darker background and vice versa (Figure 54)

Who was the artist who copied Leonardo's drawings in such a clever way? A number of clues point to Tommaso Redi (1665–1726), a Florentine draftsman much appreciated as copyist, especially of ancient

172 Holkham Hall Archive, MS 700, 33 x 24 cm, f. 68r. The name 'rifredi' is on the wrong side.

works of art.[173] Together with Giovanni Domenico Campiglia (1692–1768) he was asked to illustrate the *De Etruria Regali*, the edition promoted by Coke. Two other Tuscan artists worked on Coke's behalf in 1717 for the Livy edition and for other purposes: Giuseppe Grisoni and Giuseppe Menabuoi, both Redi's pupils.[174] It makes sense to search among these artists for the author of the drawings. The copy was realized in Florence between 1717 and 1718, which excludes some of the possible artists. Campiglia had moved to Rome by 1716,[175] where he frequented the Academy of St Luke and made drawings for the *Etruria Regali*; at the same time, he also contributed drawings to Giuseppe Ghezzi's *Le tre belle arti Pittura, Scultura e Architettura . . .* (Roma, 1716), a copy of which is in the Holkham Hall library and was almost certainly acquired by Thomas Coke.[176] Grisoni was in London from 1716.[177] Redi was actually in Florence in 1717–18 and he seems the most likely candidate. The Holkham Hall collections hold his drawings for *De Etruria Regali*.[178] Even if depicting very different subjects, the techniques and style of Redi's copies of Etruscan objects are not far from the drawings in the Ducci copy. The latter are in pen, ink, and wash, like the archaeological drawings, in which, however, pencil or black chalk is also used. The two groups of drawings share the same style and the versatility that allowed it to be adapted to the specific item to be copied.[179] Similar styles and graphic techniques were used in the archaeological drawings, which, however, lack the lively and minutely detailed quality that Redi reserved for the copy after Leonardo (Figures 56–7). A series of comparisons shows the compatibility in styles of the two sets of illustrations, for example the way in which Medusa's hair is depicted in one of Redi's drawings for *De Etruria Regali* (Figure 59) and the flames in one of the drawing in the copy of the Codex Leicester (Figure 58), or the skilful and animated use of light and shadow, and the shadow cast by the vessels (Figures 60–1).

Finally, indirect evidence of the proposed attribution can be inferred from an unpublished and damaged self-portrait, in which Redi represents himself holding Leonardo's *Treatise on Painting* (Figure 62). The title page of the *Treatise* represented in the painting corresponds, except for the alignment of the words, to the first Italian edition of the *Treatise* in 1651 (Figure 63). In any event, this is further evidence of Redi's personal interest in Leonardo. Before or after having copied a full scientific manuscript of Leonardo, he is proud to portray himself while studying a book by the great Florentine artist-scientist.

173 Monbeig-Goguel 1994; Connor Bulman 2002, p. 347; Gialluca and Reynolds 2014, p. 321.

174 Gialluca and Reynolds 2014, p. 326 n. 43 and Reynolds 2014b, p. 88.

175 Prosperi Valentini 1974. On Campiglia and the *De Etruria Regali*, see Gialluca and Reynolds 2014, p. 321.

176 Indeed, I found it mentioned in a partial list of books in Coke's London house prepared in 1727, almost certainly by the librarian Domenico Ferrari (Holkham Hall Archive, F/TC 3, f. 82v).

177 Massimi 2002. On Grisoni and Coke, see Reynolds 2014b, p. 85 and Gialluca 2014, p. 112.

178 Gialluca and Reynolds 2009, 2014.

179 In the copy of the Codex wash layers replace Leonardo's hatching, but are tonally differentiated as in the original, (Figure 54); fluid lines and rivulets in pen and wash are used for river courses (Figure 56); refined and almost miniaturist sketches interpret and give more life to the original, like in the case of two people on a see-saw (Figure 57); while linear drawing is used for the diagrammatic images.

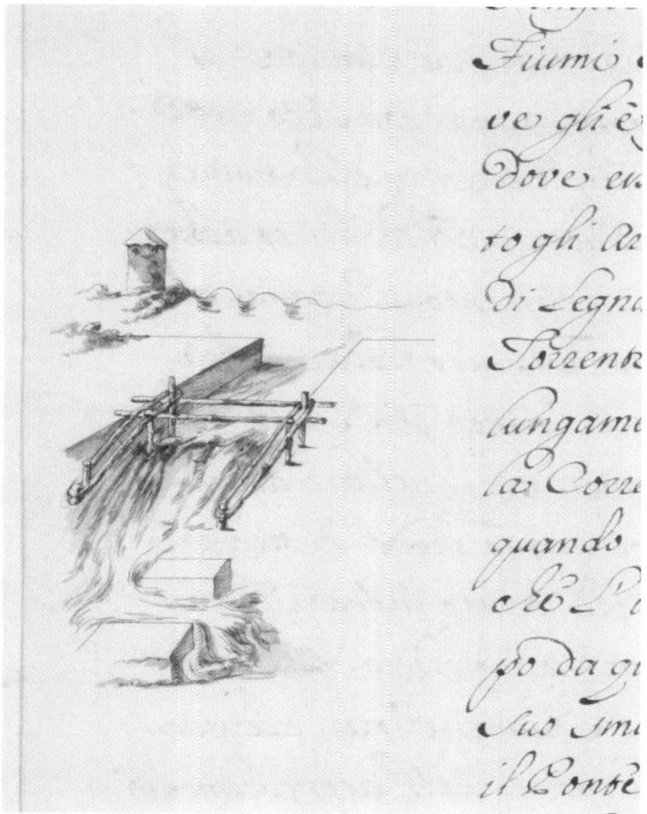

FIG. 56
Francesco Ducci (text) and Tommaso Redi (drawing), Detail of the drawing of a river. Copied after Codex Leicester, f. 9r, Holkham Hall, MS 700, f. 39v.

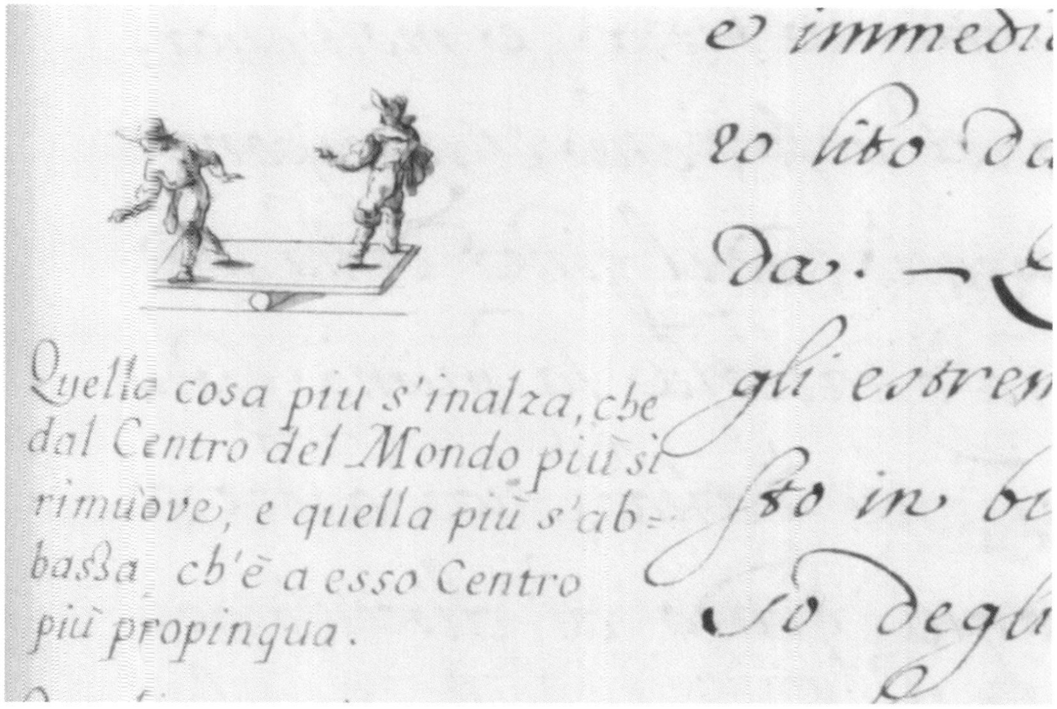

FIG. 57
Francesco Ducci (text) and Tommaso Redi (drawings), Drawing of human figures on a seesaw. Copied after the Codex Leicester, f. 8r, Holkham Hall, MS 700, f. 32r.

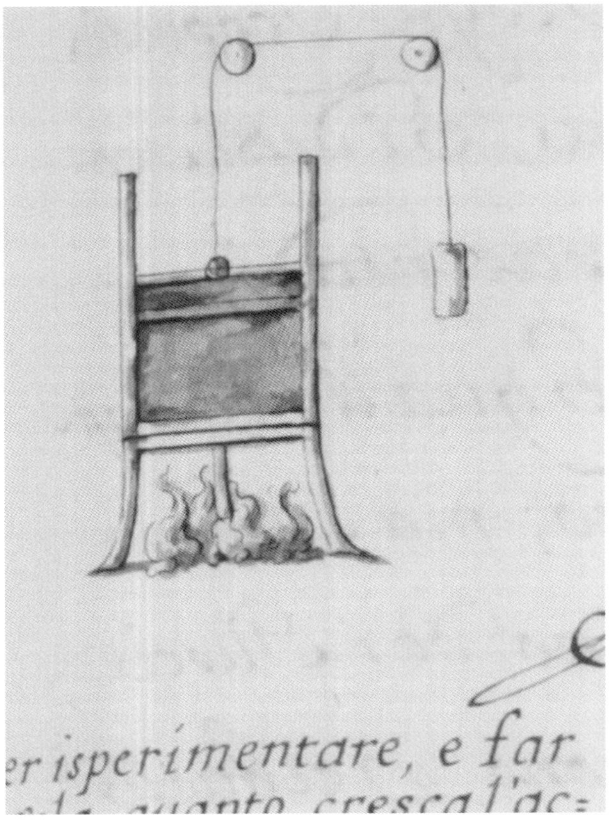

FIG. 58
Attributed to Tommaso Redi, Drawing of an experimental device. Copy of Codex Leicester, f. 10r, Holkham Hall, MS 700, f. 36r.

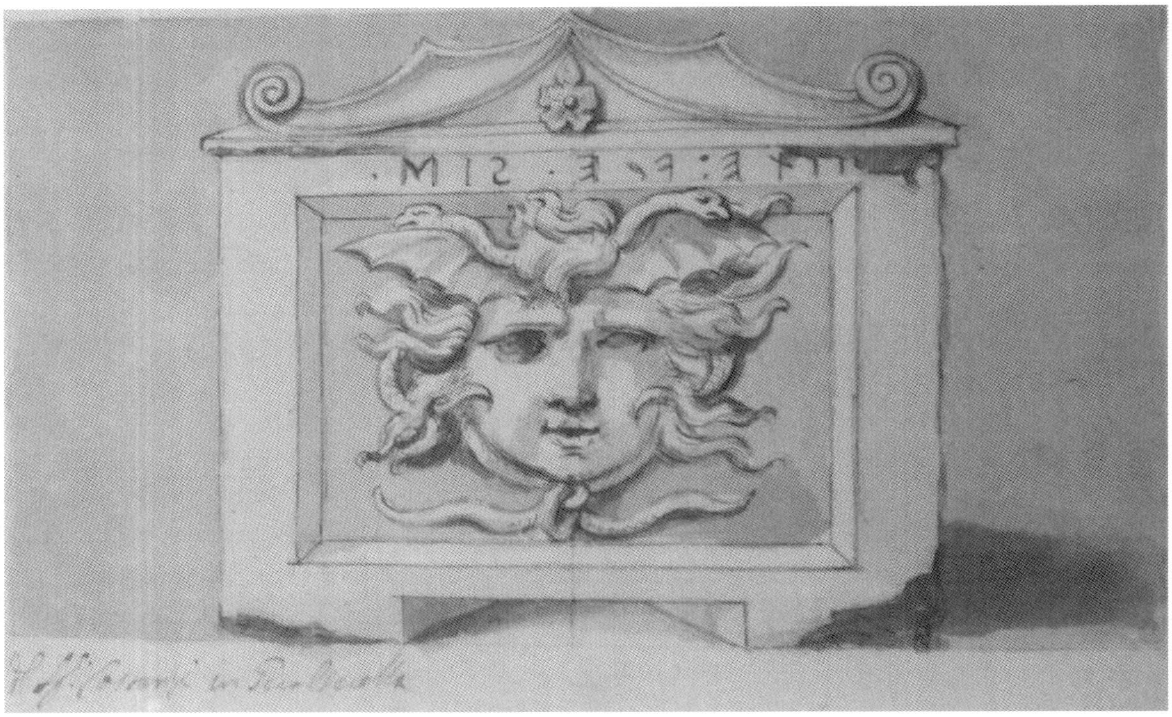

FIG. 59
Tommaso Redi, *Head of Medusa*, drawing in pen and ink and watercolor. From Preparatory drawings for *Etruria Regalis* (Florence, 1724), Holkham Hall Collections, Box 2, n. 57.

FIG. 60
Attributed to Tommaso Redi, Drawing of two vessels. Copy of Codex Leicester, f. 34v, Holkham Hall, MS 700, f. 159v.

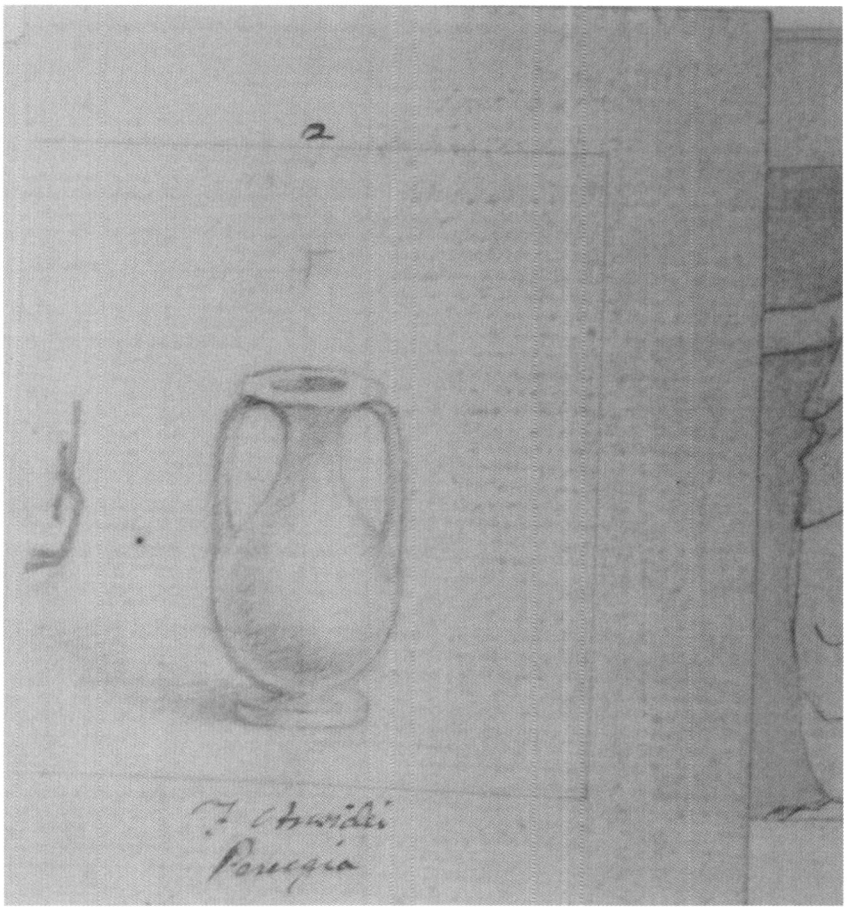

FIG. 61
Tommaso Redi, *Ancient vessel*, Drawing in pen, ink, and watercolor. From Preparatory drawings for *Etruria Regalis* (Florence, 1724). Holkham Hall Collections, Box 2, n. 69.

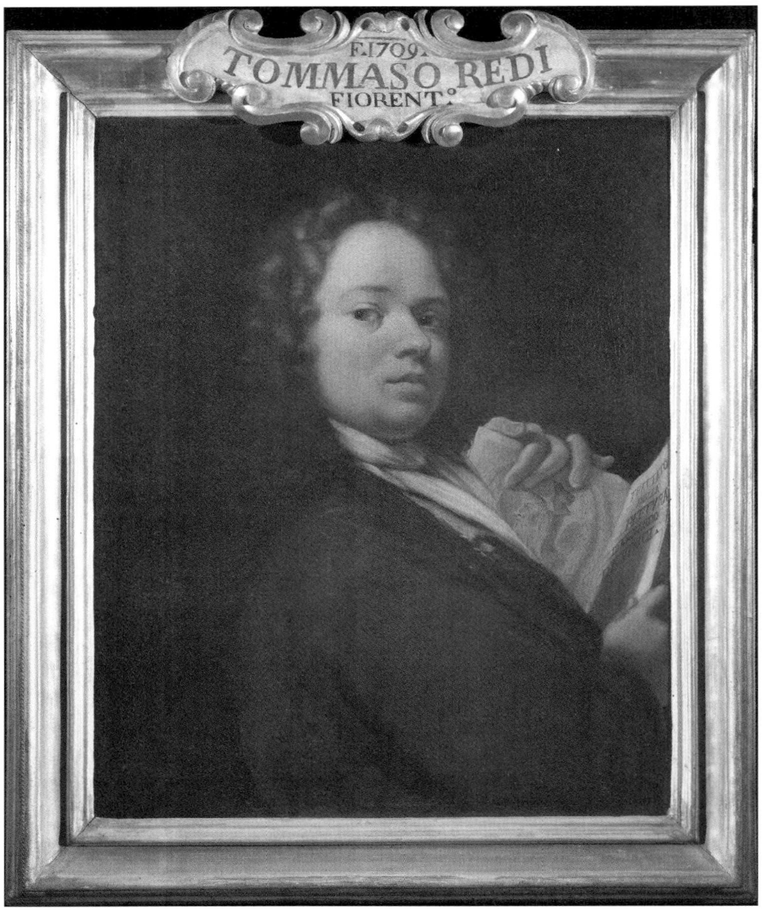

FIG. 62
Tommaso Redi, *Self-portrait*, oil on canvas. Uffizi, Florence, No. inv. 3363.

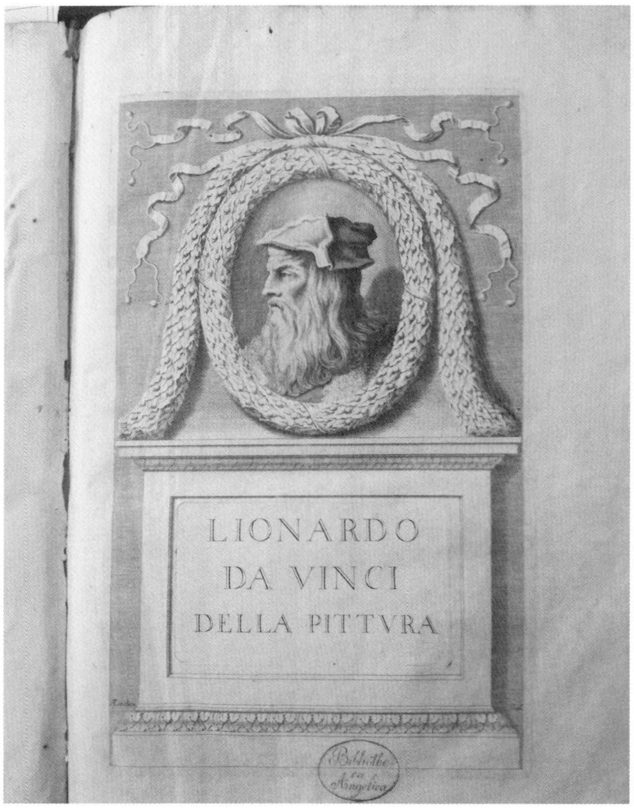

FIG. 63
Leonardo da Vinci, *Treatise on Painting*, Title page. Paris, 1656.

The Codex Leicester in the English Scientific Context in the Early Eighteenth Century

The Codex Leicester and its copy, which made Leonardo's handwriting more intelligible, remained in Thomas Coke's London house for many years. The house was attended by leading representative of English science of the period. Did some of these scientists become acquainted with the Codex? There were coincidences between the content of Codex and ideas debated in the English earth sciences of the time. At the same time, there is evidence of interest for Leonardo by English scientists in the very same period in which the Codex arrived in London.[180]

Coke left Italy in 1717 and arrived in England in 1718. As we have seen, the Codex Leicester was left in Florence so that Ducci could copy it. The Codex and its copy in clear handwriting arrived in London in 1719. An account book in Holkham Hall contains the registration of customs fees for the arrival in England of the Codex and its copy: '1719 May 21st. Payd to Mr. Hay for bringing Pictures from Abroad, Lining, Mending and Straining the same carriage. duties of the Custome House. 26:05:00. A Hebreu Bible, A Manuscript of Linoardo da binis Etc.'[181] This account book does not mention the date of arrival of the copy in England, but it is likely that it reached London in 1719 together with the original. The note dealing with the arrival seems to mention only the most precious documents: a Hebrew Bible and a manuscript by Leonardo, followed by an 'Etc.' which can include other less precious documents, such as the copy of Leonardo's manuscript.

The scientific and technological interests usually attributed to his descendant, Thomas William Coke (1754–1842), have also to be credited to Thomas Coke. In 1719, soon after the grand tour, Coke faced the problem of Holkham Hall's protection from flooding by the nearby sea. He personally inspected the situation with Nathalien Kinderley, an experienced hydraulic technician of the area, and in the same year built a bank, which would later allow the construction of the villa and its park. In a map at Holkham Hall the 1719 bank is represented by the grey bar at the centre and on the right, below the marshes and sands in the upper section of the map. Thomas Coke's name is specifically linked to this bank in the maps of the famous hydraulic engineer John Smeaton (1724–92) now in the Royal Society of London.[182]

A bank built in 1659 by Coke's family, suggests that this was a problem of which the future Earl of Leicester was aware before he left England for the grand tour.[183] Coke's direct involvement with this hydraulic engineering problem occurred the same year that the Codex Leicester arrived in England. This is of course pure coincidence. But, at the same time, that Codex—full of studies on how to drain flooded areas close to the sea—could have attracted Coke's particular interest, given that the problem of how to drain lands near the sea was not a novelty in his family.

180 On the interest in Leonardo in British scientists see the recent contributions by Barone, Field, Laurenza 2019a (including a first version of this chapter) and Mount, in Barone and Avery Quash, 2019.

181 Holkham Hall Archive, A/5, f. 67. De Ricci 1932, p. 59; Pedretti 1980, p. 17.

182 Royal Society Archive, London, JJ /6/121, vol. 5, f. 177. For example: vol. 5, f. 174 (o 171 s. l). (*A Plan of the Harbour & Haven of the Prt of Wells in the County of Norfolk* . . ., 1782): '1.Holkham Marsh Inbanked about the y.r 1719 by the Late L.d Leicester Including Creeks'. See also f. 176 (*A Plan of Weels Harbour* . . ., 1780).

183 See Hiskey 2016, pp. 79ff.

Some years later, in 1735, he became fellow of the Royal Society. In the Royal Society archives there is an 'Account of a Meteor seen near Holkham in Norfolk' that he transmitted to the Society in 1741.[184] Coke's account was then published in the Philosophical Transactions, the scientific journal of the Society, in 1742.[185] The fact that he signed the account and the article as 'Thomas Lord Lovell', one of his noble titles, explains why this submission seems to have passed unnoticed in the literature dealing with Thomas Coke and his family.

These two examples of Coke's scientific interests touch on two areas widely covered in the Codex Leicester, hydrology and astronomy. However, these sciences were at that time in advanced stage of their historical development, and Leonardo's studies would have been of a predominantly historical interest. It was different for geology: the arguments formulated by Leonardo about the history of the earth would have been of contemporary relevance for the earth sciences of the time, for which England was a major centre. In any case, the arrival in London of the Codex in 1719 coincides with various editorial initiatives involving Leonardo's work, such as the first English edition of the *Treatise on Painting* (London, 1721), whose realization involved scientific circles more than artistic ones, circles that were linked for various reasons to Thomas Coke's contacts.

When, the Codex Leicester and its copy arrived in London in 1719, the magnificent country mansion of Holkham Hall in Norfolk did not yet exist. Coke lived in his London town house, in Great Russell Street. It was therefore in his house in London that the Codex Leicester and Ducci's transcription were housed at least until 1760.[186] The Codex and its copy were therefore in central London, potentially available for perusal to the English scientific world. A search amongst Coke's contacts demonstrates that his London house was frequented by a number of significant scientists.

A key figure was Hans Sloane (1660–1753), the Coke family doctor.[187] He was also a neighbour, living on the south side of Great Russell Street and was very interested in the study of fossils. His wide-ranging collections became the seed collection of the British Museum in Great Russell Street. Another scientist who regularly attended Coke's London house was Richard Mead (1673–1754), who, in addition to having succeeded Sloane as the family doctor, was Coke's advisor for his book acquisitions and was himself a collector of paintings, drawings, and books. Payments to Mead for his medical services alternate with expenses for the acquisitions of books or drawings.[188] Collecting Bartoli's drawings of antiquities was a shared interest.[189] Perhaps another common interest was the history of the earth and the history of ancient wisdom or *prisca sapientia*, if Richard Mead is identifiable as the 'Mr. Mead' who

184 Royal Society Archive, London, L&P/1/125.

185 'An Account of a Meteor seen near Holkham in Norfolk, Aug. 1741', *Philosophical Transactions*, 42, 1742, pp. 183–4.

186 On the transport of manuscripts and books from the London house to Holkham Hall in 1760, see Murdoch 2006, p. 236. On the arrival of books and other goods from the Grand Tour into the London house, see Salmon 2014, pp. 255–6.

187 Fees to Sloane for medical services (1719–21) are recorded in Holkham Hall Archive, A/5, pp. 75, 89, 163. For letters by Coke and his wife on medical issues, see Warburton James 1929, pp. 212 and 227. For a letter on scholarly issues sent by Thomas Coke to Sloane in 1731 (British Library, Sloane MS 4051, f. 300r), see Reynolds 2015, p. 15.

188 e.g. Holkham Hall Archive, A/24: 1729, ff. 10, 14 ('Library. Paid by Mr Ferrari to Dr. Mead a Subscription for Tuani [Juani] History' and, for medical cares, 'To Dr. Mead fro my Lady'); 1731, f. 1 ('Library. To Mr Catisby in part for the Natural History of Florida. Mr Coke. To Doctor Mead for attending Master Edward Coke'); A/25: 1735, ff. 80r, 82r, 88r, 118r; 1736, f. 127r; A/28: 1742, f. 7v ('Library. To Doct:r Mead for a Book of Drawings . . . 5.5.0').

189 Modolo 2014, pp. 149 and 151.

was one of the two translators from Latin to English of Burnet's *Doctrina Antiqua* (London, 1735), which opens with an image and a reference to Burnet's famous theory of the earth. Mead was the Vice-President of the Royal Society in London, the most famous scientific society of the time, and he sponsored the admission of Domenico Ferrari, Thomas Coke's Italian librarian, into the Royal Society in 1723.[190]

Of particular significance was another of Coke's contacts, namely, Jean Theophilus Desaguliers (1683–1744). Desaguliers attended Coke's house as the teacher of 'experimental philosophy' to Coke's son, for whom he organized private scientific experiments. In February 1735, the same page of the account book with a payment to Mead ('Doctor. To Doctor Mead a fee . . . 2.2.0') also includes: 'Mr. Coke. To Dr. Disagulair for teaching Mr Coke Phylosophy . . . 11.7.6',[191] or, as is more exactly specified in another entry, 'experimental philosophy'.[192] In the same year Desaguliers is paid for a planetarium ('Extraordinary. For Mr. Coke. To Doctor Disagulier for a Planetarium . . . 10.10.0')[193] and for other scientific instruments, from microscopes[194] to what seem to be other optical instruments ('Furniture. To Dr. Desaguliers for a way Wisor [?] & other things p bill . . . 14.9.6').[195] In 1743, Desaguliers was to be Coke's consultant on hydraulic engineering issues, to which we will return later.

Desaguliers is perhaps the most representative figure in the popularization of Newtonian theories and their practical applications through experiments and illustrated texts, which is one of the main aspects of English science in the first half of the eighteenth century.[196] In 1734 Desaguliers published the first volume of an eagerly awaited work based on his popular teaching activity, *A Course of Experimental Philosophy* (London, 1734), featuring numerous illustrations, fundamental to understanding the content. What emerges in Desaguliers and in the English science of his time is a tendency to link the English Baconian scientific tradition, which paid special attention to the world of artisans and to the pursuit of science in an empirical and experimental way, with the more mathematical and theoretical character of Newtonian science.

Could the scientists who shared these ideas with Desaguliers and frequented Thomas Coke's house have become acquainted with the content of the Codex Leicester?

In London's Newtonian circles, more or less linked to the Royal Society and of which Desaguliers was one of the major representatives, there was a strong interest in figures of artists-scientists like Leonardo, as we will see in the case of first English edition of Leonardo's *Treatise on Painting* in 1721.

190 Ferrari was proposed by James Jurin, on behalf of Richard Meade. See RS, Bulloch Roll and RS online web archive ('Past fellows').

191 Holkham Hall Archive, A/25 f. 80r. 'Mr. Coke' was Edward Coke, Thomas's son. Thomas being become 'Lord Lovell', would have been mentioned with the new title. I thank Hokham Hall archivist Christine Hiskey for this clarification.

192 Holkham Hall Archive, A/25, f. 89v (April 1735): 'Mr. Coke. To Dr. Desaguliers for going through a course of experimental philosophy with Mr. Coke . . . 37.16.0.'

193 Holkham Hall Archive, A/25, f. 82r.

194 Holkham Hall Archive, A/25, f. 89v (April 1735): 'Furniture. To Dr Desaguliers for altering a fine Writing Table. D.to for a box of 7 Microscopes with . . . Appartenances.'

195 Holkham Hall Archive, A/25, f. 84v (March 1735).

196 Carpenter 2011; Fara 2004.

In 1743 Desaguliers visited Holkham Hall. Thomas Coke had started building his villa and Desaguliers was consulted on hydraulic issues.[197] It is likely that during the visit, the two also talked about the flood risk of the area located near the coast, a risk that Coke had attempted to eliminate in 1719. In 1724, Desaguliers presented to the Royal Society an experimental study on how to drain a pond near the sea, taking into account the variations of sea level at low and high tide, as represented in one of the images that accompany the study:

> An Attempt to Account for the Rising and Falling of the Water of Some Ponds Near the Sea, or Ebbing and Flowing Rivers; Where the Water is Lowest in the Pond, at the Time of High Water in the Sea or River; And the Water is Highest in the Pond, at the Time of Low Water in the Sea or River. As Also for the Increasing or Decreasing of the Water of Such Pools and Brooks as are Highest in the Dry Seasons, and Lowest in the Rainy Seasons: With an Experiment to Illustrate the Solution of the Phaenomena.[198]

The Codex Leicester contains at least two studies of the same problem, in which Leonardo considers the same variables analysed by Desaguliers:

> How one has to drain a pond that opens into the sea. . . . And the means by which the sea would not cause damage during its swelling is achieved by a device or floodgate, which will shut the pond during these refluxes of the sea; and this should not be done where the sea shows a great range in its increase and decrease.[199]

> . . . How the waters that are extracted from marshes in order to drain them have to be taken when the sea lowers its waters because of the tide.[200]

There are also conjunctions in the study of the earth. Indeed, Newtonian scientists, from Desaguliers to John Keill and William Whiston, also dealt with geology. In their studies they exhibited a strong reaction against the so-called 'theorists', the builders of abstract geological and cosmological systems, such as Burnet and Descartes. They were drawn to the study of the Earth in hydrostatic terms, that is, as landmasses floating on a liquid substrate in the context of the Newtonian concept of gravity. They favoured the use of visual diagrams of dynamic phenomena. And they were involved in polemics for and against the Biblical Deluge.[201]

Most of these ideas are very reminiscent of the geological content of the Codex Leicester. As we have seen, Leonardo's complex theory of collapse and uplift is illustrated on f. 36r.[202] The concept of 'gravity', in its pre-Newtonian sense, is a dominant theoretical notion in the Codex Leicester, both in geology and in other fields of study. There is a close comparison with Keill's geological diagrams published in his treatise before the arrival of the Codex in England (Figure 64). Keill was a leading British Newtonian geologist. In both cases, these diagrams of dynamics are intended to demonstrate, in

197 Hiskey 2007, p. 7.

198 J. T. Desaguliers, 'An Attempt to Account for the Rising and Falling of the Water of Some Ponds Near the Sea . . .', *Philosophical Transactions*, 33.384 (1724–25; published 1726), pp. 132–35, esp. 132.

199 Leicester, f. 2v.

200 Leicester, f. 17r.

201 See Porter 1977; Davies 1969; Pole 2010.

202 See Chapter 4, Figure 29, in this volume.

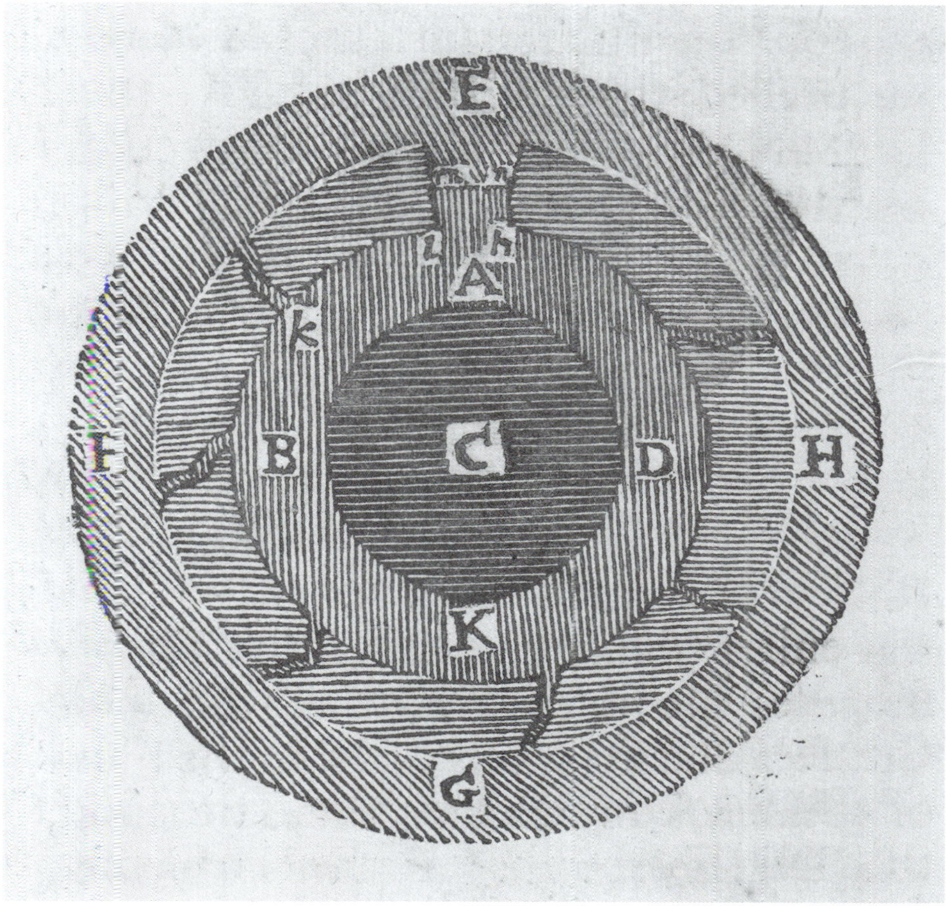

FIG. 64 Diagram with the hydrostatic movement of the Earth's crust. From John Keill, *An Examination of Dr. Burnet's Theory of the Earth . . . and A Defence of the Remarks on Mr. Whiston's New Theory*, London, 1698, p. 194.

a synthetic way, the hydrostatic movement of the Earth's crust in relation to an underground abyss occupied by water.[203]

Keill's work is mentioned in Desaguliers's articles, published in the *Philosophical Transactions* in 1724–5 and devoted to the shape of the earth.[204] The purpose was the defence of Newton's hypothesis, recently challenged by French cosmologist Jacques Cassini. Keill advocated a terraqueous globe of a somewhat squashed shape, flattened at the poles, and raised to the equator. The discourse on the earth, quoting from Keill's *An Examination of Dr. Burnet's Theory of the Earth*, uses two fundamental concepts: gravity, attracting bodies to the centre of the earth, and, since the earth was in circular motion, the application of the centrifugal force. The latter, being stronger at the equator, has lifted land from the centre

203 The date of the first edition of Keill's book (*An Examination of Dr. Burnet's Theory of the Earth . . . and A Defense of the Remarks on Mr. Whiston's New Theory*, London, 1698), where this figure first appeared rules out an influence of the Codex Leicester. However, Keill, well before the arrival of the Codex Leicester in London, was in touch with the Italian scientist Francesco Bianchini, who in Rome, as we have seen, was close to Resta and Ghezzi, former owner of the Codex. A search in the Keill and Bianchini papers might clarify this point.

204 'A Dissertation concerning the Figure of the Earth, by the Reverend John Theophilus Desaguliers', *Philosophical Transactions*, 33, 1724–5, pp. 201–22; 'The Dissertation concerning the Figure of the Earth Continued', *Philosophical Transactions*, 33, 1724–5, pp. 239–55; 'A Dissertation concerning the Figure of the Earth. Part the Second', *Philosophical Transactions*, 33, 1724–5, pp. 277–304; 'An Experiment to Illustrate What Has Been Said in the *Philosophical Transactions*, No. 386, 387, 388, concerning the Figure of the Earth', *Philosophical Transactions*, 33, 1724–5, pp. 344–5.

more at the equator than at the poles, which are flattened by gravitational attraction towards the centre of the earth.

The contribution given by these Newtonian concepts to the development of geology deserves more attention.[205] There was already a theory of uplift of the continents from the sea as consequence of rebalancing of weights within a pre-Newtonian concept of gravity, as advocated by medieval philosophers such as Buridan and as developed by Leonardo.[206] This theory re-emerged, in connection with a new notion of gravity, among some Newtonian 'geologists' in late seventeenth- to early eighteenth-century England.

Both this pre- and post-Newtonian applications of the concept of gravity to geology have to be differentiated from other apparently similar cases. For example, in Girolamo Fracastoro[207] and Robert Hooke,[208] the variation of the earth's rotation axis is favoured over gravity. Changes in the centre of gravity, as described at the beginning of Kircher's *Mundus Subterraneus*, do not affect his geodynamic theory,[209] since they are the consequence and not the cause of geological changes (collapses of the crust and the Deluge). The same position is apparent in Burnet,[210] while in Bernier,[211] who perhaps influenced Leibniz on this issue,[212] the changes only generate a passive uncovering of lands as a result of seawater displacement.

Instead, among English Newtonians, William Whiston,[213] who, as it turns out from his memoirs had some contact with Thomas Coke,[214] imagines the continents as huge columns floating in the water, and explains the level variations on their surface (valleys, mountains, etc.) by the variations in gravity that generate a greater or lesser sinking in the underlying fluid.[215] As we have seen, Keill diagrammatically

205 Newman 2009, p. 41.

206 See Chapter 4 in this volume.

207 G. Fracastoro, *Hyeronimi Fracastorii Homocentrica*, Verona, 1538. See Dal Prete 2006.

208 In Robert Hooke's *Posthumous Works* (London, 1705) the hypothesis of the shifting of the centre of gravity of the earth caused by an earthquake (p. 322) is only mentioned as one among many other hypothesis and not especially developed, contrary to the shifting of the earth's axis of rotation (pp. 346–8). See Turner 1974; Oldroyd 2006, pp. 35–8.

209 Athanasius Kirchner, *Mundus Subterraneus* [*The Subterranean World*], Bk I: *Centrographicus*, Amsterdam, 1665.

210 Thomas Burnet, *The Sacred Theory of the Earth*, 3rd ed., London, 1697, Book II, ch. 2, pp. 128ff., esp. 133–4.

211 François Bernier, *Abrégé de la philosophie de M. Gassendi*, Lyon, 1684, pp. 465–78.

212 Leibniz 2008 [1749], ch. VI, p. 15.

213 William Whiston, *A New Theory of the Earth*, London, 1696.

214 *Memoirs of the Life and Writings of Mr. William Whiston. Containing, Memoirs of Several of his Friends Also. Written by Himself*, London, 1753, vol. 1, part II, p. 324:

> When I was at Pensehurst, Dr. Hammond's old Living, during my first day's stay at Tunbridge Wells, on Saturday August 23d, 1746 (Witch Living of near 400l. a year had been offered me by the Earl of Leicester, with whom I had not the last acquaintance, above 5 years before, but was by me refused; as the Reader may find in my Letter to the Archbishop of Canterbury thereupon, printed in my Three Tracts, Page 1–12).

For the letter, see W. Whiston, *Three Tracts Being* . . ., London, 1742, pp. 5–7, included in the British Library collectanea 'Whiston Controversy', 699 g.14, 12–16 (the collectanea includes small volumes printed in 1727–48).

215 William Whiston, *A New Theory of the Earth*, London, 1708, p. 66:

> [A]n Orb of earth will be compos'd on the Surface of the Fluid, and its different Columns being made up of very different Natures and specifick Gravity . . . the Orb will sink into the Fluid in different degrees, and thereby render its Surface unequal, or distinguished into Mountains, Plains, and Vallies.

illustrated Whiston's theory (Figure 64) with floating continents in a static-dynamic equilibrium, a theory also envisaging their 'uplift' from the underlying fluid, the same concept illustrated in one of the figure in the Codex Leicester.

Indeed, in the years preceding the departure of Thomas Coke and his tutor Hobart for the Grand Tour (1712), and in the years immediately following it, a major debate in the development of the earth sciences arose in Britain, in which it is possible to distinguish two opposing lines.

The first is best represented by Thomas Burnet, whose work on the history of the earth emerged in 1680–9 in Latin and in 1691 in English. Burnet's work, apart from contributing to the conceptual connection between history and geology, features frequent use of diagrams of the earth. His starting point are the illustrations in Descartes's *Principia Philosophiae* (Amsterdam, 1644), but he enriches them in a physical-geographic sense. Descartes presented his history of the earth as valid for any celestial body.[216] Burnet, on the contrary, surpasses this abstract dimension, addressing specifically the history of 'our earth'. His purpose, common to other English geologists at the time, was to apply Descartes to the Bible. As a consequence, schematic diagrams of Cartesian origin are interweaved with geographical and physical images of 'our earth', which seem profoundly influenced by the 'physical' or qualitative approach, as well as by the figures of Kircher's *Mundus Subterraneus*.

In the alternative English, 'Newtonian', tradition, Burnet was bitterly challenged as a fanciful constructor of theoretical systems. The books by Keill and Whiston were written in order to oppose the theories of Burnet (and those of the French Descartes) in the name of Newton and in the service of a more empirical approach. Similarly, Desaguliers, in his 1724–5 articles on the shape of the earth, scorns Burnet by calling him 'the Theorist', a builder of abstract and empty conceptions. As we have seen, in this Newtonian line, the study of the earth in terms of gravity and static-dynamic equilibrium will become fundamental, contributing to the subsequent and slow overcoming of the previous notion dominant in seventeenth-century geology, that is, the idea of 'passive' uplift of the continents, caused by the breakdown of the earth's crust which set its pieces more or less in vertical or inclined positions. Other differences between these two currents in English earth sciences over the course of the two centuries were the 'deist' trends of Burnet and other scholars such as William Woodward,[217] who tended to explain the Deluge and other biblical events in a scientific and natural way, and those scientists, who, like Keill, believed that the Deluge was an event explicable only as a miracle.

This English debate, which attracted popular attention in the *Spectator* magazine,[218] could have contributed to draw Coke's and Hobart's attention to the Codex Leicester during their Roman stay and, once in England, attracting those Newtonians interested in Earth sciences who had contact with Coke. In Ducci's fine copy, the very clear titles and large characters in the final pages (e.g. *Delle terra in se*) along with the diagrams of the earth and the recurring word *gravità* (Figure 65) could easily attract attention.

See Harrison 2000.

216 Rudwick 2014, pp. 55–8.

217 Woodward 1695 explains marine fossils as the consequence of a momentary and general diminution of the cohesion of bodies during the Deluge, which allowed the shells to enter the rock formations.

218 Porter 1977, p. 87.

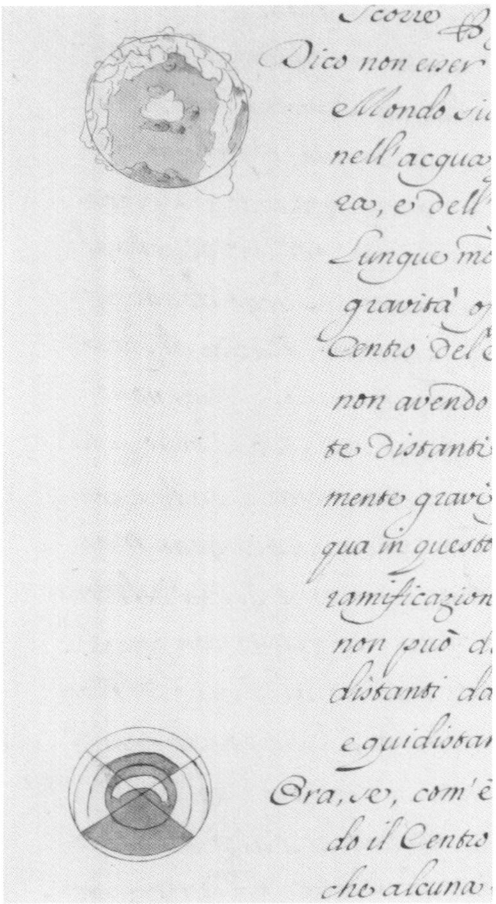

FIG. 65
Two drawings of the earth. Copy of the Codex Leicester, f. 36r, Holkham Hall MS 700, f. 167v.

Evidence in British science of an interest in Leonardo

So far, we have concentrated on conjunctions of ideas between the Codex Leicester and the English geological debates at the time when the Codex arrived in London, but there is evidence of a more general interest in the writings of Leonardo.

An article published in the *Spectator*, the well-known daily journal edited in London from 1711 to 1712 by Joseph Addison and Richard Steele, deals with 'genius' in general, of which it cites three scientists, Bacon, Boyle, and Newton, as examples, alongside the artist-scientist Leonardo. The inclusion of Leonardo in an article focused on scientists is remarkable. The article dates to 1712, the year before Thomas Coke left Britain for his Grand Tour and a few years before the first English edition of Leonardo's *Treatise on Painting*.[219]

By contrast, in the same period, in Florence, the correspondence between Buonaventuri and the scientist Guido Grandi reveals a certain detachment from, if not overt hostility toward, Leonardo and the Codex Leicester. This attitude reflects the way that, within Italian hydraulic science of the time, some Galilean scientists and mathematicians who dealt with water distanced themselves from the

219 On the article in the *Spectator*, see Craske 2013 and Mount 2019. Edition consulted: *The Spectator*, edited by Joseph Addison and Richard Steele, Edinburgh, 1785, vol. 7, pp. 305–9.

'practitioners', that is to say, artists and architects who were often involved in hydraulic works. This is very different from the British attitude towards Leonardo as a scientist in this period.[220]

The first English edition of Leonardo's *Treatise on Painting*, published in London in 1721,[221] offers clear proof of the interest of the Newtonian scientists working in London in Leonardo as an artist-scientist.[222] In the preface, Leonardo's optical theories are compared to those of Newton,[223] while the disorder of his original manuscripts are praised and compared to the *Sylva Silvarum* by Francis Bacon, philosopher of empiricism, and opposed to the methodical but empty writings of the 'System-Mongers':

> Whereas, had the whole [Leonardo's research] been woven into a Regular System, there had been no taking out, without tearing; the drawing of a few Threads, wou'd not only have disfigured the Rest, but have even endangered the unraveling of the whole Piece. 'Tis for this Reason perhaps, that my Lord Bacon's Silva Silvarum, which is written perfectly in Leonardo's manner, continues still in Use and Esteem; while the more Methodical Productions of most of our System-Mongers, are become antiquated and out of Date.[224]

The scornful definition of 'System-Mongers' recalls that used by Desaguliers in 1724–5 against Burnet the 'Theorist' within the general controversy that, as we have seen, had opposed Newtonians to the builders of abstract geological systems.

The publisher of the first English edition of the *Treatise on Painting*, John Senex (bap. 1678–1740),[225] was an active publisher of the works by Desaguliers and other Newtonian scientists active in these same years, like John Keill and William Whiston, who were all advertised at the end of the edition of Leonardo's *Treatise on Painting*. One example is the solar system chart, created by Whiston and published by Senex (Figure 66):[226] the text is arranged in a non-linear way, in individual blocks adapted to the images. Leonardo's original page with its similar mix of text and images, at least from a graphic point of view, might have not discouraged these English scientists too much.

Senex also published scientific works translated from French and other languages by the polyglot Desaguliers,[227] and I wonder whether Desaguliers could be the author of the anonymous preface and English translation of Leonardo's *Treatise on Painting* published in 1721 by Senex. In any event, behind the edition of the *Treatise* lies a scientific sensitivity. Thus, for example, in Chapter 315 concerning the static and dynamic equilibrium of the human figure, Leonardo's Italian terms *quantità* (quantity) and *somma del peso* (sum of weight), which in the 1651 French edition[228] are both generically translated

220 Laurenza 2019a and section on 'Scientists Versus Artists', p. 192 of this chapter. On this distinction in the science of waters, see Maffioli 2008 and 2010, pp. 31ff.

221 On this edition, see Woodfield 2009; Barone 2019; Mount 2019.

222 Woodfield 2009.

223 *A Treatise of Painting by Leonardo da Vinci Translated from the Original Italian*, London, 1721, 'The Translator's Preface' (n.p.): 'We see what Laudable Uses he makes, even of a defective Doctrine of Light. To what a pitch wou'd he have carried his Art, had he been acquainted with the New, the Nobel, the Newtonian Theory of Light and Colours?'

224 *A Treatise of Painting by Leonardo da Vinci Translated from the Original Italian*, London, 1721, 'The Translator's Preface' (n.p.).

225 Worms 2004.

226 1720, but possibly first published in 1712.

227 He gave public lectures in English, French, and Latin. See also Fara 2004, p. 891.

228 *Trattato della Pittura di Lionardo da Vinci*, Paris, 1651, p. 68.

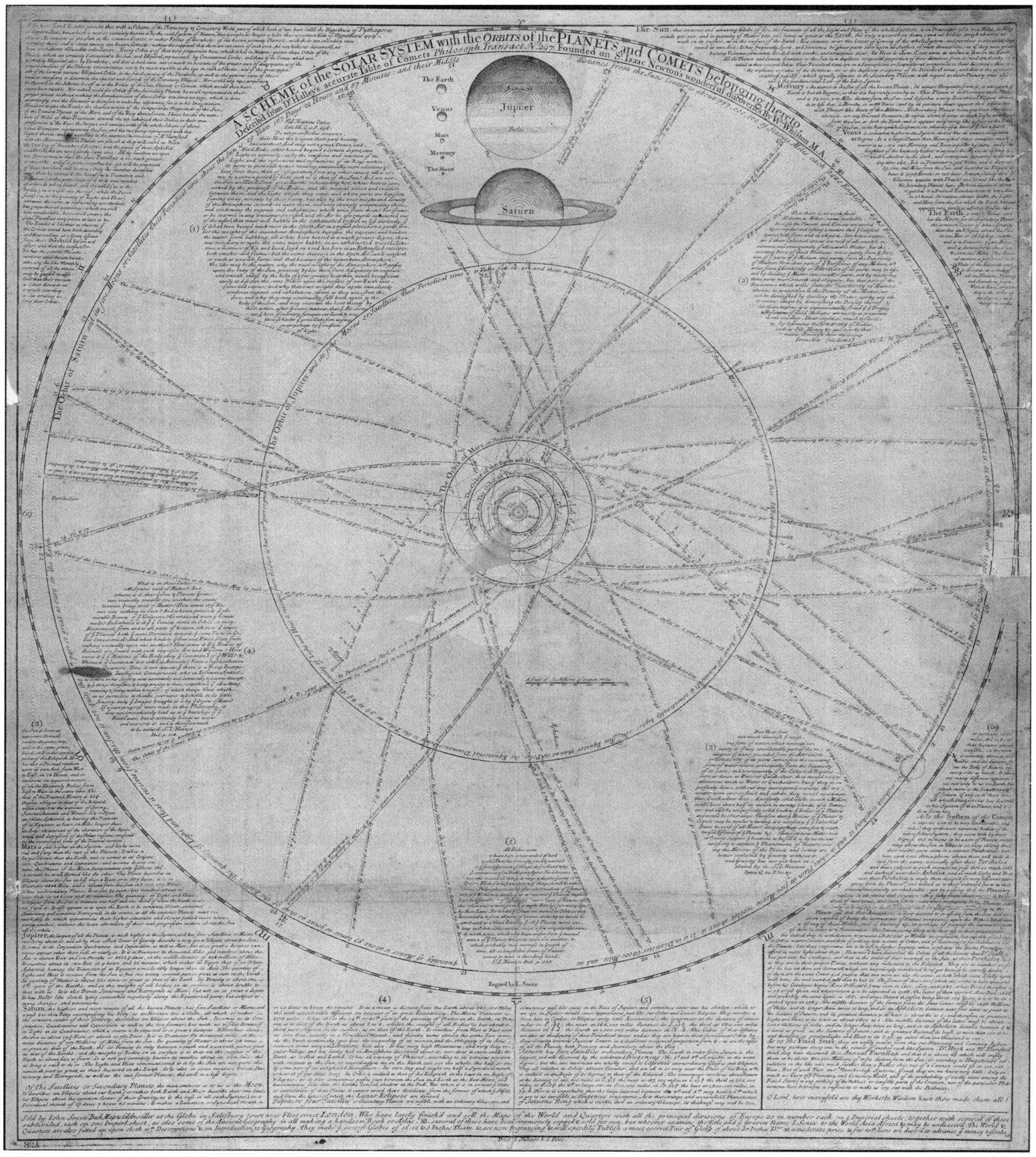

FIG. 66
From William Whiston, *A Scheme of the Solar System with the Orbits of the Planets and Comets . . . Founded on Sr. Isaac Newton's Wonderful Discoveries . . .*, London, 1712.

as *masse*, in the English edition are translated, respectively, as *magnitude* and *mass*,[229] that is using terminologies and distinctions that recall Newton's *Principia*, where *Definitio I* for the first time distinguishes the concept of 'mass' from that of 'weight'. Desaguliers studied and illustrated the static and dynamic equilibrium of the human body in his *A Course of Experimental Philosophy*, the first edition of which was published by Senex in 1735. These are certainly influenced by Borelli, but it is difficult not to compare them with similar studies in the Leonardo's *Treatise on Painting*. Indeed, in the Senex edition of the *Treatise*, particular attention is paid to the studies on the balance of the human figure. The plumb

229 *A Treatise on Painting by Leonardo da Vinci . . .*, p. 113.

lines, which in some cases is not visible in the figures of the 1651 French and Italian editions, are made very evident in the English edition. Where the line is completely missing in both the French[230] and the Italian editions,[231] it is clearly indicated in English.[232]

Studies of the kinematics of the human body in the so-called Codex Huygens in London, at that time believed to be wholly by Leonardo, were printed for the first time in two different editions around 1720.[233] Even in this case, the concepts of gravity and centrifugal force were involved, and it seems no coincidence the person who bought the Codex in London in 1690 was Constantijn Huygens (1628–97), the brother of one of the greatest scientists of the time, Christiaan Huygens (1629–95), who among other things offered the first definition of *centrifugal force*.[234]

Gravity and centrifugal force were concepts applied by Desaguliers and other English Newtonians to the study of the earth. It would therefore be perfectly understandable if the Codex Leicester, in which Leonardo applied notions such as gravity, had attracted the attention of these English scientists, in a similar way to his studies of the statics and dynamics of the human figure.

Once again it is in Thomas Coke's circle of contacts that we find interesting clues. Indeed, Coke enjoyed contacts with the people behind the English printed editions of the *Treatise on Painting* and of the Codex Huygens. The edition of some of the studies from the latter bears a dedication to Thomas Coke 'Vice Chamberlain to his most sacred Majesty King George'. This is not Thomas Coke the Earl of Leicester and the owner of the Codex Leicester, but a member of another family, the Cokes of Derbyshire (1674–1727).[235] However, they were relatives, on good terms, and kept in regular touch.[236] The publisher of the *Treatise on Painting*, John Senex, was associated with William and John Innys, book dealers and printers; and in 1745, Mary Senex, who took care of the printing house after her husband's death co-published with William Innys the second edition of Desauguliers' *A Course of Experimental Philosophy*. It is from John Innys that Thomas Coke, the owner of the Codex Leicester, will make what may be his biggest purchase in the field of printed books, the *General System of Cosmography* . . . by John Innys, a monumental collection of nearly 15,000 maps, preserved at Holkham Hall and still substantially unexplored.[237] Thomas Coke was a grand master of masonry in 1727 and, since 1735, fellow of the Royal Society. Two more sets of contacts are at work here: John Innys was a Mason, while Senex and Desaguliers were members of both organizations.

230 *Traité de la Peinture de Léonard de Vinci*, Paris, 1651, p. 69.

231 *Trattato della Pittura di Lionardo da Vinci*, p. 62.

232 *A Treatise of Painting by Leonardo da Vinci* . . ., p. 114.

233 See Laurenza 2019.

234 In the same period, Coke's family doctor, Richard Mead, adds his mechanical analysis of the muscles to the posthumous publication of William Cowper anatomical treatise (*Myotomia reformata*, London, 1724). Senex's English edition of Bernardino Genga's *Anatomy improved and illustrated with regard to the uses thereof in designing* (London, 1723) was dedicated to Mead and included, among the subscribers, Desaguliers.

235 'Coke, Thomas (1674–1727), of Melbourne, Derbys.; Melton Mowbray, Leics.; and St James's Place, London'. See Hayton, Cruickshanks, and Handley 2002.

236 Warburton 1929, pp. 216–17.

237 John Innys, 'General System of Cosmography, or Elements of Astronomy and Geography Illustrated by Maps, Plans and Views, Collected from the Most Eminent Authors Ancient and Modern', Holkham Hall Archive, London. See Mortlock 2006, pp. 33–96, who, because of the presence of maps dated 1754 and 1755, assigns a later date to this acquisition.

In conclusion, we may say that accumulation of evidence paints a picture of a rich intellectual context in which the Codex Leicester could have thrived as more than an illegible historical curiosity. Its verbal and visual demonstrations of the history of the earth were such as to have warranted the interest of major players in the reform of geology.

Scientists Versus Artists: Copies of the Codex Leicester in Eighteenth-Century Florence

Some of the copies of the Codex Leicester known in Florence in these years attracted the attention of various authorities. There was the architect and hydraulic engineer, Bernardino Ciurini (1695–1752); also Antonio Maria Niccolini (1701–69) from a noble family with extensive interests and contacts in the scientific world; and various Italian scientists, from Francesco Algarotti to Paolo Frisi. All three of the main fields addressed in the Codex—water, geology, and astronomy—were prominent subjects of attention at this time, including the denial of the Deluge. A feature shared by those interested in the Codex is their sympathy for the world of hands-on professionals and, consistently, for English science of the time. This stands in marked contrast to the more 'theoretical' and mathematical stance of leading figures in the new science of waters, such as the scientific publisher Tommaso Buonaventuri, who clearly distanced themselves from Leonardo and the Codex Leicester.

Before the Codex Leicester and the copy commissioned by Thomas Coke left Florence for London in 1718–9, at least two other complete copies were produced in Florence. One of them, now missing, belonged to the Tuscan architect and hydraulic engineer, Bernardino Ciurini, the other is now in Weimar. In 1767, on behalf of the Florentine marquis Antonio Maria Niccolini, another complete copy after one of the first two was realized plus a copy after an abridged version of the Codex that had been produced in the meantime.[238] Therefore, three complete and two abridged copies are attested in Florence in the eighteenth century

All this clearly reflects an interest in Leonardo's scientific work. But it was not an uncontested interest, since it was part of (and perhaps contributed to) a dispute between 'scientists' and 'practitioners' in the Tuscan science of the waters of the time. On one side were those who, on mainly theoretical and mathematical grounds, had inaugurated a new approach to the severe problem of water regulation. On the other were those who tackled the same problem on more empirical and observational grounds. The latter group often included architects and artists. In fact, the great works of water regulation during the sixteenth century that tried to provide a solution to the frequent floods caused by rivers such as the Arno or the Tiber, were mostly by architects and artists. Alternatively, the mathematical side of the Galilean science had generated a more theoretical approach, which supplies the basis of modern hydrology. And, as it often happens, a 'new-born' science tends to emphasize, even institutionally, its theoretical value. Thus the mathematical approach to the science of water distanced itself from that of hands-on professionals.

238 See Appendix to this Chapter.

Beyond the field of the science of water, the reconstruction of the production and circulation of copies of the Codex Leicester in Florence can be set within more general philosophical battles. Particularly prominent was the debate around one of the central geological themes of the Codex, namely the impossibility of explaining the Biblical Deluge on a scientific basis as a causal agent in the history of the earth. In any event, this serves to demonstrate the contrasting nature of the British attitude toward Leonardo as a scientist. The reception of the Codex Leicester in Florence was more uncertain. Not surprisingly, those who in the Tuscan city showed most interest in Leonardo's scientific thought were in one way or the other linked to the British context of the Royal Society.

A good place to start is in 1717, when the Codex Leicester is still in Florence in order to be copied in clear handwriting by Ducci on behalf of Thomas Coke. As we have seen, the correspondence in that year between two leading figures of the Florentine scientific culture, Tommaso Buonaventuri and Guido Grandi, deals with the Codex Leicester and the production of its copies. Grandi was a mathematician and hydraulic engineer. Buonaventuri, as the director of the Grand Ducal printing house, was involved in the first Florentine edition of Galileo's works. At the time of the correspondence concerning the Codex Leicester, Buonaventuri was overseeing an edition of authors on hydrology,[239] hence his interest in the Codex Leicester. Buonaventuri writes to Guido Grandi of having read a copy of Leonardo's Codex, and gives a very peculiar judgment: although the Codex, he writes, is only of historical interest for the science of water, it might still be worth including it in the printed collection of hydraulic authors. However, he adds, the person who had bought a copy from Ducci and could provide it to the printing house was asking for such an excessive amount of money that it was better to leave him 'hanging' (in Italian, 'bisogna lasciarlo frollare'). Buonaventuri's attitude reveals a certain detachment, if not hostility. In the end, the Codex Leicester was not included in the collection of writings on hydrology published by Buonaventuri in 1723. This story confirms Cesare Maffioli's recent conclusion that, within the Italian hydraulic science of the time, many Galilean scientists and mathematicians who dealt with water distanced themselves from the 'practitioners'.[240] In his introduction to the collection of writings on hydrology, Buonaventuri neatly distances the latter from the architects and other similar practitioners of hydraulics: only mathematicians, he writes, can establish a true science of water and certainly not the artists who claim to be involved in hydraulics on the basis of their ability in 'drawing' ('disegno'). Indeed, as Buonaventuri concludes:

> Thus Archimedes, Galileo, Castelli, Michelini, Borelli, Mariotte, and other living and dead mathematicians studied the architecture of waters without being able to draw or paint; and, on the contrary, Raphael, Titian, Correggio, Callotte, Stefano della Bella, and other perfect draftsmen did nothing in this field and many others [great draftsmen] could equally be cited to demonstrate the same point.[241]

239 *Raccolta d'autori che trattano del moto dell'acque* [*Collection of authors dealing with the motion of the waters*], 3 vols, Florence, 1723.

240 Maffioli 2008 and 2010, pp. 31ff.

241 *Raccolta d'autori . . .*, vol. 1, Florence, 1723, p. x:

> Così Archimede, il Galileo, il Castelli, il Michelini, il Borelli, il Mariotte, ed altrivivi, e morti mattematici molto ben divisarono circa l'architettura dell'acque, e de' fiumi senza niente disegnare, e dipingere; e niente in ciò operarono Raffaele, Tiziano, il Correggio, il Callotte, Stefano della Bella, ed altri perfetti disegnatori; e così si potrebbe dimostrare, ed esemplificare di molti altri.

Leonardo and other artists who, in the sixteenth century, successfully devoted themselves to the science of water and made use of their draughtsmanship, are of course omitted by Buonaventuri, since they did not fit into his overly simplistic binary equation.

Whereas in London, around 1720, Baconian and Newtonian scientific circles were interested in Leonardo because he was a practitioner and an artist who had dealt with science, in Florence in the same years and for the same reasons the contrary was the case. However, it is important to specify that this seems limited to the hydraulic science of the Galilean scientists linked to the Medici court, because in other Florentine contexts, as we will see, the work of artists and practitioners and, among them, of Leonardo was subject to a different kind of evaluation.

We do not know which of the copies of the Codex Leicester made in Florence around 1717 was examined by Buonaventuri, whether it was the copy now in Weimar, the copy belonging to Ciurini, or another, as yet unknown copy. However, it seems that most likely it was the copy obtained by Bernardino Ciurini. Ciurini was a typical representative of architects involved in the science of water, a professional class from which Buonaventuri intended to distance himself, which might explain the ironic and the irritated tone Buonaventuri uses in reference to the owner of the copy in one of his letters to Grandi.

Bernardino Ciurini's copy of the Codex Leicester and Alessandro Galilei's work: Two approaches to the science of water

Bernardino Ciurini was born in Castelfiorentino and soon moved to Florence. He is well known, but not well studied as an architect, apparently an excellent one, since in 1724 he became a member of the Academia del Disegno in Florence (later to become the Academy of Fine Arts), and in 1739, architect of the Opera del Duomo.[242] In the Accademia del Disegno he was public lecturer and teacher of mathematics and geometry, and, later on, of civil and military architecture. His courses at the Academy included the science of waters.[243] Ciurini was elected a member of the Academy alongside another architect and hydraulic engineer, Alessandro Galilei (1691–1737) of the family of Galileo Galilei.[244]

A legal document in the Florence State Archive[245] introduces us to Ciurini's work as a hydraulic engineer. The document concerns a dispute between various landowners in the valley of the Elsa River, which flows into the Arno west of Florence. One landowner, Pietro Setticelli, had work done on his property, which aggravated flood damage along the banks of the Elsa River. Setticelli had built a palisade along the embankment of the river on his own property, which altered the natural course of the river, causing it to break against the opposite bank with considerable damage to the property of the neighbours. In 1736, Bernardino Ciurini was consulted as an 'engineer'. The document, which is dated 1750, mentions a report prepared in 1736 by Bernardino Ciurini:

242 Del Pela 1893; Zangheri 1982.

243 Camerota 2008, p. 173.

244 Cusmano 1998.

245 State Archive, Florence, Decis. XIII, *Florentina Ripae Muniendae et Praetensae Alluvionis*, f. 332. See *Raccolta delle decisioni della Ruota fiorentina dal MDCC al MDCCCVIII*, vol. 1, Florence, 1845, pp. 237ff.

> The same was also observed and reported more precisely by Sig.re Bernardino Ciurini skilful engineer, who had been sent an order to scrutinize the work done . . . and who in his report, after recording the same dimensions and configuration as already mentioned, asserts therein that the said work is all planted in the river bed and carried out so voraciously and so low that *it forces the current, which inundates it from every side, to make a revolution contrary to the course of river, bumping into the opposed bank, raising and filling the side of said work, and corroding and hollowing it out on the opposite side . . .*[246]

Contents and words of Ciurini's report are evocative of the Codex Leicester. Here is one of the many examples from the Codex Leicester:

> When the circumrotating wave, reflected from the object where it hits the bottom of the river and which, in the proceeding of its *revolution*, meets the incidental wave, which passes above it, a part of it continues the *revolution* that has already begun and a part casts itself in an *opposite revolution and hits* the beach, and hence the sand of the slope in front of it descends.[247]

Following Ciurini's intervention, the works improperly constructed were demolished, 'so that the river's course would be restored to its primeval state', as stated at the beginning of the eighteenth-century document mentioned.[248] Indeed, Ciurini's report includes the same respect for the river's natural course and for works that are compatible with it that is so apparent in the Codex Leicester. In the future, the identification of Ciurini's original report, which included drawings as is mentioned in the document concerning the eighteenth-century dispute, may provide further evidence.[249]

Where is Ciurini's copy of the Codex now, if it still exists? From the only study fully dedicated to Ciurini—an article published in 1893[250]—we know that he married a Florentine noblewoman and that, in 1752, he died without heirs. In 1766, Anton Francesco Durazzini wrote that a copy of the Codex Leicester, made by Ducci, was 'in the estate' (*nell'eredità*) of Ciurini', and in another passage that 'it belonged Ciurini's heirs': 'This copy of the Ducci is presently in the estate of the well-known architect Bernardino Ciurini', where the term 'estate' (*eredità*) does not necessarily refers to the heirs, the family members.[251] Since Ciurini died without children, could his books and manuscripts have ended up in some public institution?

Ciurini was a professor of architecture and mathematics at the Academy in Florence, where we find a manuscript formerly belonging to him. It is an anonymous treatise on 'architecture' (*Breve discorso e pratica di Architettura*), including parts of geometry, perspective, etc., written in Florence in the

246 State Archive, Florence, Decis. XIII, *Florentina Ripae Muniendae et Praetensae Alluvionis*, f. 332 (emphasis added):

> L'istesso fu osservato e riferito anche più esattamente dal Sig.re Bernardino Ciurini Perito Ingegnere, il quale era stato mandato là a riconoscere il lavoro d'ordine . . . e che nella sua relazione, dopo riportate l'istesse misure e l'istessa figura di già accennata asserisce ivi *che il detto lavoro è piantato tutto nel letto del fiume e lavorato talmente ingordo e basso che obbliga la corrente, che per ogni parte lo bagna, a formare una revoluzione contraria al corso del fiume, ad urtare nell'argine contrapposto, a rialzare e riempire dalla parte di detto lavoro, e corrodere e incavare dalla parte apposta.*

247 Leicester, f. 13r.

248 Durazzini 1770, p. 237: 'onde il corso del fiume medesimo venga ad esser rimesso nel suo stato primiero'.

249 On Ciurini's expertise concerning the Bagnolo River in 1734–48, see the *Relazione fatta a S. M. C. Granduca di Toscana*, Decision XXXVIII, in *Selectarum Rotæ florentinæ decisionum thesaurus* . . . , vol. 1, Florence, 1772, pp. 513ff.

250 Del Pela 1893.

251 For full quote see later in this Section and Appendix to this chapter.

seventeenth century. When Ciurini bought the manuscript, he added a note on the title page: 'Of Bernardino Ciurini from Castelfiorentino bought the day of April 2, 1717'.[252]

The year corresponds to the period in which Ciurini, perhaps after Francesco Ducci's death (1718), also acquired one of the copies of the Codex Leicester made by Ducci. The most important part of the Academy's library was taking shape in the early nineteenth century through the acquisition of the libraries of architects and engineers, from Giuseppe Salvetti to Giuseppe Martelli and Giovan Battista Nelli. These libraries were a source of many scientific books and manuscripts in the Academy library collections, including texts on hydrology and geology, as well as a recently rediscovered group of printed and manuscript Galilean texts, including some in Galileo's own hand.[253] The presence of these scientific texts is not surprising, since the Academy's teachings embraced areas that we would now define as engineering. The manuscript once belonging to Ciurini could have arrived at the Academy directly or indirectly from his collection, as in the case of other books from architects' libraries. At the moment, the manuscripts of the Academy's library are only partially accessible, and there is no updated catalogue. Only a future search can clarify whether Ciurini's copy of the Codex Leicester passed into the Academy's possession.

The Accademia del Disegno was a place of great cultural vivacity. In 1732 it was Ciurini who introduced into the Academy the collector Jean-Pierre Mariette, the author of a letter on Leonardo's manuscript on mills.[254] In 1734, the Academy hosted Martin Folkes,[255] who was known for his antiquarian interests and for his atheist convictions, and who was one of the main British Newtonian scientists of the Royal Society, and thus a member of the English community that was so interested in Leonardo.

Alessandro Galilei, Ciurini's colleague at the Academy, was also linked to the English context. He had lived for a long time in London and written a bilingual treatise on architecture in Italian and English.[256] His manuscript papers in Florence State Archives include scientific arguments, ranging from a compendium of cosmography to a copy of Galileo's *Bilancetta*; from drawings of the dynamics of the human body[257] to hydraulic studies, dealing, for example, with the shape of the buttresses of the Ponte Vecchio in Florence.[258] The papers on the human body are clearly linked to the illustrations of Desaguliers in *A course of Mechanical and Experimental Philosophy* (London, 1725) and include a printed sheet from it.[259]

252 'Di Bernardino Ciurini da Castelfiorentino comprato il dì 2 Aprile 1717'. On this manuscript, see Camerota 2010, p. 42 n. 71. The name of the seventeenth-century author was erased.

253 Tongiorgi Tomasi et al. 2009, pp. 336–7 and plates pp. 84–6.

254 State Archive, Florence, *Accademia del Disegno*, 23, c. 51: 'Addì 14 marzo 1732 . . . Monsù Pietro Mariett' di Parigi prop.o dal Sig. Bernardino Ciurini.'

255 State Archive, Florence, *Accademia del Disegno*, 23, c. 55.

256 Malservisi 1991.

257 State Archive, Florence, *Fondo Galilei*, filza 18 ins. 2, ff. 389–90.

258 State Archive, Florence, *Fondo Galilei*, filza 18 ins. 2, ff. 85ff.

259 State Archive, Florence, *Fondo Galilei*, filza V (21), ins. 1, ff. 408–9.

It has been noted that in his treatise on architecture Galilei does not credit the authors from whom he borrows.[260] In relation to the Ponte Vecchio buttresses, only Leon Battista Alberti is credited. Galilei dealt with hydraulics at the same time when copies of the Codex Leicester circulated in Florence, and one of them belonged to his colleague Ciurini. It is very likely that he knew one of these copies. The shape of the buttresses of a bridge is considered in Codex Leicester, f. 7r, including a drawing in which the pillars are not far from Galilei's drawing. More research on his papers may demonstrate further overlaps. Certainly, Galilei's interest for Leonardo is demonstrated by the fact that when he moved from Florence to Rome he took a drawing by Leonardo with the intention to sell it at high price.[261]

The Niccolini copy: favouring a hands-on approach and an anti-diluvian stance

Two more decisive, and better documented, episodes in the circulation of the Codex Leicester in Florence date to 1765–9. Again, many of the people involved were close to the Royal Society and shared a Baconian and pre-Enlightenment sympathy towards the knowledge of hands-on professionals. All three areas addressed in the Codex Leicester attracted interest: water studies, astronomy, and geology. Interest in the latter, was particularly important because this was a 'new' science, involving pre-Enlightenment arguments that emphasized anti-Diluvial aspects.

One episode in the story is the production in 1767 of two copies of the Codex Leicester for Marquis Antonio Maria Niccolini, one complete, the other abridged (Figures 67–8).[262] The learned *abate* Giuseppe Bacherini,[263] librarian of the Niccolini family, made the first copy and verified the correctness of the second. Niccolini was, according to the eighteenth century definition, a 'virtuoso'. He studied law in Pisa, but his skills ranged from antiquarianism to science. Difficulties in accessing the private archives of the Niccolini family explain the lack of knowledge of his intellectual stature.[264] If the attribution of Jansenism[265] and 'deistic and libertine sympathies'[266] to Niccolini has recently been judged as excessive,[267] his sensibility was certainly that of an enlightened man with anti-Jesuit and anti-inquisitorial beliefs. An unpublished text by Niccolini contains a historical reconstruction of the drainage works begun in the sixteenth century in the Foligno marshes in Umbria, where Niccolini possessed lands.[268] Written in 1765 in the form of a letter addressed to Cardinal Alessandro Albani in Rome, this text can help us to understand why two years later, in 1767, Niccolini commissioned copies of the Codex Leicester. Niccolini informs Cardinal Albani about important hydraulic works undertaken

260 Malservisi 1991.

261 Gaburri's letter to Mariette, 1732. Bottari 1754–73, vol. 2, pp. 282–3. The drawing represented, according to Bottari, 'la circoncisione di nostro Signore. . . . Vi sono ventiquattro figure con una bellissima architettura . . .'.

262 Laurenza 2005.

263 See *Supplemento d'Istorie Toscane*, Firenze, 1769, p. 21: 'dotto Sig Abate Giuseppe Bacherini, Bibliotecario de Sigg Marchesi Niccolini'.

264 I thank marquis Niccolini for allowing me to get access at the archive and dr. Rita Romanelli for her help during the research. For Niccolini, see Scopetani 1770; Prosperi 1771; Pasta 2013.

265 Rosa 1980.

266 Ferrone 1982, pp. 230 and 318.

267 Pasta 2013.

268 Niccolini Archive, Florence, Fondo Antico 51, fasc. 58.

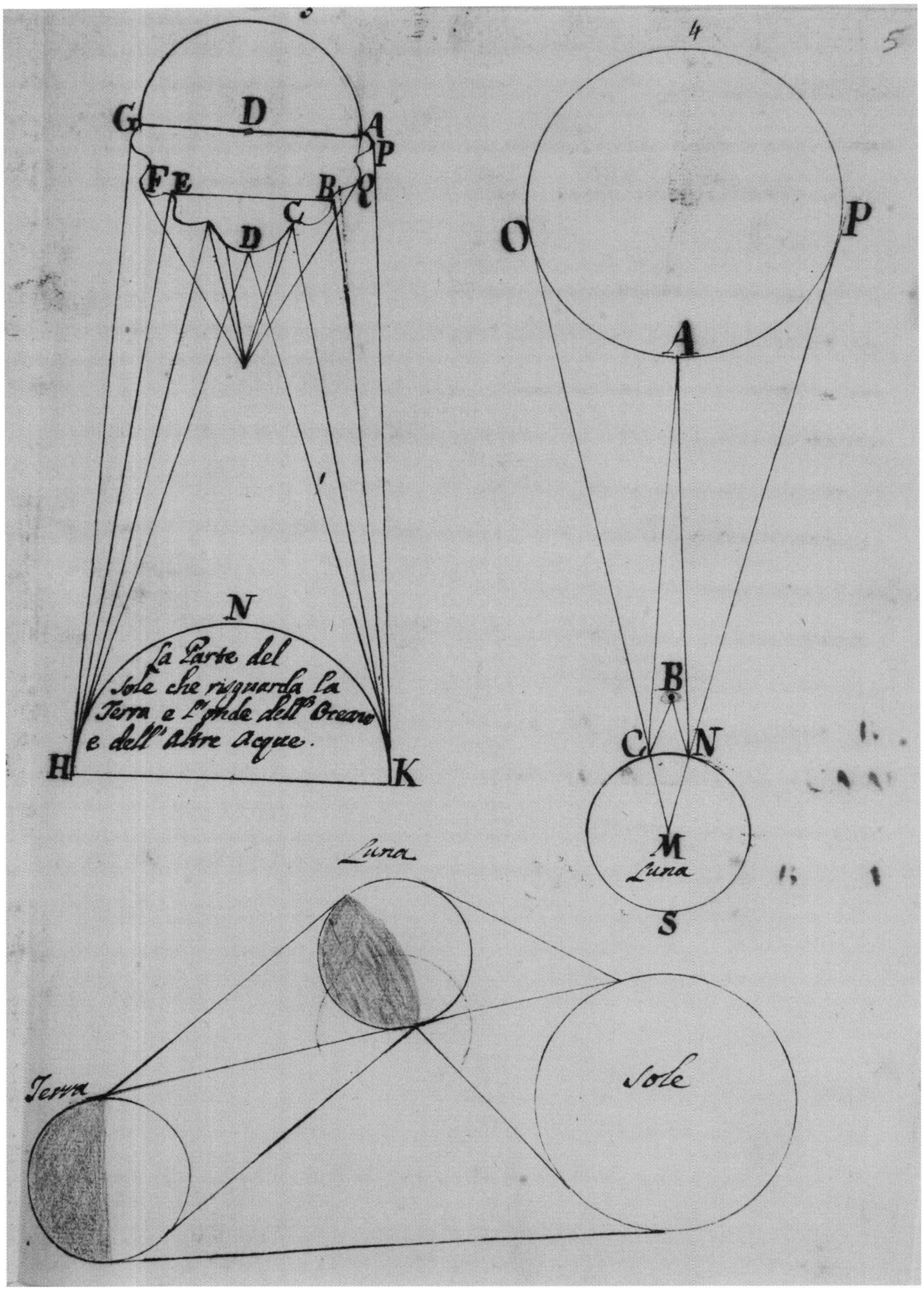

FIG. 67
Astronomical diagrams. Giuseppe Bacherini in the Niccolini copy of Codex Leicester, ff. 1r and 2r. National Library, Rome, MS Vitt. Em., 1767, f. 5.

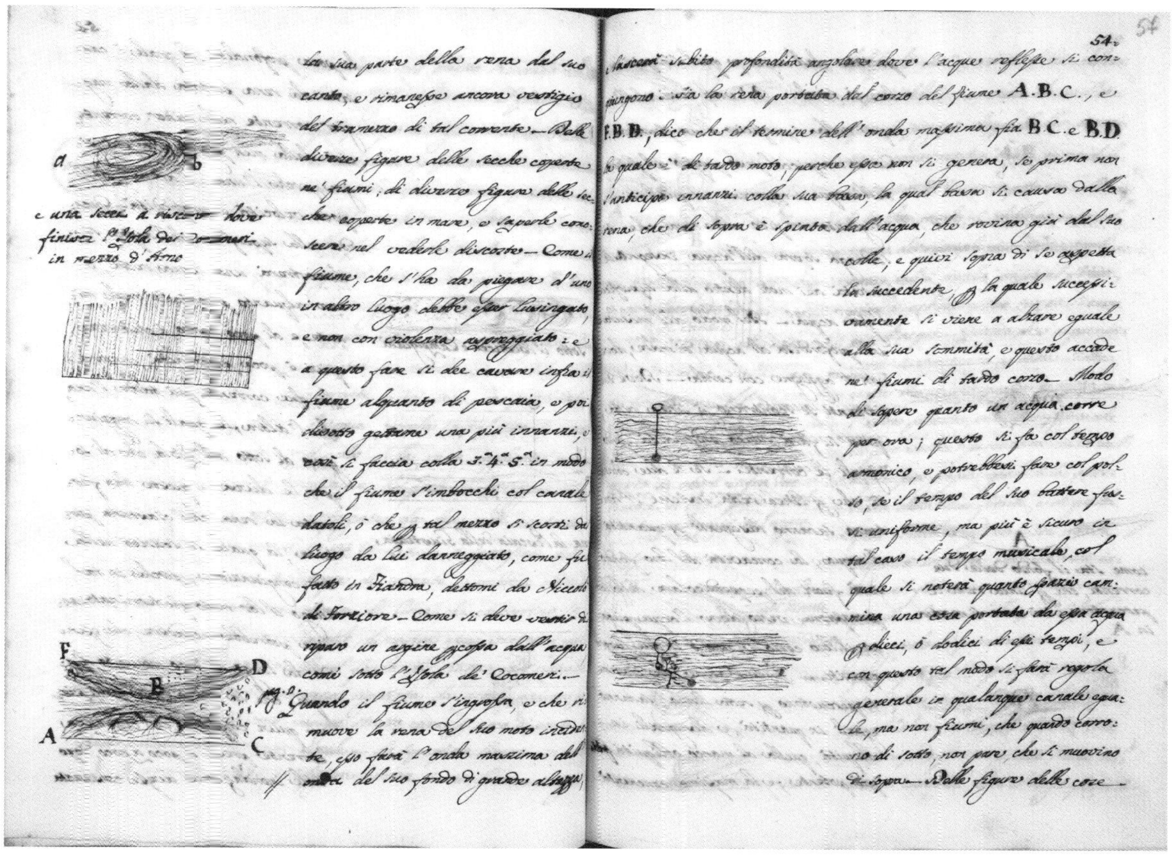

FIG. 68
Studies of water and other topics. Giuseppe Bacherini in the Niccolini copy of Codex Leicester, ff. 13r and 13v. National Library, Rome, MS Vitt. Em. 1767, ff. 56v–57r.

in Umbria in order to drain one of his lands near the River Tiber. Alongside the description of the on-going works, Niccolini adds a long historical survey describing, from the Renaissance to his time, all the hydraulic works carried out in the same area by famous 'architects'. Hydraulic works in Foligno, carried out by artists and architects of the past—Antonio da Sangallo, Baldassare Peruzzi, and above all Carlo Maderno—are highly praised by Niccolini. The text reveals Niccolini's technical knowledge of hydraulic matters, which is also testified by the correspondence exchanged with the scientist Paolo Frisi.[269]

Niccolini's many correspondents include scientists and prominent figures in European culture of the time, from Morgagni to Manfredi, from Montesquieu to Haller, to name a few.[270] During his long 'reverse' grand tour in Northern Europe (1745–50), Niccolini spent two years in England (1746–8), where he became a member of the Royal Society, presented by Martin Folkes, with whom he had long been in contact, probably since the time of Folke's Florentine stay in 1734 when he was received at the Accademia del Disegno. During his stay in England, Niccolini meet Thomas Coke, the owner of the Codex Leicester. As well as expressing his gratitude for the help given by Niccolini with the invoicing

269 e.g. Letters of Paolo Frisi to Antonio Niccolini, 1761, 1763, 1769. Niccolini Archive, Florence, Fondo Antica, segnatura 279, inserto 32.

270 Prosperi 1771; Pasta 2013.

of works of art from Rome, an unpublished letter sent by Coke to Niccolini also mentions their meeting in England.[271]

Given the content of Niccolini's text on Foligno, it is understandable that he would also have been interested in Leonardo, an author who had been excluded as a 'practitioner' from Buonaventuri's collection of hydraulic authors.

In a catalogue of the Niccolini library compiled in 1795, when it was evaluated by an antiquarian to be divided amongst the different heirs of the family, the copies of the Codex Leicester appear:

> ...
>
> Grandi. Institut. Mechanicae. Chart. in 4° .1.6.8
>
> Vinci. Dell'Acqua etc. Cart. in f° . Tom .2 .6.13.4
>
> Geometria e Fortificazioni. Cart. in f° 1.
>
> ...[272]

The title 'Dell'Acqua' (*On Water*) is a short form of the title page of the Codex Leicester (*Libro della Natura peso e moto delle Acque* ...). This document seems to suggest that the two copies were originally separated and formed two different volumes ('Tom. 2'), whereas they are now bound in a single volume. They could have therefore circulated separately. The numbers on the far right are the estimated value of the books. For each entry in the catalogue we find from one to three numbers corresponding to three monetary units, scudi, soldi, and denari. The two copies of the Codex Leicester are estimated at 6 scudi, 13 soldi, and 4 denari, which is far more, for example, than the other books, including the one by the Tuscan scientist Guido Grandi, valued at 1 scudo, 6 soldi, and 8 denari. This could be evidence of the importance assigned to Leonardo's scientific work, particularly given that these were copies and not original manuscripts.[273]

Of the two copies produced on behalf of Niccolini, the abridged version is particularly interesting because it makes an attempt to group Leonardo's texts around two themes: water science and geology. Of its two parts, the first is a selection of texts dealing with rivers and waters, and it is titled: *Discorso sopra la Natura, moto, e peso dell'acque et osservazioni sopra il corso de' Fiumi* (*Discourse on the Nature, Motion, and Weight of Water and Observations on the course of the Rivers*). Then, starting with folio 55v (f. 203v, according to modern numbering), the cosmological and geological texts are collected. In

271 Niccolini Archive, Florence, segnatura 283, inserto 4, Letter of Thomas Coke Earl of Leicester to Antonio Niccolini. Undated. 'I fear we are long divided from meeting, that nevertheless will never diminish my attention.'

272 Niccolini Archive, Florence, Library catalogue, Ms 1795, f. 59.

273 A painting attributed to Leonardo is included in an inventory of the Niccolini goods. Niccolini Archive, Florence, Inventari per l'interesse di Monsig.re Niccolini [1738], 114 fasc. 13, f. 71:

> Un quadro in tavola con ritratto di Donna, alto c.ca braccia 1 [?]. Ornamento dorato, e antico, e sua custodia davanti, nella quale v'è dipinto una maschera col motto 'Sui cuique Persona' di mano da Leonardo da Vinci.

See also *Serie degli uomini più illustri* ..., vol. 2, pp. 29–41, esp. p. 31:

> Il Signor Marchese Niccolini di Firenze tra le altre pitture, e sculture antiche, possede un'eccellente ritratto di femmina in mezza figura al naturale lavorato da Leonardo con stupendo Artifizio.

The portrait of the woman with the motto, 'to each his [or her] mask' is likely to be by a follower and is untraced.

particular, from folio 71r (f. 219r, modern numbering) onward, there are texts on fossils and the Deluge, grouped under the general title *Del Diluvio e de' Nichi Marini* (*On the Deluge and Marine Shells* [fossils]; after Codex Leicester, f. 8v). Compared to the first known example of a 'critical' anthology of Leonardo's manuscripts, made on behalf of Cassiano del Pozzo in the mid-seventeenth century, a novelty here is the selection of brief texts on geology, a science that in Cassiano's time was still in a seminal phase of its modern development.

On the basis of Barberini's statement at the beginning of the abridged copy, we know that it was based on another already existing copy.[274] It is therefore difficult to say how much of the 'critical' arrangement of Leonardo's texts is due to Niccolini's initiative or to the copy on which the arrangement was based.

However, it is likely that the inclusion, at the end of the abridged copy,[275] of texts copied from the work of another author, Filippo Maria Bonini's *Il Tevere incatenato* (*The Tiber Shackled*, Rome, 1663), was due to Niccolini. From 1661, Bonini lived between Palestrina and Rome[276] and wrote a treatise that, in addition to the possible regimentation of the Tiber, contains many theoretical sections on hydrology and, in a more limited form, on geology. The texts included in the Niccolini copy of the Codex Leicester deal with theoretical and hydrological subjects and show clear affinities with Leonardo's texts. They concern the macro-microcosmic analogy between water in the earth and the circulation of blood;[277] the study of the inverse relationship between water velocity and turbidity;[278] and finally a long, general text on the distinction between rivers, streams, lakes, and ponds.[279] This appears to be the first attempt at direct comparison of Leonardo's scientific texts with those of another author. A previous example of such comparison with regards Leonardo as a writer on art is the inclusion of Vignola's texts on perspective in manuscript copies of Leonardo's *Treatise on Painting*.[280]

Some sections of Bonini's treatise deal with the Deluge, which Bonini judges to be naturally impossible, as did Leonardo.[281] This was a rather original stance in the period, as even those, like Steno and Scilla, who defended the animal origin of fossils, did not deny the Deluge as their cause or one of their causes. Indeed, the recognition of the organic origin of marine fossils increasingly shared by scientists, was often reconciled with a scientific account of the Deluge. This stubborn adherence to 'Diluvialism' will last until the nineteenth century.[282]

Another aspect of Bonini's work that might be reminiscent of Leonardo is the use of pictorial images (Figure 69) as an integral part of the empirical, qualitative approach to the science of water, in contrast to the more mathematical, quantitative approach inaugurated by the new science of water of

274 See Appendix to this chapter.

275 National Library, Rome, Ms. Vitt. Em. 1767, ff. 80r–82v (ff. 228r–230v, modern numbering).

276 Buonora 2003; Carré Brunenghi 1996.

277 Bonini, *Il Tevere incatenato*, Bk I, ch. 3 ('L'Oceano è fonte e miniera di tutte l'acque è al mondo quel che all'uomo il fegato . . .'). Cf. Leicester, f. 34r.

278 Bonini, *Il Tevere incatenato*, Bk I, chs. 3 and 16 ('Perchè l'acqua mista d'atomi terrestri sia più lenta . . .'). Cf. Leicester, f. 20v.

279 Bonini, *Il Tevere incatenato*, Bk I, ch. 4.

280 Fiorani 2003; Farago 2009.

281 Bonini, *Il Tevere incatenato*, Bk I, ch. 5, pp. 28–32 (emphasis added): 'Dico dunque il Diluvio universale essere stato *naturalmente impossibile*, non solo per il tempo, nel qual successe, ma per la quantità dell'acque. . . .'

282 Rudwick 2014, pp. 120ff.

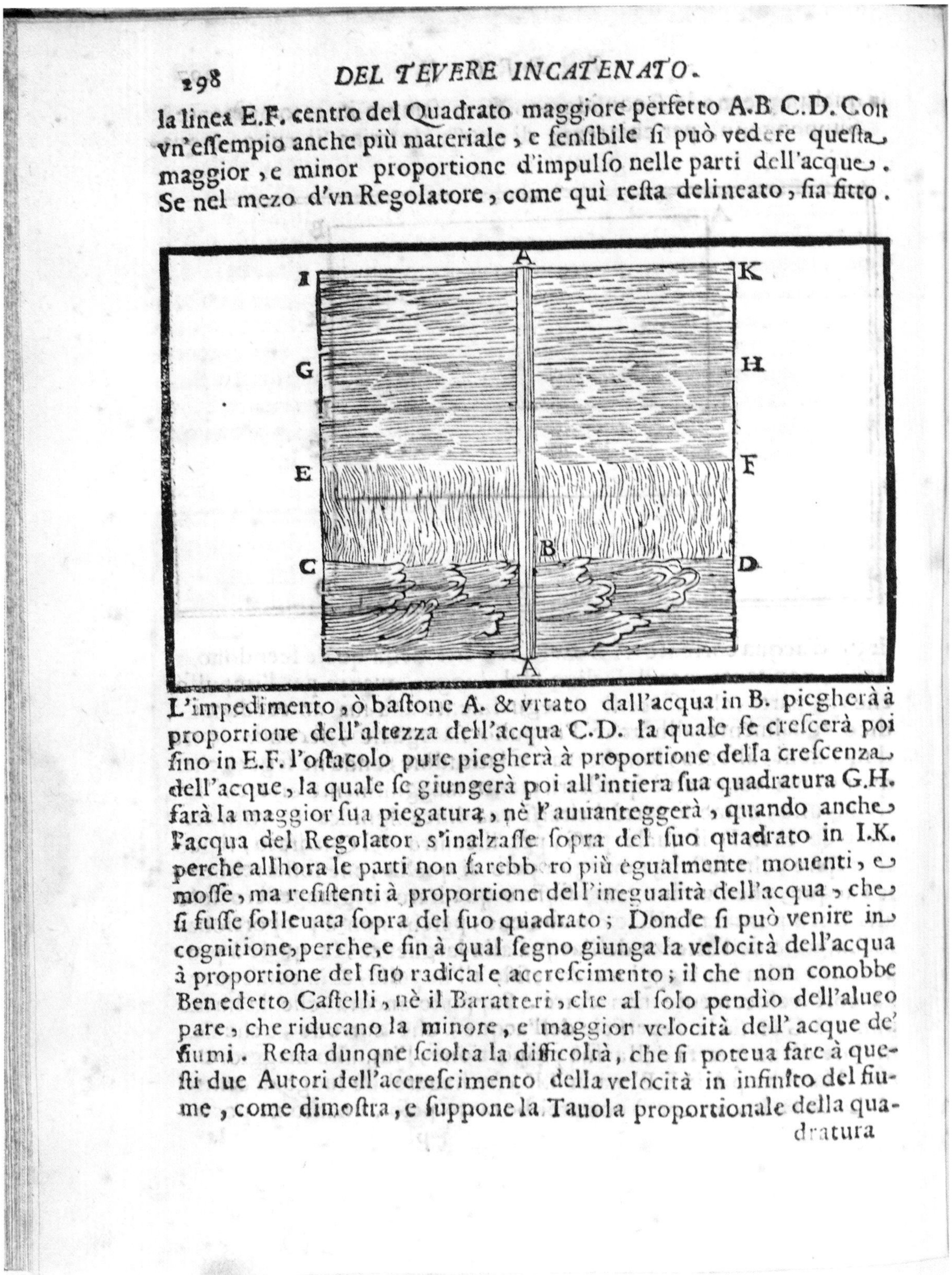

298 DEL TEVERE INCATENATO.

la linea E.F. centro del Quadrato maggiore perfetto A.B C.D. Con vn'eſſempio anche più materiale, e ſenſibile ſi può vedere queſta maggior, e minor proportione d'impulſo nelle parti dell'acque. Se nel mezo d'vn Regolatore, come qui reſta delineato, ſia fitto.

L'impedimento, ò baſtone A. & vrtato dall'acqua in B. piegherà à proportione dell'altezza dell'acqua C.D. la quale ſe creſcerà poi fino in E.F. l'oſtacolo pure piegherà à proportione della creſcenza dell'acque, la quale ſe giungerà poi all'intiera ſua quadratura G.H. farà la maggior ſua piegatura, nè l'auuantaggerà, quando anche l'acqua del Regolator s'inalzaſſe ſopra del ſuo quadrato in I.K. perche allhora le parti non ſarebbero più egualmente mouenti, e moſſe, ma reſiſtenti à proportione dell'inegualità dell'acqua, che ſi fuſſe ſolleuata ſopra del ſuo quadrato; Donde ſi può venire in cognitione, perche, e fin à qual ſegno giunga la velocità dell'acqua à proportione del ſuo radicale accreſcimento; il che non conobbe Benedetto Caſtelli, nè il Baratteri, che al ſolo pendio dell'alueo pare, che riducano la minore, e maggior velocità dell'acque de' fiumi. Reſta dunque ſciolta la difficoltà, che ſi poteua fare à queſti due Autori dell'accreſcimento della velocità in infinito del fiume, come dimoſtra, e ſuppone la Tauola proportionale della quadratura

FIG. 69
Filippo Maria Bonini, Hydraulic studies. From *Il Tevere incatenato*, Roma, 1663, p. 298.

Benedetto Castelli and other Galilean authors, who used more schematic images. These aspects of Bonini's book recall the hydrological sections in Kircher's *Mundus Subterraneus*, a work published in the same period in Amsterdam (1664–66), but conceived earlier in Rome.[283] Leonardo's images could have served as a powerful precursor in this field. It is worth considering the possible influence on both authors of the drawings in *Del moto et Misura dell'Acqua*, the copy after Leonardo's manuscripts made around 1634–43 on behalf of Cassiano del Pozzo for the Roman library of Cardinal Francesco Barberini.

For various reasons, Bonini's work fell well within Antonio Niccolini's scientific and intellectual horizon, and helps us understand, given the comparison with Leonardo, how Niccolini and other open-minded intellectuals in mid-eighteenth-century Florence perceived Leonardo as a scientist. Bonini, like Leonardo, belonged to those hands-on professionals who addressed the science of water and for which Niccolini, in his historical text of Foligno's marshes, had shown great appreciation. Furthermore, Bonini took libertine and anti-Jesuit positions in his theological, political, and historical books, some of which were included in the *Index Librorum Prohibitorum*.[284] As for the Deluge, Florence is in the mid-eighteenth century one of the few places in Europe where the possible scientific explanation of the Deluge is viewed with suspicion by some geologists,[285] including Antonio Maria Matani.[286]

Leonardo and the ashen light of the moon *c.*1765–71

In the years when the Niccolini copies were made, Leonardo's theory of the ashen light of the new moon contained in the Codex Leicester became a matter of note. Between 1765 and 1769, it is discussed by Italian and foreign scientists and erudite writers, serving to draw attention to other aspects of Leonardo's science as well.[287]

This trend is exemplified by the first substantial print publication of Leonardo's scientific ideas, well before Giambattista Venturi's *Essai* of 1779. This publication was triggered by the Codex Leicester, in particular by the Giurini copy. The series of publications mentioning Leonardo's ideas seems to begin with the posthumous publication of Francesco Algarotti's works in 1765. Algarotti was a scientist and

283 Regarding the Deluge, Bonini ambiguously mentions Kircher.

284 Marré Brunenghi 1996.

285 Vai 2003a.

286 Matani 1762, p. [illegible]:

> Therefore setting aside what is only supposed to have happened in the time of the universal Deluge, to which perhaps attribute too much importance, it should be suspect that in the past the marine waters covered the Cutigliano Mountains, where I was told that some shells were found. . . .

Niccolini's interest in geological issues can be confirmed by the fact that he was among the three fellows who sponsored the admission to the Royal Society in 1747 of Giovan Battista Passeri (1694–1780). Passeri studied geology, fossils, and marbles, as mentioned in the presentation of Passeri to the Royal Society. See Passeri 1753–4, 1755–9, 1775. Passeri also practised antiquarianism and was the author of an appendix to Thomas Dempster's Etruscology treatise, printed on behalf of Thomas Coke. See Passeri 1767.

287 Algarotti 1765; Frisi 1768, p. 98 and 104, 1775, p. 31, 1778a, pp. 33–4, 1778b, p. 6, 1958 [1783], p. 354; Lalande 1769, pp. 310–12, 1771, p. [illegible]; *Serie degli uomini più illustri* . . . 1769–76, vol. 3; Durazzini 1770, p. 130; Giovanni Targioni Tozzetti, *Selva di notizie spettanti all'origine dei progressi e miglioramenti delle scienze fisiche in Toscana* National Library, Florence, MS Targioni 189, vol. [illegible]

polymath who spent the last years of his life in Tuscany, where he died in Pisa in 1764 after having taken care of the edition of his work. The latter includes a description of Leonardo's theory, which is worth mentioning in full since it even escaped Ettore Verga's detailed *Bibliografia Vinciana*:

> Some very trustworthy men claim that there is in Tuscany a manuscript by Lionardo da Vinci, in which he explains the true cause of that secondary light which is visible in the new moon and thanks to which its whole body is visible despite its being a crescent moon. He says that this secondary light corresponds to the lunar hemisphere that, in its night time, is illuminated by the earth. And, he adds, do you want to convince yourself of this? Look how that secondary light is more alive when the moon is old and rises slightly before the sun than when it is new and almost sets in the company of it. The moon receives little light from the parts of the earth which are to the west of us and are covered by the water, and more by the Eastern parts, which extend into a vast continent; things that perfectly agree with the truth and with what was shown by Galileo thanks to the telescope.[288]

Algarotti's text, which was to be mentioned by the Florentine naturalist Giovanni Targioni Tozzetti in his *Selva*, the monumental manuscript history of the sciences in Tuscany,[289] is a paraphrase of Leonardo's theory which highlights the 'geographical' tenor of many passages in the Codex Leicester (ff. 1r, 2r, 7r) and in particular f. 1r:

> The face of the moon as it first enters has so much light as it derives from our seas, which reflect the sun there, and this light is so much greater as the ocean receives greater light from the sun, and this is when the new moon that enters rises a little after the sunset of the sun and this is less [apparent] when the moon is older.

Algarotti thus had indirect knowledge of Leonardo's text through 'trustworthy men', who had read a 'manuscript of Lionardo da Vinci' present in Tuscany, and obviously corresponding to one of the copies of the Codex produced in Florence. The date of publication of Algarotti's text (1765) eliminates the Niccolini copy for this role, since it was made two years later, in 1767. The source can therefore be identified as one of the other two copies that we know were in Florence in these years: the copy now in Weimar or the lost version that belonged to Bernardino Ciurini. The latter is explicitly mentioned by one of the other authors who cite Leonardo's theory at the time, namely Anton Francesco Durazzini. In his *Raccolta d'elogi d'uomini illustri . . .*, he offers praise of Leonardo:

> [Leonardo] was the first of all mortals to discover, with admirable skill, the true origin of that opaque light, thanks to which the dark side of the lunar disc becomes visible in the new moon and he attributed it to the reflexion of the sun light, which, when the moon is 'silent', fully illuminates the surface of the earth.[290]

288 Algarotti 1764–5, vol. 7 (1765), pp. 186–7:

> Asseriscono uomini degnissimi di fede, che vi abbia in Toscana un manuscritto di Lionardo da Vinci, nel quale egli spiega la vera causa di quel lume secondario, che si scorge nella luna quando è nuova, e per cui si vede tutto il corpo di essa benchè falcata. Quel lume secondario mostra, dic'egli, l'emisfero della Luna illuminata nel suo notte-tempo dalla Terra: E volete convincervi, egli aggiuge, che sia così? Vedete come quel lume secondario è assai più vivo quando la Luna è vecchia e si leva poco tempo innanzi al Sole, che quando ella è nuova, e quasi tramonta in compagnia di lui: Dalle parti, rispetto a noi, Occidentali della Terra, le quali sono coperte d'acqua, poco lume riceve la Luna, e maggiore ne riceve dalle parti Orientali, che si estendono in un vasto continenente; cose che concordano a un puntino col vero, e con quanto insegnò il Galilei armato del Cannocchiale.

289 Targioni Tozzetti, *Selva. . . .*, Florence, National Library, MS Targioni 189, vol. 5, ff. 2343–5.

290 Durazzini 1770.

Durazzini adds a note that 'this discovery is recorded in the Manuscript belonging to Ciurini heirs'.[291]

Algarotti shared the same scientific vision as other figures, including Ciurini and Niccolini. These figures were in fact very close to the English and Newtonian scientific world. Algarotti had for a long time lived in England where he came into contact with the 'post-Newtonian' British circles, which, as we have seen, form the background of the first English edition of Leonardo's *Treatise on Painting*. He enjoyed a relationship with Desaguliers and travelled with his son.[292] Like Desaguliers, Algarotti was engaged in the popularization and dissemination of Newton. Indeed, he is the author of *Il newtonianismo per le dame ovvero dialoghi sopra la luce e i colori* (Naples, 1737), a work that enjoyed great fortune at that time, with many editions and translations in English (*Sir Isaac Newton's Philosophy Explain'd for the Use of the Ladies in Six Dialogues on Light and Colour*). It was also translated into French.[293]

Algarotti wrote on art, and in his *Saggio sopra la Pittura* (*Essay on Painting*) compared the Newtonian theory of white light as composed of various colours to Leonardo's theory, a comparison that became another topos for various authors interested in this cultural conjunction.[294] Algarotti's radical Newtonianism was to be condemned by religious authorities,[295] recalling the reactions to Niccolini's 'progressive' position. The two men were in written contact with each other, as shown by a letter of 1763 concerning a new edition of the Crusca Italian dictionary.[296]

Niccolini's correspondence allows us to reconstruct a direct relationship between some of the protagonists of this early episode of the circulation of Leonardo's scientific work. Paolo Frisi was another scientist who, in that period, mentions in his books the theory of ashen light and other aspects of Leonardo's scientific research. Frisi, in a letter to Niccolini written in Pisa and dated 1763, mentions as common friends Algarotti and Giuseppe Pecis. The latter is known to Leonardo scholars, because in

291 Durazzini 1770, pp. 127–37, esp. 132:

> [I]l primo di tutti i mortali con ammirabile sagacità ritrovata la vera origine di quella luce abbagliata, per la quale la parte oscura del disco lunare si rende visibile nel Novilunio, attribuendola alla riflessione del lume Solare, dal quale nel tempo che la Luna tace, viene appieno illuminata la superfice terrestre . . .
>
> [In a footnote:] Questa scoperta è registrata nel MS. degli Eredi Ciurini.

The Ciurini copy is also mentioned at p. 130 n. XXX. See Appendix to this chapter. In the *Serie degli uomini più illustri* . . . , vol. 3: *L'Elogio di Leonardo*, pp. 29–41, the Codex Leicester is the first of Leonardo's manuscripts to be mentioned, with the title in the title page:

> Non sappiamo precisamente che contenessero questi libri [di Leonardo]: è noto però che egli uno ne scrisse della natura, peso e moto delle acque; alla qual'opera diede impulso l'impresa del Naviglio di Martesana. (p. 39 n. 1).

The same book (p. 3[illegible]) also mentions a Leonardo painting in the Niccolini palace.

292 Carpenter 2011, pp. 69–70, 241.

293 Mazzotti 2004; Porter 2001, pp. 130ff.

294 *Saggio sopra la Pittura*, in *Saggio del Conte Algarotti sull'architettura e sulla pittura*, Milano, undated, but probably 1756 (reprinted in Livorno, 1762), p. 74: 'Il qual bianco non è colore per se, come disse espressamente quasi precursore del Neutono il dottissimo Leonardo da Vinci, ma è ricetto di qualunche colore'. The source mentioned is ch. 104 of the *Treatise on Painting*. The same comparison can be found in Durazzini 1770, p. 132; Frisi 1778, 1783; Lalande 1769.

295 Mazzotti 2004.

296 Niccolini Archive, Florence, Segnatura 265, inserto 13, Letter of Francesco Algarotti to Antonio Niccolini, Pisa, 10 January 1763.

a letter of 1771 he mentioned a Leonardo manuscript about the Martesana canal in the Imperiali library in Rome.[297]

Paolo Frisi, like Niccolini and Algarotti, had frequent contacts with the English scientific world and was very much interested in the role played in the history of science by hands-on professionals such as artists and architects. The first work in which Frisi deals with Leonardo was the second edition of his hydrological treatise *Del modo di regolare i fiumi . . .* (*On how to regulate the rivers. . . .*, Lucca, 1768). In it, he denies that Leonardo was the architect of the Martesana canal, but recognizes that:

> The invention of the locks and supports, made in the fifteenth century in the Paduan region, has prepared Leonardo da Vinci's later union in Milan of the two canals of the Adda and of the Tesino. The latter served as a model and, as a rule, to many other navigable channels and most in Languedoc.[298]

It is worth noting that all these references to Leonardo's work as a hydraulic engineer, which are notable for their sense of history, were absent from the first edition of Frisi's treatise in 1762. Frisi lived in Tuscany from 1755 to 1764, and therefore might have had access to one of the copies of the Codex Leicester that circulated in Tuscany. It is possible that he had access to the original Codex in London and/or its copy there. Indeed, in 1766 Frisi was travelling between Paris and London. One of the most important representative of this episode in Leonardo's legacy as a scientist, the French astronomer Lalande, mentions Leonardo's theory of the ashen light, adding that he was informed about it by Paolo Frisi and that the theory was included in a Leonardo Codex in London.[299] As we will see, Lalande's interest in this theory contained in the Codex Leicester will have marked consequences for the later fortune of Leonardo as a scientist.

Frisi included other aspects of Leonardo's work in his later books, printed between 1775 and 1783. In his *Elogio di Galileo*,[300] Frisi mentions the ashen light and various other subjects, including Leonardo's

297 Motta 1908–9.

298 P. Frisi, *Del modo di regolare i fiumi . . .*, Lucca, 1768, Introduction, n.p.:

> L'invenzione delle chiuse, e sostegni, che nel secolo decimo quinto s'è fatta sul Padovano, ha preparato l'unione, che Leonardo da Vinci ha poi fatto in Milano dei due Navigli dell'Adda, e del Tesino: e quest'unione ha servito di modello e di norma a tant'altri canali navigabili e a quello massimamente di Linguadoca.

See p. 68 on the Martesana canal.

299 Lalande 1771, p. 192:

> [I]l y a des Italiens qui attribuent cette reparque [i.e. la spiegazione della vera causa della luce cinerea della luna] à Leonardo del Vinci célèbre peintre Toscan mort en 1518, & le P. Frisi m'a assuré qu'elle se trouve dans un des ses manuscrits sur les Rivieres [*sic*], que l'on conserve à Londres.

Lalande 1790, vol. 1, p. 294:

> On conserve à Londres un manucrit sur les rivières, par Leonardo da Vinci, où le P. Frisi m'a assuré qu'on trouve la première explication de la lumière cendrée de la lune, quand elle est nouvelle, quoiqu'on en ait fait honneur à Maestlinus.

300 Frisi 1775, pp. 31–2:

> Leonardo da Vinci, e il Mestlino avean congetturato anche prima che quella fosse un testo della luce mandata dal Sole alla Terra, e dalla Terra ripercossa poi nella Luna, e della Luna nuovamente rimandata alla Terra.

Frisi further notes:

> In un manoscritto, che adesso ritrovasi in Inghilterra, incominciò a trattare dell'acque correnti, e congetturò, che la luce secondaria nasca da una doppia riflessione della terra, e della Luna.

optical theory of white light; these subjects reprised Frisi's works of 1778 and 1783.[301] In the *Elogio di Maria Teresa Imperatrice* (1783) he associates, as the greatest representatives of the new age of reason, Locke, Newton, Leonardo da Vinci, and Michelangelo.[302] We are here opposed to the position expressed by Buonaventuri, who, in the early eighteenth century, excluded Leonardo from his *Collection* of hydraulic authors.

In conclusion, the focus on the scientific contribution of hands-on professionals is a common characteristic of people and scientists interested in Leonardo and the Codex Leicester. It is symptomatic of the divergence in the scientists who came into contact with Leonardo's legacy, between Anglophile and Baconian thinkers, and those who were Italiaphile and Galilean. Finally, the theme of 'precursors', another aspect common to these authors, was a key feature of the historical vision current in the Enlightenment—even if it is not a prime concern in modern history of science.

A Copy of the Codex Leicester in Naples: Interest in Leonardo and the Uplift of Land as a Key Theory in New Geology

Between the end of the eighteenth and the early nineteenth century, a new phase began in the history of the legacy of the Codex Leicester and Leonardo as a scientist. Leonardo's geology attracted attention, in particular his theory of an uplift of continents, exploiting a mechanism that involved a pre-Newtonian notion of gravity. It was a 'theory' formulated at a time when geology, at the end of the eighteenth and in the first half of nineteenth centuries, was dominated by empirical research, including field studies and stratigraphy. However, many geologists at that time became increasingly aware that it was not possible to proceed without some recourse to the detested 'systems' or 'theories'.

On the one hand Charles Lyell (1797–1875) stated that, in order to understand the past, it was necessary to limit oneself to the 'empirical study of current geological phenomena', assuming substantial 'uniformity' between geological processes in the present and the past (the so-called *uniformitarianism*). On the other hand in Paris George Cuvier (1769–1832), in addition to promoting field research, felt it necessary to 'hypothesize' that in the earth's distant past, catastrophic changes occurred, events that were on a much vaster scale than those empirically visible in the present. A key idea, important also from the point of view of what happened later in the direction of the theory of plate tectonics, was active uplift, involving more or less sudden and catastrophic vertical movements. These notions were elaborated according to the theories of volcanism and in relation to cosmological considerations, including the notion of gravity and the problem of the overall shape of the earth.

Paris contended with London as the key place for the new geology and, not by chance, Leonardo's legacy as a scientist figured mainly in the French context, even if the phase of its circulation in Naples provides the starting point for our reconstruction of this later period in the history of the Codex Leicester.

Starting with the disastrous eruption of 1631, Vesuvius resumed an intense phase of activity that would last for three centuries and made Naples during the eighteenth and nineteenth centuries one of the major laboratories for the new earth sciences. In particular, the years between the end of the eight-

301 Frisi 1778a, pp. 3–4, 1778b, p. 8, 1783.

302 Frisi 1958 [1783] p. 354.

eenth century and the beginning of the next represented a notably flourishing period. In 1776–9, William Hamilton, English ambassador to the Bourbon capital, important antiquarian, and enthusiastic volcanologist, published his *Campi Phlegraei*, probably the best-illustrated geological treatise of all time.

In 1779, the German geologist Leopold von Buch visited Vesuvius, coming back to Naples in 1805 with the great Alexander von Humboldt. Von Buch was later to elaborate the theory known as the 'craters of elevation' which, by giving importance to vertical movements in the earth's crust, contributed to break the dominant *neptunist* theory of a passive uncovering of the mountains through the lowering of the sea level.

Several manuscript copies of Leonardo's scientific works circulated in Naples at the time. A copy of the Codex Leicester was in the library of Duke Luigi Serra di Cassano (1747–1825) while two abridgements based on other manuscripts, were owned by Vincenzo Corazza, a Bolognese scholar who, from 1772, lived in Naples (Figures 70–2).[303]

The copies attracted the attention of scientists and scholars. Two are key figures: Giuseppe Bossi, a Milanese artist who spent some time in Naples in 1810, and Vincenzo Corazza. An interest in Leonardo's geology arose from their personal feedback and in the context of their contacts with scientists who were representative of geology studied in Naples at the time. This circle included William Hamilton, Alberto Fortis, Teodoro Monticelli, and Scipione di Breislak.

Giuseppe Bossi and the Serra di Cassano copy of the Codex Leicester

In a letter written in Naples and addressed to the sculptor Antonio Canova, Giuseppe Bossi notes:

> I discovered various things about this man [i.e. Leonardo] that set him at the head of the fabricators of geological systems: his system is similar to those elaborated by the best physicists who were born just over a century ago.[304]

The term 'systems' designated at the time theories of the earth, that is, the attempts to go beyond empirical and experimental data and propose hypotheses about the origins of the earth and its transformations. The priority given either to explanatory systems or to empirical collection of data had been the subject of harsh controversies, but some geologists felt the need to use 'systems' in order to draw conclusions from the growing amount of empirical data collected in the field. Informing Canova about the Codex Leicester, Bossi mentions only its geological content, and in particular Leonardo's 'system' or theory of the earth, considering it as the most important and innovative aspect of the Codex. It is a clear sign of a new kind of interest in Leonardo during the golden age of geology. In previous years, it was hydrology that attracted most attention.

Bossi's letter is dated 23 July 1810. A few days before on July 6 Bossi took possession of the copy of the Codex Leicester from the library of the Duke Luigi Serra di Cassano, as Bossi informs us in a note added at the beginning of the manuscript: 'Naples, July 6, 1810. Kindly donated to me by the Duke

303 On Corazza copies, see Buccaro 2011, pp. 111ff.

304 Ciardi 1982, vol. 2, p. 667: 'Intorno a quest'uomo [i.e. Leonardo] ho scoperto varie cose che lo mettono alla testa de' fabricatori di sistemi geologici: il suo è simile ai riconosciuti dai migliori fisici, e ch'ebbero nascimento poco più d'un secolo fa.'

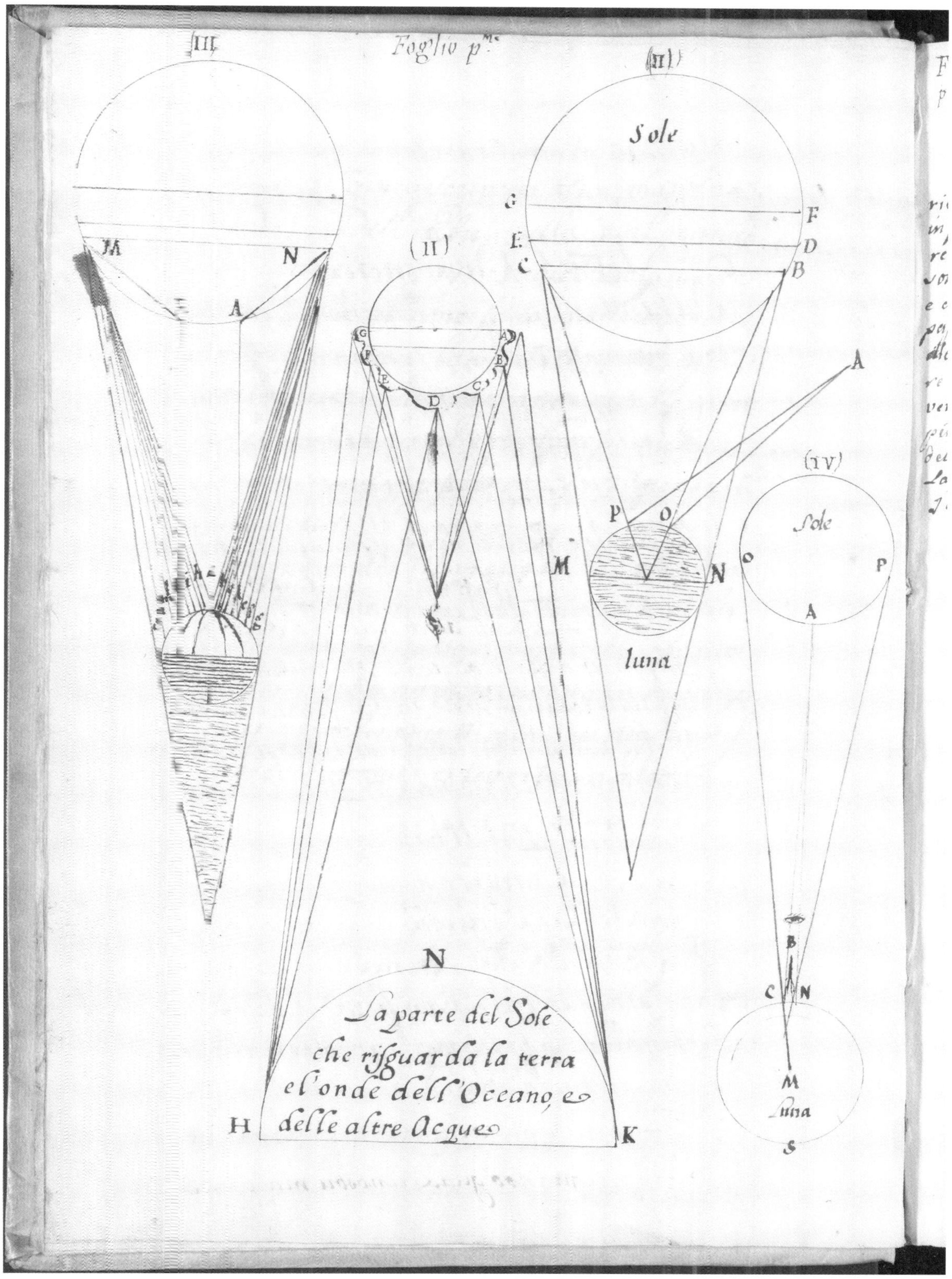

FIG. 70 Astronomical diagrams. Attributed to Francesco Ducci, Weimar copy of the Codex Leicester. Herzogin Anna Amalia Bibliothek, Weimar, f. 26, unnumbered 'Folio p.mo'.

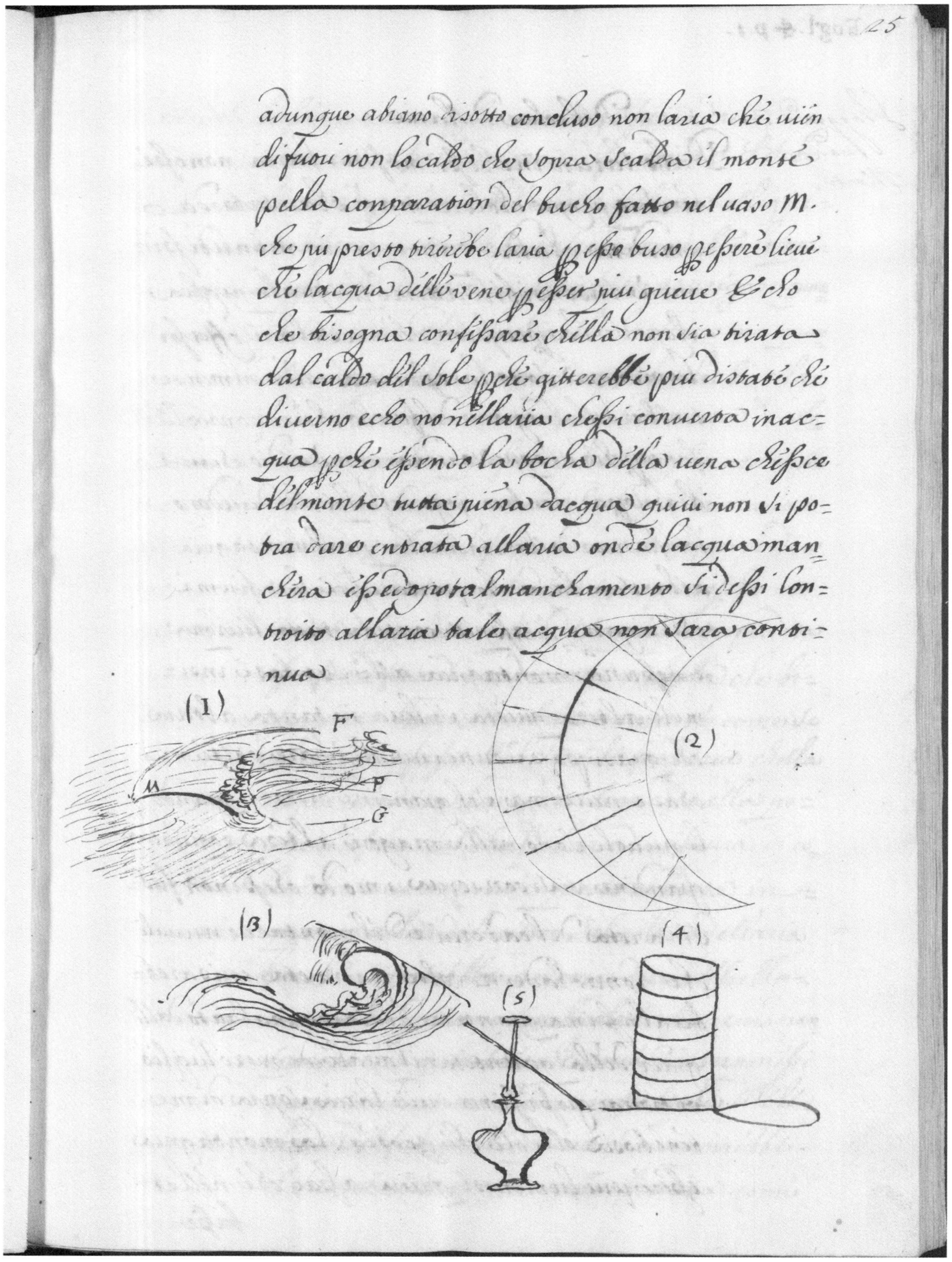
25

adunque abiano disotto concluso non laria che uien
di fuori non lo caldo che sopra scalda il monte
pella conparation del bucho fatto nel uaso M.
che piu presto tirerebbe laria pesse buso pesere lieue
che la acqua dello ueno pesser piu grieue & che
che bisogna consessare chella non sia tirata
dal caldo del sole pche gitterebbe piu distate che
diuerno e che no nellaria chessi conuersa in ac=
qua però essendo la bocha della uena che esce
del monte tutta piena dacqua quiui non si po=
tria dare entrata allaria onde lacqua man
chera esse dopo tal manchamento si dessi con=
tinuo allaria tale acqua non sara conti=
nua

FIG. 71
Attributed to Francesco Ducci. Weimar copy of Codex Leicester, f. 4v. Herzogin Anna Amalia Bibliothek, Weimar, f. 326, f. 25.

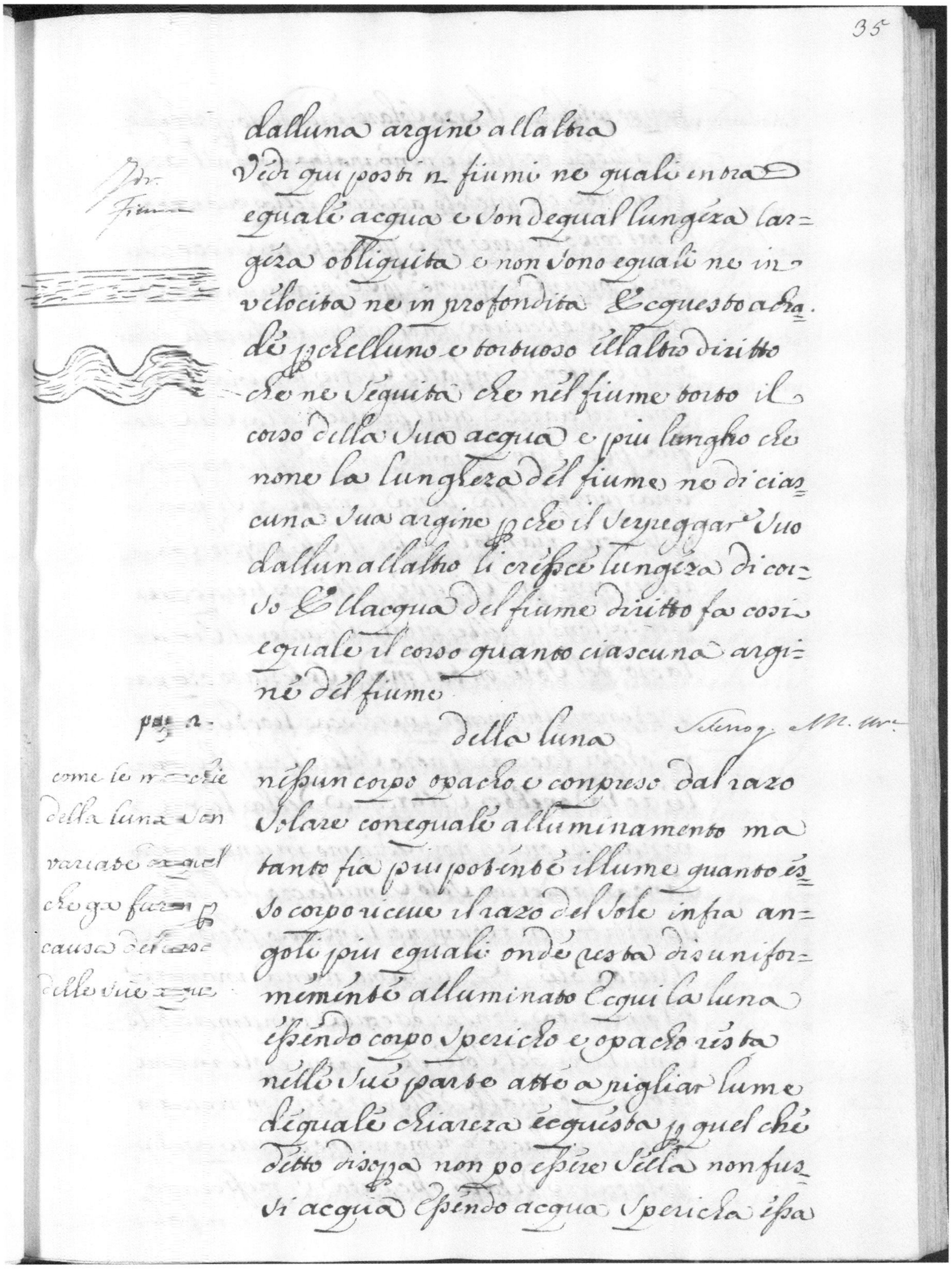
35

dalluna argine allaltra
Vedi qui posti 2 fiumi ne quali in tra
equali acqua e son dequal lungeza lar=
geza obliquita e non sono equali ne in
velocita ne in profondita E questo a ra
de perelluno e tortuoso illaltro diritto
che ne seguita che nel fiume torto il
corso della sua acqua e piu lungho che
none la lungheza del fiume no di cias=
cuna sua argine per che il serpeggiar suo
dallun allaltro li crescie lungeza di cor=
so E llacqua del fiume diritto fa cosi
equale il corso quanto ciascuna argi=
ne del fiume
della luna
nessun corpo opacho e conpreso dal razo
solare con equale alluminamento ma
tanto fia piu potente il lume quanto es=
so corpo riceve il razo del sole infra an=
goli piu equali onde resta disunifor=
memente alluminato E qui la luna
essendo corpo spericho e opacho resta
nelle sue parte atto a pigliar lume
d'equale chiareza e questo per quel che
detto disopra non po essere sella non fus=
si acqua essendo acqua sperica essa

FIG. 72 Attributed to Francesco Ducci, Various studies with Giuseppe Bossi's marginal annotations. Weimar copy of Codex Leicester, f. 5r. Herzogin Anna Amalia Bibliothek, Weimar, f. 326, f. 35r.

of Cassano, Great "Hunter" of S. M. G. Bossi Painter.'[305] Bossi arrived in Naples on June 30 and Serra di Cassano is amongst the first people to whom he paid a visit. On July 2, he visited Palazzo Serra di Cassano and noted in his diary:

> 3 of the said month [July]. Yesterday morning I met Tassoni . . . then the Duke of Cassano, where I viewed the stupendous collection of 15th century authors and other rare books for three hours. *I also saw Leonardo's treatise. It is a copy made perhaps a century ago; but is nevertheless important and I do not despair of getting it. It deals with the ebb and flow of the sea caused by the moon's attraction. etc.*[306]

This is Bossi's first mention of the copy of the Codex Leicester. The focus of attention is no longer, as in previous years, on the studies of water and rivers but on its more cosmological and geological content. This emphasis is reflected in a passage of poetry by Bronzino inserted by Bossi at the beginning of the manuscript containing the copy of the Codex:

> The depths of the sea were once above the hills
> and one sees its mark; and they will come back there,
> unless the fire will drive it out.
> Agnolo Bronzino in the second Chapter, 'on his Excuses'.[307]

Bossi was not a geologist, but he was a painter and a cultivated man. He is known as one of the first scholars of Leonardo and is the author of a famous book on the Last Supper,[308] which was to influence Goethe. The quote from a poem by the sixteenth century Florentine painter Agnolo Bronzino,[309] is significant for various reasons. First of all, it places emphasis on the geological content of the Codex. It is a comparison between Leonardo and a geological passage from a poem by another author, similar to that with Bonini's book on the River Tiber in the 1767 Niccolini copy. But now, the focus of the comparison is geology, not hydrology. Bossi began to study the received copy, identifying three thematic areas and inserting the relevant names in the margin of the manuscript near the corresponding passages: astronomy (*selenologia e miscellanea uranologica*), hydrology (*idrologia*), and geology (*geologia*) (Figure 74).[310]

Bossi seems to have been planning a version that organized texts around the same themes, somewhat like the abridged copy commissioned by Niccolini in Florence in 1767. He would have later transcribed Leonardo's passages in an anthology thematically arranged. This anthology, if ever compiled, is now

305 Weimar copy, Weimar, Herzogin Anna Amalia Bibliothek, manuscript f. 326, 2nd flyleaf verso: 'Napoli 6 luglio 1810. Donatomi cortesemente dal Duca di Cassano, Gran Cacciatore di S. M. G. Bossi Pittore'. See Appendix to this chapter. Both Duke Luigi and his son Giuseppe were granted the honorary title of 'Great Hunter' by His Majesty the King Ferdinand IV. See Russo 2007, p. 47 and Augurio and Musella 1999, p. 426.

306 Ciardi 1982, vol. 2, p. 733 (emphasis added):
3 detto [luglio]. Jeri mattina fui da Tassoni . . . indi dal Duca di Cassano, dove vidi la stupenda raccolta de' quattrocentisti, ed altri rari libri per tre ore. Ho veduto anche il Trattato di Leon. È una copia fatta forse un secolo fa; ma non lascia di essere importante, e non dispero ottenerla. Vi si parla del flusso del mare cagionato dalla attrazione della luna etc.'

307 Weimar, 1st flyleaf verso. 'Già fu il profondo del mare in su poggi / E se ne vede il segno; e torneravvi /se prima il fuoco non fa ch'ei diloggi. Agnolo Bronzino nel Cap.° 2.° delle Scuse.' See Appendix to this chapter.

308 Giuseppe Bossi, *Del cenacolo di Leonardo da Vinci*, Milan, 1810.

309 Manuscript in the Marciana Library, Venice. See Agnolo Bronzino, *Li capitoli faceti editi ed inediti di Mess. Agnolo Bronzino . . . ora per la prima volta raccolti e pubblicati. . . .*, 1822, ch. 2 ('Delle scuse', a burlesque poetic composition included in the manuscript). See p. 89 for the 'geological' verses. Cf. Petrucci Nardelli 1988.

310 The notes in the Weimar copy stop after f. 38.

missing.[311] From Bossi's marginal notes, we know that the hydrological section included a chapter on the waves and another on the rivers, while the geological section's '2nd article' dealt with the Deluge.[312] These are new terms and concepts. The name geology, in the modern sense in which Bossi used it, had only been introduced since the end of the eighteenth century.[313]

During his Neapolitan stay, Bossi also made an excursion to Vesuvius, which was still active. Above all, he found himself in one of the liveliest centres of geological studies of the time, a context that certainly directed his attention towards Leonardo's geological ideas. The Jacobin historian Vincenzo Cuoco, among Bossi's Neapolitan friends,[314] opens his history of Italy with a geological reconstruction inspired by Boulanger's and d'Holbach's theories of geological catastrophes.[315] This exemplifies the pervasiveness of geology at the time. On 4 July, two days after the examination of the Serra di Cassano copy of the Codex Leicester, Bossi met Teodoro Monticelli, a leading geologist and one of the major experts on Vesuvius.[316] He noted in his diary: 'I met Cavalier Monticchi, Secretary of the Academy of Science',[317] a clear *lapsus calami* since the 'cavalier Monticchi' can only be Monticelli who, since 1808, was the secretary of the Academy of Sciences in Naples, an institution that included the best Neapolitan scientists of the time and which placed geology among its central goals. Monticelli, well known for his studies of volcanic sediments, was, among the Neapolitan volcanologists, one of the most open to new ideas that explored the phenomenon of Vesuvius as a key to broader issues.[318] For example, he exchanged letters with Alexander von Humboldt about the theory of the 'craters of elevation' and Christian Leopold von Buch's observations on Vesuvius.[319] The general goal at the centre of such reflections was to understand whether it was possible to derive a general 'system' or theory applicable to the genesis of mountains from the continuous variations in the Vesuvius volcano (in its growth and subsidence). As we have seen, uplift was Leonardo's key idea contained in the Codex Leicester. Leonardo viewed uplift as 'gravitational' whereas many geologist of his era saw it as volcanic.

It is possible that Serra di Cassano introduced Bossi to Monticelli, since Monticelli was imprisoned during the repressive Bourbon campaign of 1793–6, together with Giuseppe Serra, Duke Luigi's son, because of their shared Jacobin convictions.[320] As we shall see, on at least one occasion, Bossi discussed Leonardo's scientific texts with Giuseppe Serra. The Serra di Cassanos were one of the noble Neapolitan families who supported the famous anti-Bourbon, Neapolitan, Jacobin revolution of 1799. After the failure of the revolution and the restoration of the Bourbon regime, Gennaro Serra, Giuseppe's brother, paid dearly for his participation in the revolution and was executed in August 1799, while the family

311 See Appendix to this chapter.

312 Weimar copy, Herzogin Anna Amalia Bibliothek, Weimar, manuscript f. 326: 'Geologia. Questo può aprire il 2° articolo' ('Geology. This can open the second article'); f. 21 referred to a text on the Deluge.

313 Rudwick 2005, pp. 134–5.

314 On Cuoco and Bossi, see Nenci 2004, p. 97 and Ciardi 1982, vol. 2, p. 733.

315 Ciancio 2009, p. [illegible]4; Andreoni 2006, pp. 267ff, 1998.

316 De Ceglia 201[illegible]; Nazzaro 2001, pp. 88–92.

317 Ciardi 1982, vol. 2, p. 734: 'Conobbi il Cavalier Monticchi, Seg.° dell'Accade.a delle Scienze'.

318 Nazzaro 2001, p. [illegible] n. 60.

319 *Opere dell'abate Teodoro Monticelli*, vol. 2, Naples, 1841, p. 93 n. 1, quoted in Nazzaro 2001, p. 222.

320 According to D'Ayala 1873, pp. 99ff., Monticelli and Serra were both imprisoned for political reasons in 1795. See Augurio and Musella 1999.

was forced into exile for several years. When Bossi visited Naples, the situation had changed again, since in 1806 the French had re-conquered Naples, initiating the so-called 'revolutionary decade' with a government led by Joseph-Napoléon Bonaparte and then by Joachim Murat. In the same year, the Serra came back to Naples and shortly afterwards Duke Luigi, father of Giuseppe and Gennaro Serra, became a minister in the new government. It is within the new progressive climate of the French 'decade' (1806–15) that Bossi's Neapolitan journey occurred. Again, as in Florence in the mid-eighteenth century, interest in the Codex Leicester comes from very progressive minds, now inspired not by England but by revolutionary France. Duke Luigi sent his sons Gennaro and Giuseppe to study in France, and the Serra family house was attended by the most progressive thinkers in Naples. Luigi was a bibliophile and avid collector, especially of fifteenth-century printed books, for which his library was famous.[321] His correspondence with leading figures in the book market demonstrates the quality and intensity of his acquisitions,[322] which included the copy of the Codex Leicester. But when was the copy of the Leonardo's Codex purchased? The long period covered by Luigi's acquisitions opens up many possibilities. However, two dates seem most likely. The copy was made in Florence at the beginning of the eighteenth century by Francesco Ducci, librarian of the Laurenziana.[323] In 1787, Duke Luigi Serra was in Florence to buy books, as we know from a letter dated 2 July 1787 sent by the erudite Florentine Angelo Maria Bandini to the philologist Cesare Lucchesini.[324] Bandini was also a librarian at the Laurenziana in Florence, so he could have been aware of the existence of a copy made at the beginning of the century by his predecessor, Ducci. Therefore, 1787 is a possible date for Serra's acquisition of the copy. Alternatively, the manuscript may have been purchased later, during in the family's exile from Naples post-1799, when they were residing in Tuscany.[325]

The library was sold in 1818–19, but in the archive of the Serra di Cassano there is an 1837 inventory of the family property, which includes books. Among the scientific books, many are devoted to geology and cosmology, and these, apart from Scipione di Breislak's *Introduzione alla geologia* (*Introduction to Geology*, Milan, 1811), are almost all by French authors or by geologists linked to the French world of science, such as Pierre Simond Laplace's *Système du monde* (Paris, 1808), Jean-Claude de La Métherie's *Théorie de la terre* (Paris, 1795), Jean-André De Luc's *Traité élémentaire de Géologie* (Paris, 1809), and Phillipe Bertrand's *Nouveaux Principes de Géologie* (Paris, 1797) and *Renouvellements périodiques des continents terrestres* (Geneva, 1803). We will look later at some of the traits shared by these authors: their theoretical inclinations, the importance of gravitational measurements in the study of the earth, and catastrophic hypotheses. In his *Introduction to Geology*, Breislak, a geologist long active in Naples, also supports the topical character and usefulness of some of the older geological systems.[326] In the Neapolitan context, dominated by the empirical description of Vesuvius's eruptions and of their aftermath, these works brought a breath of fresh air, as did the political ideas coming from Paris.

321 Many books were sold in 1819–20 to George Spencer. See Trombetta 2002, pp. 263–7.

322 Serra di Cassano Archive, Naples, Palazzo Serra di Cassano. See Carnevale 2002.

323 See Appendix to this chapter.

324 Lucca,Biblioteca Statale, Ms. 1360, fols. 132r–133v. See Paoli 1987.

325 Augurio and Musella 1999.

326 Breislak 1811, p. 243: 'Perhaps we will come back to the systems of Buffon, Leibniz, Moro and many ancient schools, but we will come back to them through a way worthy of the present century, namely through observations'. First published in his *Voyages physiques et lythologiques dans la Campanie*, Paris, 1801, p. xiv.

It is difficult to say whether Serra di Cassano's choices of books were motivated by clear awareness of the orientation of the geological theories they contained. Certainly, the young Giuseppe Serra had studied in France, sharing his French political ideas with the geologist Teodoro Monticelli, and was therefore probably well informed about the main issues in the fashionable science of geology. Certainly, the small collection of geological treatises put together by the Serras consisted of texts in which geology was dealt with according to concepts shared by Vincenzo Corazza in Naples and Giambattista Venturi in Paris—two men who were interested in Leonardo's theory of the earth.

Corazza's copies of Leonardo's manuscripts and two geologists interested in Leonardo: William Hamilton and Alberto Fortis

It was Giuseppe Serra di Cassano who, in a conversation about the copy of Codex Leicester, informed Bossi of the existence in Naples of other manuscript copies after Leonardo:

> The young marquis Serra di Cassano informed me about Leonardo's things that were in Corazza's hand. He gave these writings and drawings to the Royal Prince, son of Ferdinand IV. It is not known what happened to them.[327]

In reality, the manuscripts were in the Royal Library of Naples, and Bossi was copying one of them,[328] without knowing its origin. The learned *abate* Vincenzo Corazza (d.1799) owned two copies of Leonardo's writings based on manuscripts other than the Codex Leicester: a seventeenth-century anthology, probably one of those made in Rome on behalf of Cassiano del Pozzo[329] and a copy by Corazza himself after the seventeenth-century anthology *Del moto et Misura delle Acque*.[330] Both copies contained a few cosmological and geological passages, and Corazza, in his collection of *Termini di arte negli scritti di Lionardo da Vinci ed altri* (*Terms of Art in the Writings of Lionardo da Vinci and others*) drew upon them.[331] He commented, for example, on Leonardo's term 'world' ('mondo'):

> World. S° With this term Lionardo always means the terraqueous globe we live in. The following example suffices: 'high things do not remain at their height, but over time they will all descend, and so with time the world will remain spherical and consequently it will be all covered with water, and underground veins will remain immobile.[332]

> World. 'The centre of the water sphere is the true centre of the roundness of our World, which is made of waters and earth in round shape' (We have alleged other examples of World as Earth; this suffices

327 Ciardi 1982, vol. 2, p. 739. 'Il M.sino Serra di Cassano mi ha istruito intorno alle cose di Leonardo ch'erano in mano di Corazza. Egli regalò questi scritti e disegni al Principe Reale f.° di Ferdinando I. Non si sa che ne avvenisse.'

328 See Bossi's diary in Ciardi 1982, vol. 2, and Buccaro 2011.

329 National Library, Naples, Ms. XII.D.79.

330 National Library, Naples, Ms. XII.D.80. See Buccaro 2011 and Rascaglia 2000.

331 National Library, Naples, Ms. XII.D.81.

332 National Library, Naples, Ms. XII.D.81, f. 10:
Mondo. s° Con questa V [oce:] Lionardo intende continuamente il globo terracqueo che noi abitiamo. Basti il seguente es.°: 'le cose alte non restano in loro altezza , ma col tempo tutte discenderanno , e così col tempo il Mondo resterà sferico e per conseguenza fia tutto coperto dell'Acqua, e le vene sotterranee resteranno immobili'.
Leonardo's text is from MS F, f. 84r. For a full transcription, see Buccaro 2011.

for all the others and must be considered in order to understand a very important concept of our [author] that contains the germ of the Newtonian system of universal attraction).[333]

Speaking of Leonardo's geological conceptions, Corazza emphasizes the changes in the earth's shape and their connection with the notion of gravity, which Corazza presents as a precursor to Newtonian ideas. Though generic and inaccurate, the connection with Newton is significant, because it evokes the debate on the shape of the earth conducted by English Newtonian scientists in early eighteenth-century England from Desaguliers to Keill. The theory of an earth squashed at the poles and raised at the equator included two notions interesting in terms of our reconstruction of Leonardo's impact, namely gravity and the uplifting of land. Towards the end of the eighteenth century, the issue was still on the agenda. Travels were undertaken to use a pendulum to measure gravity at various points on the planet, and, eventually, these experimental tests proved Newton to be right.

Corazza was well aware of these debates and was also interested in geology and the history of the earth. From documents now in the Archiginnasio library in Bologna, we know that he was planning to write a *Geografia pratica* (*Practical Geography*)[334] and an *Indice di geografia antica* (*Index of Ancient Geography*);[335] while among his papers now in Naples there is a text dealing with the problem of the shape of the earth,[336] a sketch or summary of a history of the earth,[337] and the *Riflessioni sulla figura della terra* (*Reflections on the Shape of the Earth's*)[338] by P. Andrej, an engineer in the French navy, who discusses all these issues, and mentions new measurements of gravity that the author had commissioned from various French ships. The text on the history of the earth is by Corazza himself and includes subjects such as the Deluge, the formation of the Mediterranean after the opening of Gibraltar, the emergence of islands in the Aegean, and other themes that span geology and historical geography.

All of this explains Corazza's interest in some specific aspects of Leonardo's geological theories, which he could share with other representatives of the Neapolitan cultural and scientific world of the time. Indeed, Corazza, who was a teacher of history and geography to the young royal heirs from 1783,[339] was close to the court. He was also a regular visitor at Belforte's house, which was frequented by many scientists and literary men active in Naples, including two geologists, Alberto Fortis (1741–1803) and Scipione di Breislak (1750–1826). In a letter of 1780, Fortis recalls his visits at the Belforte house together with Corazza.[340] The same Duke of Belforte was also the author of a poem on the eruption of Vesuvius in 1779,[341] while Breislak, as we have seen, also manifested interests in the old geological 'systems'.

333 National Library, Naples, Ms. XII.D.81, f. 15:

Mondo 'Il centro della sfera dell'Acqua è il centro vero della rotondità del nostro Mondo, il quale si compone in fra acqua et terra in forma rotonda' (si sono recati altri esempli di Mondo per Terra; questo evidentiss. dee bastar per tutti, e deb'esser presente per intender bene un importantiss. luogo del nro. [ie: nostro] aut. che contiene il germe del Sist. Newtoniano su la universale Attraz.ne).

334 Bologna, Archiginnasio library, MS FS Biografie Cart I, n. 5

335 Bologna, Archiginnasio library, Coll. Aut. XXI, 5849-6101, Letter of 1796.

336 National Library, Naples, MS X.AA.28 / 5, n.p.

337 National Library, Naples, MS X.AA.28 / 6, ff. 56–7.

338 National Library, Naples, MS X.AA.28 / 5, n.p.

339 Buccaro 2011, pp. 111ff.

340 Toscano 2004, p. 28.

341 Vetrano 1925, pp. 105–111.

In general, uplift of volcanic origin, whether local or global, was an idea circulating among geologists active in Naples. Thus Breislak, considering the opening of the Strait of Gibraltar, suggested a lifting of the bottom of the Mediterranean ('qualche straordinaria escrescenza del Mediterraneo'),[342] while Alberto Fortis, after his stays in Naples, considered the hypothesis of volcanic uplift alongside the theory of changes attributable to a shift in the inclination of the terrestrial axis.[343] The theory of the formation of volcanoes by active uplift, along with their growth through the accumulation of erupted matter, was also discussed by William Hamilton, English ambassador to Naples since 1764 and author of volcanological studies, including the beautifully illustrated *Campi Plegraei* (Naples, 1776 and 1779).[344] Leopold Von Buch, also thanks to his visits to observe phenomena of the Vesuvius, would later build a theory dominated by the idea of uplift and vertical movements.

When, for economic reasons, Corazza tried to sell his two copies of Leonardo in 1779, William Hamilton was among those interested in the purchase. The intermediaries for a possible sale were the Roman architect and friend of Corazza Giacomo Quarenghi and Hamilton's nephew, Gavin Hamilton, the Scottish painter and pioneering archaeologist.[345] It seems that initially William Hamilton was contacted as a possible intermediary to help sell the manuscripts to the King of England who, as stated in a letter of 1779 from Gavin Hamilton to sir William Hamilton, already possessed Leonardo's anatomical manuscripts.[346] But William Hamilton himself was potentially interested, as attested by Quarenghi's letter of 1779[347] and by another document dated 1802, after Corazza's death when his son Sebastiano was monitoring the sale. The 1802 document involves another geologist of the time, Alberto Fortis, who, as we have seen, was active in Naples. In a letter of May 28, Alberto Fortis, in Bologna at the time, was informed from Naples about copies of Leonardo's manuscript, in response to a query he had made of his Neapolitan friend:

> His [Corazza's] writings by Vinci do exist, and are presently in the hands of S.A.R our Royal Hereditary Prince. The Duke [of Gravina] is confident to get them for you, but I have to let you know that Corazza's son, who is in misery, does not want to give them for less than a hundred *ounces*. But I have to prevent you and let you know that these texts were for a period in the hands of an Englishman who said he wanted to buy them, but later he gave them back. So the Englishman can have made a copy before giving them back.[348]

342 Breislak 1798, pp. 39–40.

343 Ciancio 1995, pp. 95ff.

344 Hamilton 1776, p. 53:

> I imagine the subterraneous fires to have worked in this country, under the bottom of the sea as Moles in a field, throwing up here and there a hillock, and that the matter thrown out of some of these hillocks, formed into settled Volcanos, filling up the space between one and the other, has composed this part of the continent, and many of the Islands adjoining.

345 For Giacomo Quarenghi's letters, see Böhmig 2008 and Buccaro 2011, pp. 71ff.

346 Irwin 1962, p. 100; Böhmig 2008, p. 149.

347 Letter of Giacomo Quarenghi to Vincenzo Corazza of 28 May 1779, in Böhmig 2008, p. 155: '. . . questo S.r Hamilton [Gavin Hamilton] mi ha dato buone nuove e spera si efetuerà questa facenda, o con quelo [William Hamilton] o in Inghilterra [col re].'

348 Bologna Archiginnasio, MS A 1224, item 14, f. 5r:

> I suoi scritti [di Corazza] sul Vinci esistono, e al presente li possidede S.A.R il nostro Regal Principe Ereditario. Il Sig. Duca [di Gravina] non diffida di averli, ma nel caso che l'abbia, il figlio del Sig. Corazza, il quale è in miseria, non li vuole

Given the 1779 document, it is reasonably certain that the 'Englishman' was William Hamilton. As for the author of the letter, who signs only as '*Ciro*', he may be identified as Ciro Minervini, an erudite historian in Naples. Minervini was interested in geology, as proved by his correspondence with Fortis dealing with geological and historical issues of Mount Vulture, an extinct volcano in Basilicata, in which Hamilton was also interested.[349]

In any case, the document confirms interest in Leonardo's scientific writings on the part of two geologists active in Naples: Fortis and Hamilton. The latter, who had in his collection a painting attributed to Leonardo,[350] retained the two manuscript copies for inspection. Of course, Fortis's and Hamilton's interest in Leonardo's writings could be of a general nature, and, in the case of Hamilton, related to his interest in scientific images.[351] However, given the context and contacts, it seems likely that they were also interested in Leonardo's uplift theory, to which Giambattista Venturi was also alert, as were his fellow scientist in Paris at a time when both Fortis and Breislak[352] were in the French capital.

At the Centre of the Scientific World: Leonardo's Manuscripts in Paris (*c.*1796–7) and the Role of the Codex Leicester

Two interlinked events of fundamental importance to Leonardo's legacy as a scientist took place in 1796–7. The first is the transfer by Napoleon of Leonardo's manuscripts from the Ambrosian Library in Milan to Paris in 1796. The second is the publication of Giambattista Venturi's *Essai sur les ouvrages physico-mathématiques de Léonard de Vinci* in 1797. Both these events took place in Paris, at the time the centre of the scientific world.[353] Venturi's book is famous because it is usually considered as the first instance of the circulation of Leonardo's science within the scientific community. In reality, as we have seen, copies of the Codex Leicester and other manuscripts ensured that Leonardo's scientific ideas had long since been circulating. Indeed, as we will see, it was the circulation within the eighteenth-century scientific community of the contents of the Codex Leicester that catalysed the transfer to Paris of Leonardo's manuscripts and the essay by Venturi. The latter was also the author of a partial copy of the Codex.

In 1796, Napoleon's army conquered Milan and vast areas of Northern Italy, with the aim of spreading the ideals of the French Revolution. In the name of these ideals, artistic works, books, manuscripts, scientific instruments, and naturalist collections were transferred to Paris. The aim was to create in Paris a world centre of the new politics, science, and art—a new Athens to which foreign artists and scientists were encouraged to move. Among the works transferred to Paris were Leonardo's manuscripts, at that time housed in the Ambrosian Library of Milan, namely the Codex Atlanticus, subsequently returned

dare meno di cento once. Debbo però prevenirvi che quegli scritti furono per un pezzo in mano d'un Inglese, il quale disse volerseli comperare ma poi li restituì; ecco dunque che l'Inglese se ne avrà fatta fare una copia e poi li ha restituiti.

See Buccaro 2011, pp. 71ff.

349 Andreoni 2006, pp. 167–81.

350 Jenkins and Sloane 1996, pp. 280–2.

351 Hamilton's interest in anatomical images is testified by the correspondence exchanged with the anatomist William Hunter in connection to the latter's interest in Leonardo's anatomical drawings. See Kemp 1976.

352 Gennari 1972.

353 Rudwick 2014, p. 108.

to the Ambrosiana, and twelve smaller manuscripts later designated by the letters from *A* to *M* and still in Paris, owned by the Institut de France.

The circulation of the Codex Leicester and its copies played a direct role in the French choice of Leonardo's manuscripts on the initiative of an influential French scientist of the time, the astronomer Joseph Jérôme de Lalande. Lalande travelled to Italy in 1765–66,[354] at a time when the theory of the ashen light of the moon in the Codex Leicester was spreading in the scientific community. Lalande became aware of this discovery,[355] and in order to understand it better promoted a search for Leonardo's manuscripts both in Italy and abroad. He also suggested to the French that they should seize Leonardo's manuscripts in Milan. In a 1797 review of Venturi's *Essai* it is mentioned that Lalande asked Venturi about the ashen light of the moon and that 'he [Lalande] has been searching especially in Italy for Leonardo's manuscripts because of this subject'.[356] Thus Lalande, before the transfer of the manuscripts to Paris and Venturi's work on Leonardo, had already been interested in Leonardo's manuscripts and not only in Italy. When he speaks of those 'especially in Italy', he was probably also thinking of those in London, which Lalande knew since 1769 as the location of the Codex Leicester with its discussion on the ashen light.[357]

In order to better understand this theory, Lalande later promoted the transfer of Leonardo's manuscripts from Milan to Paris. His role in the affair was suggested in 1881 by Ravaisson Mollien and reprised by later authors.[358] It can now be confirmed with certainty on the basis of two explicit and unnoticed statements by Lalande in his little-known *Bibliographie astronomique avec l'histoire de l'astronomie depuis 1781 jusqu'à 1802* (Paris, 1803). Both appear in the margin of Lalande's quotation from Venturi's essay. One of the two reads:

> 1797. Paris, in 4th Essai sur les ouvrages physico-mathématiques de Léonard de Vinci, by J. B. Venturi. On page 11 it is possible to see how this famous painter understands the cause of the ashen light. He died in 1519. Because of this text in Leonardo's book I asked, at the time of Bonaparte's conquest of Milan, that the manuscripts were sent to us.[359]

Lalande was a distinguished French scientist, well known for his role as a promoter of scientific initiatives of various kinds as well as a researcher. His manuscripts at the National Library of Paris testify to the huge collections of information, scientific and historical-scientific from other authors.[360] His

354 On Leonardo's manuscripts in the Ambrosiana, see Lalande 1769, vol. 1, pp. 310–11.

355 The Italian scientist Paolo Frisi is mentioned as a source in Lalande 1771 and 1790, while in 1769, Lalande writes: 'l'on m'a assuré' ('I was told') Lalande 1769, vol. 1, p. 311 (emphasis added): 'On conserve à Londres un manuscrit sur le rivieres [*sic*], par Leonardo del Vinci [*sic*], où *l'on m'a assuré* qu'on trouve la premiere explication de la lumiere cendrée de la Lune . . .'.

356 'Il [ie Lalande] avoit meme demandé specialment en Italie les manuscrits de Vinci à cause de cet objet'. 'Review of the book *Essai sur les Ouvrages Physiquo-Mathématiques de Léonard de Vinci . . . par J. B. Venturi . . .*', *Magasin encyclopédique, ou Journal des sciences, des lettres et des arts*, 3.2, 1797, pp. 145–6. For a different interpretation, see Marcuccio 2012, pp. 23–4.

357 Lalande 1769, p. 159.

358 Ravaisson Mollien 1881, p. 10; De Toni 1974; Marani 2003 pp. 385–8, esp. 385; Marcuccio 2012, esp. p. 24.

359 Lalande 1803, p. 641 (emphasis added):

> Paris, in 4° Essai sur les ouvrages physico-mathématiques de Léonard de Vinci, par J. B. Venturi. On voit, page 11, que ce peintre célèbre comprit le premier la cause de la lumière centrée. Il mourut en 1519. *C'est pour ce passage du livre de Léonard, que j'avois demandé, lors de la conquête de Milan par Bonaparte, qu'on nous envoyat les manuscrits.*

See also Lalande 1803, p. 790.

360 e.g. Paris, National Library, Ms. Fr. 12274.

original research in astronomy made a significant contribution to the calculation of the orbit of the sun–earth–moon system, a theme that chimes with his interest in the Codex Leicester. He is the author of a historical and scientific work on hydrology, which mentions Leonardo.[361] He also deals with issues related to geology, ranging from Maskelyn's measure of the gravitational attraction of mountains[362] to the catastrophic impact of comets.[363] Many aspects of the Codex Leicester would therefore be of interest to him.

At any rate, his influence on the decision to transfer Leonardo's manuscripts to Paris, as well as the existence of new documents that allow us to integrate Ravaisson Mollien's fundamental historical reconstruction of the affair published in 1881, raises the issue of the role played by the scientists in the transfer.

After Napoleon's entry into Milan on 15 May 1796, a commission appointed to select works to be transferred to Paris swung into action. The commission included artists, such as the painter Jacques-Pierre Tinet, as well as scientists, such as the mathematician Gaspard Monge and the chemist and geologist Claude Louis Berthollet.[364] The artist Tinet was already in Italy and was the first to start working in Milan. Other commission members, including scientists like Monge, president of the commission, arrived a little later, on 5 June.[365] It is possible that this short interval resulted in two phases in the collection of Leonardo's manuscripts: the first dominated by artists, in particular Tinet, and the other by scientists, particularly Monge.

In the first phase, two documents concern the Codex Atlanticus: one, mentioned above, dated 24 May and signed by the painter Tinet;[366] the other, of 13 July, was signed by the Commissioner of the Direttorio in Italy Saliceti.[367] Although the Codex is mostly technological, some of the more spectacular drawings may have attracted the painters' attention. It is indeed listed among works of art, that is paintings and drawings by various artists. By contrast, a document dated 14 June 1796 (26 Prairial of Year IV, according to the French Revolutionary calendar), for the first time includes 'Twelve manuscript volumes by Leonardo da Vinci' ('Dodici volumetti manoscritti di Leonardo da Vinci') at the end of a list of books. This previously unknown document was signed by the artist Tinet and by the scientists of the Commission: Monge, Berthollet, Thouin, and La Billardière.[368] The two phases in the transfer of manuscripts probably correspond to two successive withdrawals from Ambrosiana Library

361 *Des canaux de navigations et spécialement du canal de Languedoc*, Paris, 1778, pp. 453ff. See also Frisi 1768, preface on Leonardo's works and Languedoc.

362 See Boistel, Lamy, and Le Lay 2010.

363 Ampollini 2014.

364 Sadoun-Goupil 1974.

365 Institut de France, Paris, MS 2192, c. 3 (copy of a letter sent by Monge from Milano apparently to his wife, 9 June 1796). See also the map of Monge's travels in Italy in Taton 1951.

366 Ravaisson Mollien 1881, p. 8.

367 Archives Nationales, Paris, AF/III/72, Dossier 295, *Objets d'arts transportés en France An IV et An V*, unnumbered pages. That the item 'Un carton d'ouvrages de Leonard de Vinci' ('A cardboard binder of Leonardo da Vinci's works') corresponds to the Codex Atlanticus is confirmed by later documents in which that Codex, with that name, is sent to the National Library in Paris.

368 Archives Nationales, Paris, AF/III/72, Dossier 295, *Objets d'arts transportés en France An IV et An V*, unnumbered pages. The date is on a piece of paper with a note included among these documents. See also the document dated June 20 in Ravaisson Mollien 1881, p. 7.

on 19–24 May and 10–25 June.[369] We know that Monge worked in Milan in June.[370] It is likely that we owe to him or his fellow scientists, all of whom arrived in Milan only at the beginning of June, the withdrawal of a further twelve smaller manuscripts by Leonardo, which predominantly dealt with science and technology. It is likely that this choice was influenced by Lalande, with whom Monge was in frequent correspondence, as is evident in Monge's private letters.[371] In his letters, Monge mentions Leonardo several times, including his works of hydraulic engineering in Lombardy for the king of France, François I.[372] Monge might have learnt of this from Lalande, author of a treatise on navigable channels which took note of Leonardo's work.

Further confirmation of the two phases is the fact that the Codex Atlanticus was given to the Bibliothèque Nationale in Paris, while the twelve manuscripts were destined for the Institut National, which under the name of the Institut de France still holds them today. The Institut National des Sciences et Arts was one of the most important scientific institutions of the time. Lalande played a prominent role there, and it was also the place, shortly after the arrival of the manuscripts, where the best French scientists of the time listened to two communications by Giambattista Venturi about Leonardo's scientific work.

Giambattista Venturi in Paris (1796–7) and Leonardo's geological theories

Leonardo's manuscripts arrived in Paris on 25 November 1796. A direct consequence of this was their study by an Italian scientist, Giambattista Venturi (1746–1822), who had arrived in Paris shortly before. Venturi was among the foreign scientists attracted to the French capital in these years of political and cultural ferment. He went there first of all for diplomatic reasons, directly related to the French campaign in progress in Italy, but, once he had accomplished his aims, he decided to stay there for scientific reasons.[373]

On 25 April 1796, in a public session of the Institut National des Sciences et Arts in Paris, Venturi read the first notice of his ongoing research on Leonardo's scientific manuscripts,[374] promoted by Lalande[375] and soon after printed.[376] On 19 June 1797, Venturi's recently published *Essai sur les ouvrages*

369 Galbiati 1951, pp. 61–4; Ballarini 2000; Ravasi 2000.

370 Institut National, Paris, Biblioteque, MS 2192, a nineteenth-century copy of a letter written by Monge in Milan to his wife, 24 June 1796 (6 Messidor of Year IV).

371 A copy of a letter of Monge to Lalande of 7 November 1797 is in the 'Dossier Monge' among the papers of the French science historian René Taton, now at the École des Hautes Études, Paris, Centre CAPHES, RT 15.4.1, n. 1; a copy of a letter to his wife (Rome, 23 August 1796), in which Lalande and the Institut are both mentioned, is at the Institut National, Paris, Bibliothèque, Ms. 2192, pp. 17–18.

372 Institut National, Bibliothèque, Paris, MS 2192, pp. 6–7, copy of a letter of Monge to his wife, Bologna, 10 July 1796, therefore written shortly after the withdrawal of Leonardo's manuscripts in Milan: 'C'est François Ier qui, dirigé par Leonard de Vinci, a fait realizer [?] le premier canal, au moyen duquel on tire du Tessin une énorme quantite d'eau qui se distribue ensuite dans toute la Lombardie . . .'.

373 June 1796 to October 1797. Spiaggiari 1984, p. 66; De Toni 1924, pp. 29ff.; Marcuccio 2012, pp. 15–17.

374 See manuscripts with the reports of the public sessions at the Institut de France, Paris, Archives de l'Academie des Sciences, *Pochettes de séance*. See also *Procès-verbaux* 1910 [1795–1835], vol. 1, pp. 205–8 and 225–7.

375 Venturi 1797, p. 6 'Le C. Lalande, cet astronome si zélé, m'a encouragé à vous en offrir maintenant quelques essais.'

376 Venturi 1797. A work on optics by Flaugergues, presented during the same session, mentions Leonardo in connexion with the coloured shadows in the setting and rising sun.

physico-mathématiques . . ., was presented by the author during another public session of the Institut National. The Institut was founded in 1795, and the Museum of Natural History two years earlier. Both originate from the revolutionary transformation of former royal institutions, and many scientists became active both at the Institute and at the Museum. Scientific renewal followed the political revolution, and Paris became a major centre of the scientific world, a lighthouse attracting a wide range of international intelligences.

In the work on Leonardo presented by Venturi to the members of the Institut National, geology is the only field that is approached not as a historical precursor but as a topical theory, capable of influencing the ongoing scientific debate. The title of the relevant chapter is *De l'état ancien de la Terre*, and Venturi cited passages from the Leonardo's manuscripts recently arrived in Paris.[377]

In the commentary the following aspects of Leonardo's theory are highlighted:

1| Leonardo's precocious recognition that most of the surface of the continents was originally at the bottom of the sea;

2| his supposition that the mass of water is equal to, or greater than, that of the earth in relation to recent measurements of the gravitational attraction of mountains and the ideas of Bouguer, Maskelyne, and Laplace on the internal structure/density of the terraqueous globe;

4| his hypothesis that every continent is a small part of the earth's mass, and that the matter eroded and discharged into the sea at some point became heavier than the continents and caused them to elevate further above the water (slow uplift) or generate imbalance towards one side;

5| his theory that, as a result of erosion and the change in the balance of weight, a great shock suddenly disrupted the situation of a continent or uplifted new masses from the bottom of the sea (catastrophic uplift).

Venturi emphasized that Leonardo's ideas were such as to exercise a contemporary impact, given the continuing difficulties in explaining major geological changes in accordance with gravitational theory.

The object of Venturi's commentary is Leonardo's theory of the origin of dry land by a static or 'gravitational' mechanism that is mainly developed in ff. 35v and 36r of the Codex. Much less of this theory emerges from the passages from manuscripts E and F mentioned by Venturi. We will return to this issue later. It should be emphasized that Venturi prefers the most theoretical part of Leonardo's geology. By contrast, Charles Lyell, writing in the historical section of his *Principles of Geology*, was to highlight Leonardo's demonstration of the organic origin of fossils, which at that time was only a historical curiosity.[378] As we have noted, Lyell was the most important representative of 'uniformitarianism', while

377 MS F, f. 80; MS E, f. 4v; MS F, f. 11v; MS N, f. 124 (i.e. Codex Atlanticus). Venturi 1797, pp. 12–15.

378 Lyell 1840, p. 34. See also Verga 1931, p. 157; Vai 2003a, pp. 233–4. Shortly afterwards, the English 'lay' scientist and geologist William Whewell (1794–1866), who first termed the opposing geological theories of the time as catastrophist and uniformitarian, opened a section 'Of the Doctrine of Geological Catastrophes' in his *History of the Inductive Sciences* with Leonardo, even if, perhaps under Lyell's influence, he stressed the 'uniformitarian' aspects of Leonardo's theory:

> Leonardo da Vinci, whom we have already noticed for his early and strenuous assertion of the real marine origin of fossil impressions of shells, also maintained that the bottom of the sea had become the top of the mountain; yet his mode of explaining this may perhaps be claimed by the modern advocates of uniform causes, as more allied to their opinion, than to the doctrine of catastrophes.

in Paris, at the end of the eighteenth century, a conception of the past of the earth gained hold that included catastrophic events. Many French geologists saw theories or systems as a necessary instrument alongside empirical collection of directly observable data. In order to understand Venturi's comments, we need to take into account the geological ideas that were being debated in Paris at the time.

From the detailed reports of the Institut's public sessions, we learn that some of the greatest scientists active in Paris, including Cuvier, Desmaret, Dolomieu, Hauy, Laplace, Lamarck, Fourcroy and, of course, Lalande, attended the two sessions in which Venturi presented and discussed Leonardo's scientific theories. Some of them were representative of the new geology.[379]

A few months before Venturi's arrival in Paris, at another public session at the Institut, on 26 January 1796, George Cuvier had presented his epochal discourse. He used his study of the comparative anatomy of the fossilized bones of a Mammoth and the skeleton of a living elephant to show that the mammoth was not a sort of elephant, but belonged to an entirely different animal species, which had disappeared from the face of the earth. The biological concept of 'extinction', which for the first time was decisively introduced, immediately called for a geological explanation. On the same occasion, Cuvier had already used the concept of 'geological catastrophes'—or, in his own words, 'quelques révolutions de ce globe', which completely destroyed an animal species—to explain its extinction.

Among the other scientists attending Venturi's presentation at the Institut was René Just Haüy (1743–1822), a great mineralogist, the founding father of crystallography, Nicolas Desmaret, and Déodat de Dolomieu, who were among the greatest geologists of the time. Dolomieu (1750–1801), from whom the name of the mountains Dolomites derives, contributed to the development of a theory of the earth, including uplifts by lateral thrusts and on a non-volcanic basis.[380] His conclusions were based on travels and field research, and lead him to underline the need for a new theory capable of explaining the collected data.[381] His ideas were presented to the Institut on November 1798.[382]

Venturi's audience also included the great Jean Baptiste Lamarck (1744–1829), who supported the idea of 'transformation' of animal species over time, one of the many pre-evolutionary theories. Lamarck was the author of a complex theory of the earth, presented at the Institut in 1799 and later published under the title of *Hydrogéologie* (Paris, 1802). He denied Cuvier's geological catastrophes, instead advocating a continuous and cyclical inversion of oceans and continents. He gave much emphasis to water and not so much to volcanic fire as a key factor in these processes.[383] Lastly, Pierre-Simon Laplace (1749–1827) was a great mathematician and astronomer, whose fundamental *Exposition du système du monde* (Paris 1796) dealt with issues relevant to geology such as the shape of the earth and its internal density.

In these years two successive editions (1795 and 1797) of Jean-Claude de La Métherie's *Théorie de la terre* were published in Paris. The work, after a long empirical and mineralogical section, introduces

Whewell 18[illegible]7, vol. [illegible], p. 606–7. See Verga 1931, p. 152.

379 *Procès-verbaux* 1910 [1795–1835], vol. 1, pp. 205–8 and pp. 225–7.

380 Dolomieu 179[illegible], p. 425. See Gohau 1991, pp. 117–18; Ellenberger 1994, pp. 285ff.

381 Dolomieu 1798, pp. 429–30.

382 *Procès-verbaux* 1910 [1795–1835], vol. 1, p. 309.

383 Corsi 1988, pp. [illegible]03–4; Rudwick 2005, pp. 388–92.

the main 'systems' or geological theories.[384] La Métherie assesses the volcanological hypothesis as unlikely from a global point of view, both in the cyclical and anti-catastrophic versions of James Hutton and John Playfair[385] and in Anton Lazzaro Moro's more catastrophic one.[386] La Métherie devotes much space to the 'gravitational' theories of the shape of the earth and its interior composition or density, mentioning Pierre Bouguer's and Nevil Maskelyn's measures of gravity and Laplace's theory of the uneven density of the interior of the earth.[387] Finally, introducing his preferred system, he considers the problem of the ratio of water to earth, stressing the need for a greater mass of water and large underground caverns, particularly in the second edition (1797)[388] after a passing mention in the first (1795).[389]

What emerges from the reconstruction of this chain of events is that in Parisian geology at the end of the eighteenth century there was a faction predisposed to reject volcanic hypotheses and to emphasize catastrophic, gravity-based hypotheses. All this helps us to understand Venturi's presentation of Leonardo's geology, which, in this context, could serve as more than a merely historical-scientific curiosity. Not surprisingly, in a review of Venturi's *Essai* that appeared in the *Annales de Chimie* of 1797, Leonardo's geological texts were among the few mentioned, along with some chemical and methodological matters. Even in what was a chemistry periodical, Leonardo's geological texts opened the short review. Among the citations from Venturi's *Essai* were passages dealing with catastrophes in the history of the earth, but not the texts on the animal origin of fossils.

The title page of Venturi's book reads: 'Essay . . . read in the first course of the Institut National des Sciences et Arts'. We do not know how much of the final content of the *Essai* was actually read in front of the members of the Institut. In any event, everything was soon available in print. Whether, upon Venturi's initiative, some geologists of the Institut addressed Leonardo's theory and texts remains an open question. But, regardless of Leonardo's immediate legacy in Paris, Venturi's emphasis on Leonardo's particular theory of the earth in the context of the lively debate is a significant episode in a more theoretical and less well-known current in the history of geology, which we have already encountered in considering the English context of the late seventeenth to the early eighteenth centuries.

Venturi does not mention the Codex Leicester either in relation to Leonardo's theory of the earth or elsewhere in the *Essai*.[390] This is not surprising, since the work aimed to present to the scientific world excerpts from the manuscripts transferred to Paris. However, Venturi's comment on Leonardo's geological theory implies knowledge that seems to go beyond the three cited passages from manuscripts E

384 Rudwick 2005, pp. 340–1

385 La Métherie 1797, vol. 5, p. 488.

386 La Métherie 1797, vol. 5, pp. 472ff.

387 La Métherie 1797, vol. 4, pp. 4ff.

388 La Métherie 1797, vol. 5, pp. 523ff., esp. pp. 523 and 532 (respectively):

> D'après tous le sentiments des géologues, que nous venons de rapporter, on voi que le très-grand nombre pense que la masse intérieure du globe est solide, et qu'elle renferme seulement quelques cavités plus ou moins considérables.
>
> Il faut donc supposer dans l'intérieur de la terre des cavernes bien plus vastes que celles que nous connoissons. Mais comment auront-elles été produites? C'est peut-être une des questions les plus difficiles de toute la géologie.

389 La Métherie 1795, vol. 3, p. 441: 'Dans tous ces systemes on suppose des cavernes intérieures plus ou moins considerable.'

390 At p. 36, Ducci's copy of the Codex Leicester is wrongly considered as a copy of another work by Leonardo.

and F and the Codex Atlanticus. Only the Codex Leicester, ff. 35r–36v, contains a presentation of Leonardo's theory, in which the ratio of earth to water provides the starting point, and only these folios are comparable for completeness and clarity to the content of Venturi's account of Leonardo's ideas. Had he read a full or partial copy of the Codex Leicester while working on his *Essai* in Paris, or even before that? Is it possible that Lalande, who, as we have seen, tried to verify the ashen light theory in the Codex Leicester, had passages copied from the Codex or from one of its transcriptions?

In fact it is certain that Venturi had later access to one of the copies of the Codex Leicester and began copying it. The existence of this partial copy[391] has escaped all studies on the history of Codex Leicester.[392] This is a further small tassel of the circulation of Leonardo's manuscript and therefore of his thought (Figure 73). Venturi copied two opening excerpts from folios 1r and 2r of the Codex Leicester, the former dealing with the light reflected by the moon to the earth,[393] and the latter describing the moon as a heavy body like the earth.[394] The copy of the folio 2r excerpt quickly breaks off and is completed by a free paraphrase of the remaining text. Then a comment follows:

> We see that Da Vinci did not believe that solid bodies illuminated by the Sun make a very luminous reflection. In this same work he attributes the weak light between the new moon's horns to the reflection from the water of our Ocean, which proves what I have just said.[395]

This corresponds to Leonardo's explanation of the ashen light, in which Lalande was so interested. Venturi's note on the copy he was using follows: 'The work went to Germany to the Duke of Saxe-Weimar; neither had I time to transcribe more from it.'[396]

Venturi used the copy of the Codex Leicester now in Weimar, formerly in Naples at the time of Giuseppe Bossi in Milan. This is demonstrated by the number 'III' marking one of the copied passages ('Here it is shown [. . .]', 'Qui si dimostra [. . .]') (Figure 73) corresponding to the Weimar copy (Figure 70) and lacking in the original Codex and in its other known copies (Figure 67). It is also confirmed by the drawing derived from Codex Leicester, f. 1r, which appears in similar form both in Venturi and the Weimar copy (Figures 73 and 70). In the upper left corner, on the recto of the sheet containing the copy, Venturi's notes: 'written in 1815, or 16'.[397] This uncertainty indicates that the note was added later, as was the note on the transfer of the copy to Weimar, which took place in 1817. We know that Venturi already had access to Bossi's papers on Leonardo in 1810 in Milan and perhaps on that occasion he could have seen the copy of the Codex Leicester.[398] In 1817–18 he tried but failed to

391 Panizzi Library, Reggio Emilia, MS Regg. A 35/3, ff. 276r and v.

392 The only mention known to me is a footnote in De Toni 1924, p. 59.

393 Panizzi Library, Reggio Emilia, MS Regg. A 35/3, f. 276r: 'Here it is demonstrated how the moon, not having any light. . . .'

394 Panizzi Library, Reggio Emilia, MS Regg. A 35/3, ff. 276r and v: '[Since we have proved . . .] at present it must be proved. . . .'

395 Panizzi Library, Reggio Emilia, MS Regg. A 35/3, ff. 276v: 'Si vede che il da Vinci non credeva che i corpi solidi illuminati dal Sole facciano molto lume riflesso. In quest'opera stessa attribuisce il lume debole della Luna nova fra le corna al riflesso dell'Acqua del nostro Oceano, lo che prova ciò che ho detto pur ora.'

396 Panizzi Library, Reggio Emilia, MS Regg. A 35/3, ff. 276v: 'L'opera è andata in Germania al Duca di Saxe-Weimar; nè io ho avuto tempo di trascriverne di più.'

397 Panizzi Library, Reggio Emilia, MS Regg. A 35/3, ff. 276r: 'scritto del 1815, o'16'.

398 Panizzi Library, Reggio Emilia, MS Regg A 35/3, sez. Q, ff. 31–2. See also De Toni 1924, p. 59.

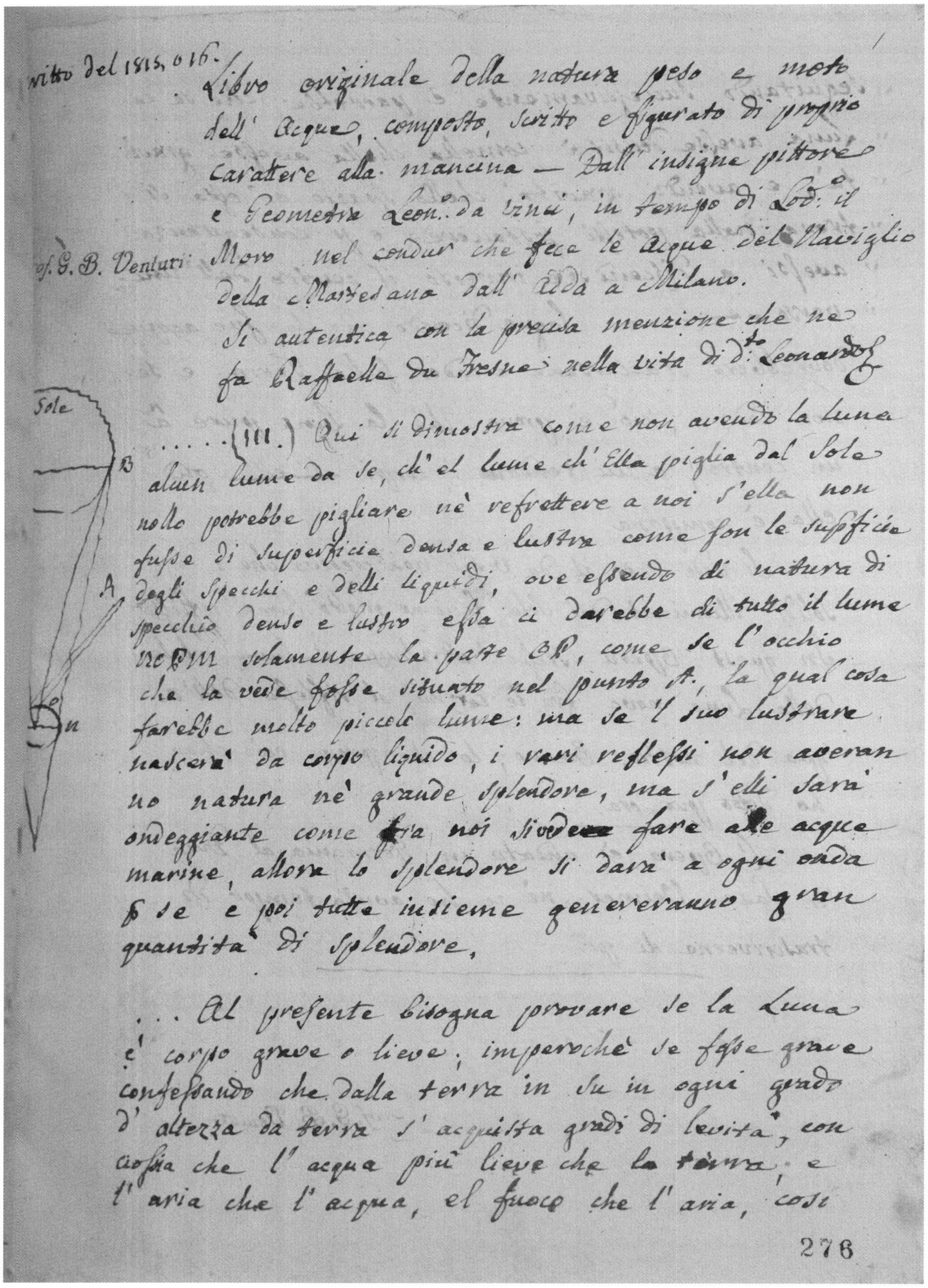
scritto del 1815, o 16. Libro originale della natura peso e moto dell' Acqua, composto, scritto e figurato di proprio carattere alla mancina — Dall' insigne pittore e Geometra Leon.o da Vinci, in tempo di Lod.o il Moro nel condur che fece le Acque del Naviglio della Martesana dall' Adda a Milano.

Prof. G. B. Venturi

Si autentica con la precisa menzione che ne fa Raffaelle du Fresne nella vita di d.to Leonardo

Sole
B
A
n

....(III.) Qui si dimostra come non avendo la luna alcun lume da se, ch'el lume ch'Ella piglia dal Sole nollo potrebbe pigliare nè refrettere a noi s'ella non fusse di superficie densa e lustra come son le superficie degli specchi e delli liquidi, ove essendo di natura di specchio denso e lustro essa ci darebbe di tutto il lume non solamente la parte CB, come se l'occhio che la vede fosse situato nel punto A, la qual cosa farebbe molto piccolo lume: ma se 'l suo lustrare nascerà da corpo liquido, i vari reflessi non averanno natura nè grande splendore, ma s'elli sarà ondeggiante come fra noi si vede fare alle acque marine, allora lo splendore si darà a ogni onda per se e poi tutte insieme genereranno gran quantità di splendore.

...Al presente bisogna provare se la Luna è corpo grave o lieve, imperochè se fosse grave confessando che dalla terra in su in ogni grado d'altezza da terra s'acquista gradi di levità, con ciosia che l'acqua più lieve che la terra, e l'aria che l'acqua, el fuoco che l'aria, così

276

FIG. 73
Giambattista Venturi, Manuscript copy of Codex Leicester, f. 1r. Biblioteca Panizzi, Reggio Emilia, MS Regg. A 35/3, f. 276 r.

complete the copy before it left Italy and in this connection wrote letters to Gaetano Cattaneo, who was in charge of its sale in Milan.[399]

It is possible that Venturi's interest in the Codex Leicester during his later years was also motivated by the preparation of one of his geological works, *Memoria intorno ad alcuni fenomeni geologici* (*Memorandum about some geological phenomena*, Pavia, 1817), which had been presented in public in 1816.[400] To date, there has been no specific study of Venturi as a geologist. However, Alexandre Brongniart, Cuvier's companion in his geological explorations of the area surrounding Paris, quotes Venturi alongside noted geologists, Deluc, Dolomieu, and Von Buch, in relation to one of the central problems of geology of the time, the explanation of 'erratic blocks'[401]—that is to say, big isolated blocks of rock extraneous to the surrounding area.[402] Venturi's *Memorandum* was precisely about this problem. In addition to this printed work, the Panizzi library in Reggio Emilia holds unpublished manuscripts by Venturi on geological subjects. One of them is very interesting because it emphasizes the arguments developed in his comment on Leonardo's theory of the earth.[403] Once it was established that marine fossils in the mountains imply a prior presence of the sea, Venturi writes that geologists proposed various explanatory theories, all still hypothetical.[404] His presentation begins with the theory of uplift, and is followed by the others, which include the neptunist hypothesis of a lowering of the sea level.[405] Above all Venturi presents his study of the inner structure of the earth in the light of gravitational theories and measurements, including those of Maskelyn, who advocated a greater compactness within the earth.[406] Maskelyn's conclusions were mentioned in Venturi's commentary on Leonardo's geology of 1796–7, where he also cites Laplace's opposing hypothesis and links it to Leonardo's idea of the greater proportion of water to earth, implying the existence of underground reservoirs, which was the starting point for Leonardo's theory of the earth. It is possible that Venturi's geological manuscript dates to the years around the 1797 *Essai*,[407] and therefore bears witness to the first phase of Venturi's geological studies, characterized by a more theoretical slant, which changed at least in part by the time of the 1816–17 *Memorandum*, which was dominated by more empirical approach with no hypotheses of 'furious cataclysms'.[408]

399 Panizzi Library, Reggio Emilia, MS Regg. A 15/34, ff. 1–2, Letter of Gaetano Cattaneo to Giambattista Venturi, 8 April 1818. See De Toni 1924, p. 126 for a letter sent by Venturi to Cattaneo on 17 March 1818.

400 De Brignoli 1835.

401 Brongniart 1829, p. 77.

402 Rudwick 2005, pp. 581–5, 2014, pp. 116–20.

403 Panizzi Library, Reggio Emilia, MS Regg. A 75/2.

404 Panizzi Library, Reggio Emilia, MS Regg. A 75/2, ff. 33r–34v.

405 Panizzi Library, Reggio Emilia, MS Regg. A 75/2, f. 33v:

> Ossia che una Rivoluzione del Nostro Globo abbia sollevati dal fondo del mare i continenti d'oggidì e sprofondando in loro vece gli antichi. Ossia che il livello dell'Oceano fosse un tempo veramente più alto, ma che poi siasi ribassato perchè l'acqua si abbia aperto l'adito e siasi ritirata in caverne sotterranee. . . .

406 Panizzi Library, Reggio Emilia, MS Regg. A 75/2, ff. 72ff., esp. f. 113r: '. . . Therefore the earth is not a great hollow with an immense vacuity in the midst of it. Therefore the attraction of this mountain is in the same inverse ratio of the squares of the distance. See Maskelyn in *Philosophical Transactions*, 65, 1775, pp. 495–9 and Todhunter 1873, pp. 459–60.

407 *c.*1780–1820, according to Marcuccio 2001, p. 186.

408 *Memoria intorno ad alcuni fenomeni geologici* (*Memorandum on certain geological phenomena*), Pavia, 1817, p. 8: 'furiosi cataclismi'. According to Di Brignoli 1835, Venturi refers to Breislak.

A Copy of the Codex Leicester in Weimar: Leonardo and Goethe as Artist-Geologists

In 1818, a copy of the Codex Leicester arrived in Weimar, Germany, where it is still today. It came from Milan and was the copy formerly belonging to the painter and scholar of Leonardo, Giuseppe Bossi, and, before him, to the Serra di Cassano's family in Naples. Bossi's copy and other books sold by his heirs, were purchased by Grand Duke Karl August of Saxony-Weimar. Behind the purchases of the Grand Duke, not least the copy of the Codex Leicester, was Johann Wolfgang Goethe, the great German poet and scientist. From geology to optics, from botany to anatomy, Goethe's scientific interests were expansive in scope and entertained a profound relationship to his poetry, comparable in European culture only to Leonardo's science and art centuries earlier. Goethe was also a fairly good draftsman, just as Leonardo was a fairly good man of letters. The German episode in the life of this copy of the Codex Leicester provides an opportunity for a comparative analysis of the geological theories of these two geniuses of European culture.

On 2 January 1818, Grand Duke Karl August sent to Goethe a copy of the sale catalogue of Bossi's books. The catalogue included, among the manuscripts, the *Libro della natura, peso e moto dell'acque* (*Book of Nature, Weight and Movement of water*), that is a copy of the Codex Leicester, correctly identified as a volume from the eighteenth century.[409] Goethe had been in Jena and, since the end of November 1817, he had been immersed in a study of Leonardo, focusing on the *Last Supper*. At the beginning of November, Bossi's drawings, sketches, and tracings of the *Last Supper* and its copies arrived in Weimar, along with a copy of Bossi's pioneering book on the *Last Supper* published in 1810. These documents, acquired by Karl August who had learnt of them during his stay in Milan during the previous summer,[410] stimulated Goethe's account of the *Last Supper*, finished between January and February 1818 and published in the same year.[411]

Thus Goethe received the sale catalogue of Bossi's library just as he was immersed in his Leonardo studies, and, not surprisingly, he selected for purchase the copy of the Codex Leicester and a seventeenth century copy of the *Treatise on Painting*.[412] It seems that the latter was not purchased, while the copy of the Codex Leicester was. Choosing a scientific manuscript rather than an artistic one—if it really was a choice—may be significant. It was in 1817–18 that Goethe had resumed the geological studies that he had begun in 1776, when Karl August had appointed him the administrative supervisor of the reactivation of Ilmenau's mines in Thuringia.[413]

This was the 'epoch of geology', perhaps the most rejuvenated and popular science of the time. The 1817–18 correspondence between the Grand Duke, Goethe, and Gaetano Cattaneo, who was handling the sale of Bossi's books in Milan, is full of references to the despatch of specimens of rocks and minerals. We also learn of contacts with Italian geologists and the despatch of their texts, including those by the

409 *Catalogo della libreria del fu Cavaliere Giuseppe Bossi*, Milan 1817, p. 235.

410 Blank 1992a; Mildenberger 1999; Mildenberger and Zanaboni 2016; Venuti 2005; Zanaboni 2016.

411 Goethe 1818.

412 Herzogin Anna Amalia Bibliothek, Weimar, Dd 8 680, includes marks by apparently three different hands. The librarian Vulpius also contributed to the selection: see Letter of Karl August to Goether of 2 January 1818.

413 Vaccari 2005, p. 417; Fink 1991, p. 14.

'neptunist' Gian Battista Brocchi (1772–1826)[414] and his rival the 'plutonist' Scipione di Breislak,[415] two scientists representative of the debate between those who prioritized the theory of the marine origin of the rocks (the name 'neptunists' comes from Neptune, the ancient Roman god of the sea) and those who favoured the theory of the volcanic origin (the 'plutonists' are named after Pluto, the god of the underworld).

Goethe was directly involved in the great dispute that divided geology. According to the neptunists, the main rocks, including granite, had been deposited at the bottom of the oceans and then emerged passively, following the slow and continuing lowering of the oceans. According to the plutonists, many rocks, such as granite or basalt, were of volcanic origin and had been lifted up actively by the inner volcanic forces of the earth. The main school of the neptunists was based in Germany, where the leading figure was Abraham Gotlob Werner (1749–1817). It is no coincidence that Goethe was a neptunist, although, as we will see, of a particular kind. In the neptunist system, the key role was played by water; in the plutonist, by fire. Goethe and Karl August's interest in the copy of the Codex Leicester can therefore be set within this geological context, even if we do not know whether Goethe or the Grand Duke knew before the arrival of the copy about its content, other than what was indicated by its 'hydrological' title in the Bossi sale catalogue. Bossi's books, which included the copy of the Codex Leicester, finally arrived in Weimar on 16 June 1818, along with some medals that had been purchased, as recorded by Goethe in his diary[416] and in a letter to Cattaneo at the end of June.[417]

Goethe's interest in Leonardo soon extended beyond the artistic area of the *Last Supper* to embrace scientific issues, and this happened at the same time as the re-emergence of his geological studies. In a more general chapter on Leonardo in his essay on the *Last Supper*,[418] Goethe describes Leonardo's extraordinary capacity for rational analysis and his exposition of natural phenomena. After his mention of Leonardo's visual studies of proportions and of 'monsters', he turned to Leonardo's writings, first mentioning Leonardo's geological studies:

> To his sharp and rational conception of the world we owe the expressive richness with which he makes us aware of the most violent movement of complicated earthly phenomena using words as if they were paintings.[419]

A 6 December 1817 entry in his diary includes, next to notes concerning his study of the *Last Supper*, a 'geological' reference: 'Leonardo's Last Supper. . . . I have read again Bossi. Leonardo da Vinci on colours. *Leonhards' great Mineralogy*. Later I have corrected the article on Bossi.'[420] The reference is to the *Bedeutung und Stand der Mineralogie* (*The Importance and State of Mineralogy*) published in 1816

414 Blank 1992b, p. 92 ('alcuni libri, che Brocchi mi ha pregato di far pervenire a V. E.') and p. 117.

415 Blank 1992b, pp. 68 (19 November 1817), 134 (October 1818).

416 Goethe 1975, p. 217.

417 Blank 1992b, p. 122.

418 Goethe 1818, pp. 167ff.; *Uno sguardo su Leonardo*, in Venuti 2005, pp. 50–3.

419 Goethe 1818, p. 168:
Dieser scharfen, verständigen Weltanschauung verdanken wir auch die grosse Ausführlichkeit, womit er verwickelter Erdenbegegnisse heftigste Bewegung wit Worten vorzufühern weiss, eben als wenn as Gemälde werden könnten.

420 Goethe 1975, p. 144 (emphasis added).

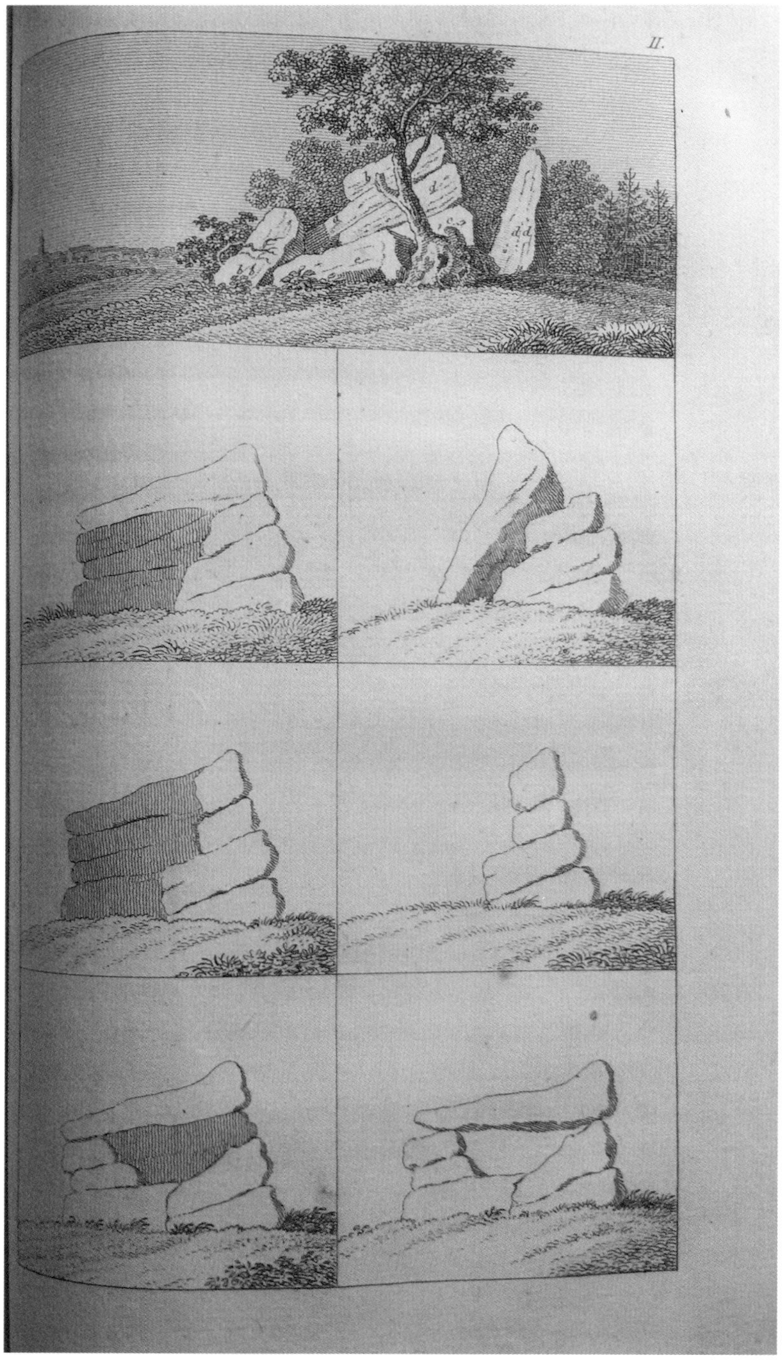

FIG. 74
After Johann Wolfgang Goethe, *Die Luisenburg bei Alexanders-Bad*. From Werke: *Vollständige Ausgabe letzter Hand*, Stuttgart, 1833, p. 167. First published in 1820.

by Karl Cäsar von Leonhard, who corresponded with Goethe. It is interesting that Goethe should chose to think about Leonardo's pigments in the context of mineralogy.

In the same period, among other geological writings, Goethe wrote a short essay on the morphology of granite, *Die Luisenburg bei Alexanders-Bad*, which is included in his *Naturwissenschaftliche Hefte* (1817–20). The essay is accompanied by an illustration engraved by Carl Lieber after drawings made by Goethe that date back to 1820 (Figure 74).[421]

421 Femmel 1958–69, vol. 5b, nos. 213, 214, 214a.

The purpose of the drawings is to demonstrate dynamic geological processes, that is, the before and after of an event. In particular, Goethe tries to understand how the morphology of four groups of granite changed over time and how the original block was crushed. The top illustration in a pictorial mode is labelled with letters, and demonstrates 'after', that is, the present situation: the blocks labelled '*bb*' and '*dd*' were originally part of the blocks indicated respectively with '*b*' and '*d*'. The images below illustrate the same process in three other cases of granite groups, but they do so by showing 'before' (left) and 'after' (right) by shading the blocks that will become lost.[422] As a neptunist, Goethe privileged a harmonic conception of natural processes that excluded catastrophic changes. Granite was, according to neptunists, the primitive rock formed by crystallization on the bottom of the ocean. Once it emerged, following the lowering of the sea level, it underwent metamorphic processes by atmospheric agents. These slow, continuous, non-catastrophic processes were the subject of Goethe's figures. This illustrative technique is unique in Goethe's geological drawings and was devised around 1820, shortly after the arrival in Weimar of the copy of the Codex Leicester. Codex Leicester, f. 36r, contains a figure that corresponds, conceptually and graphically, to Goethe's chiaroscuro illustrations of Luisenburg granite. Leonardo describes and represents the dynamic process through which continents and dry land emerged. The process described in words in the opening text of the page (*Of the Earth Itself*) is illustrated by a chiaroscuro technique at the bottom drawing in the right margin, where the underground mass affected by the dynamic process is represented as a shaded arc in its original position and as a blank arc in its final position. Leonardo represents a continental and catastrophic geological process; on a smaller scale, Goethe illustrates a non-catastrophic change. Nevertheless, the dynamic content and the chiaroscuro technique are very similar. However, this drawing is missing from the copy that arrived in Weimar, and therefore any direct influence is out of the question. We seem to be witnessing parallel conceptions of how to represent geological processes graphically.

Goethe's geology, along with other aspects of his natural philosophy, found a particular poetic transfiguration in his *Classical Walpurgis Night*, outlined in 1826 and written in 1830,[423] which is one of the most vivid sections of the second part of *Faust*, the dramatic poem written by Goethe in successive phases. Some parts of the *Classical Walpurgis Night* are like a scientific drama in which the later natural philosophy of Goethe reverberates, just like Leonardo's late *Deluge* drawings are artistic transfigurations of his research in fluid dynamics and geology.

Set in Greece on the Aegean's coast, the scene imagined by Goethe includes figures from ancient mythology and philosophy, and it is suffused by various themes that cut across poetry and science. We find echoes of the geological dispute between neptunists and plutonists. Goethe evokes the motion of the waves and the origin of life from the waters. He speaks of the harmonic metamorphosis of natural forms, and of the watery nature of the moon and its reflection in the waters of the sea. All these themes are reminiscent of the content of the Codex Leicester. However, it is a fit subject for further study and here we can only suggest the following comparisons.

In the *Classical Walpurgis Night*, two Greek philosophers exemplify the geological disputes: the plutonist Anaxagoras and the neptunist Tales. Provoked by Anaxagoras, Tales enters the scene pronouncing the following words:

422 Fink 1998, esp. p. 277.

423 Amoretti 1991, p. 775 n. 1.

The waves respond to every whim of air,
but stay away from rugged rocks like this.[424]

Reefs, Anaxagoras claims in the following verses, have been created by the 'vapours of fire', that is, they have volcanic origin, like all things in the earth. But Tales rebuts this and says, 'life originated in the waters'. The verses on Tales, supporter of the primacy of water, begin with the marine wave in a way that recalls the numerous aphoristic descriptions of the behaviour of waves in the Codex Leicester. For example, 'waves are of two natures, one of which is created by the wind and this kind breaks and ruins its tops toward the place where it runs' (f. 22r); and 'how the waves of the sea continually consume their promontories and rocks' (f. 22v). It seems that Goethe's Tales has been endowed with a Leonardesque character.

Another theme developed by Goethe in the *Classical Walpurgis Night* is the origin of life from the water. To dramatize this process, Goethe uses the character of Homunculus who is born in the waters and the waves of the sea (Tales's proposal), rather than in underground flames and caves (Anaxagoras's proposal).[425]

Thus, according to Goethe's (and Tales's) vision, not only the mountains originated through slow and harmonic processes in marine waters, but even life was born in water. This analogy between the origin of the continents and the birth of animal life again evokes Leonardo. The years of composition of the Codex Leicester are close to Leonardo's famous embryological studies.[426] Concepts, terms, and visual formulas Leonardo uses to explain the generation and emergence of the body of the foetus from its bath of fluids recall those used in the Codex Leicester in order to explain the emergence of dry land from the sea in the body of the earth.[427]

This was also an outcome of the general macro-microcosmic analogy, which, in new forms, will still influence Goethe's theory of the earth. It should be remembered that Cattaneo's letter to Goethe, dated 2 May 1818, at the time of correspondence between Milan and Weimar dealing with the purchase of Bossi's books and the copy of the Codex Leicester, includes a discussion of Leonardo's anatomical drawing of two figures in coition.[428] Cattaneo informs Goethe that he had examined the facsimile of the drawing with Professor Blumenbach of Göttingen and suggests that Leonardo aimed to represent 'the best posture in both sexes of the Human Species to facilitate conception'.[429] Goethe was therefore interested in Leonardo's studies on animal generation, an aspect worth considering in view of the still-unresolved problem of the presence in Weimar of an anatomical drawing by Leonardo, including reproductive organs.[430]

424 Johann Wolfgang von Goethe, *Faust I and II*, ed. and trans. Stuart Atkins, Princeton, NJ, 2014, Act II, lines 7853–54. In the original: 'Die Welle beugt sich jedem Winde gern. Doch hält sie sich vom schroffen Felsen fern'.

425 *Faust*, Act II, in Amoretti 1991, pp. 449–51.

426 Windsor 19101–3.

427 Laurenza forthcoming c.

428 Windsor 19097v.

429 Blank 1992b, pp. 116 and 118.

430 Laurenza 2016b, pp. 186–87.

Conclusion

This study of an unexpected number of copies of the Codex in their contexts of production and circulation, represents the first sustained effort to study Leonardo's contribution as a scientist. Over the last fifty years, it has been increasingly acknowledged that, as an artist-engineer, Leonardo was not an isolated genius, but part of a very specific historical juncture. Likewise, the reconstruction of his legacy proposed here indicates that interest in Leonardo's scientific work between the sixteenth and the early nineteenth centuries was due not only to a few eccentric collectors curious about scientific manuscripts by a legendary artist who wrote from right to left, but was an manifestation of a specific cultural and social context of sciences during that period. Leonardo gave cultural legitimacy to the scientific ambitions of practitioners, artists, and architects who, between the sixteenth and eighteenth centuries, practised the sciences, above all the 'science of water'. These figures included Guglielmo della Porta, Giovanni Ambrogio Mazenta, Lodovico Antonio David, and Bernardino Ciurini. Leonardo, who had sought in Archimedes a model that would legitimize his cultural emancipation as a scientist, rather than a mere technician, in turn became a model for later artists-engineers. In the British scientific world of the eighteenth century, dominated by the Baconian tradition and by the modern Newtonian paradigm, as well as in the Italian context linked to British science, Leonardo's work served as a great exemplar for the contribution of practitioners to science. This attitude continued in the Enlightenment, and in particular with Venturi's work and the arrival of Leonardo's manuscripts in Paris at the end of the eighteenth century. Moreover, at that time and occasionally in the previous two centuries, Leonardo's geological ideas attracted the attention of a relatively young science.

The issue of Leonardo's influence on the history of science is more complicated than his impact on the history of art, and it is unlikely that Leonardo's legacy as a scientist will come to be seen as comparable to that in art. In judging his position within science, we need to allow for the possibility that similar conclusions can arise independently. We also need to take into account the issue of uncredited 'borrowing', which is very common in science. One of the best-known cases of plagiarism in the history of science involves the early-eighteenth-century British geologist John Woodward, who, as his contemporaries became rapidly aware, 'borrowed' much of his work from the artist-geologist Agostino Scilla without mentioning his name. Even so, we have seen some credible cases of the impact of Leonardo's geological ideas contained in the Codex Leicester. Future research may well disclose other instances.

In any event, we hope to have presented sufficient evidence to indicate that the Codex Leicester in particular and Leonardo's science in general played a far more prominent role than anyone had previously allowed.

APPENDIX

PART 3

Eighteenth- to Early-Nineteenth-Century Manuscript Copies of the Codex Leicester

DURING THE COURSE of this study, we have seen evidence of at least nine copies of the Codex Leicester, almost all realized in the eighteenth century. None of these copies have been previously the object of any specific study. Some of them were unknown or assumed missing. Below is an overview of these copies with some philological notes.

Codex Leicester 18th-early 19th century copies. Stemma Codicum
(Whereabouts unknown: *)

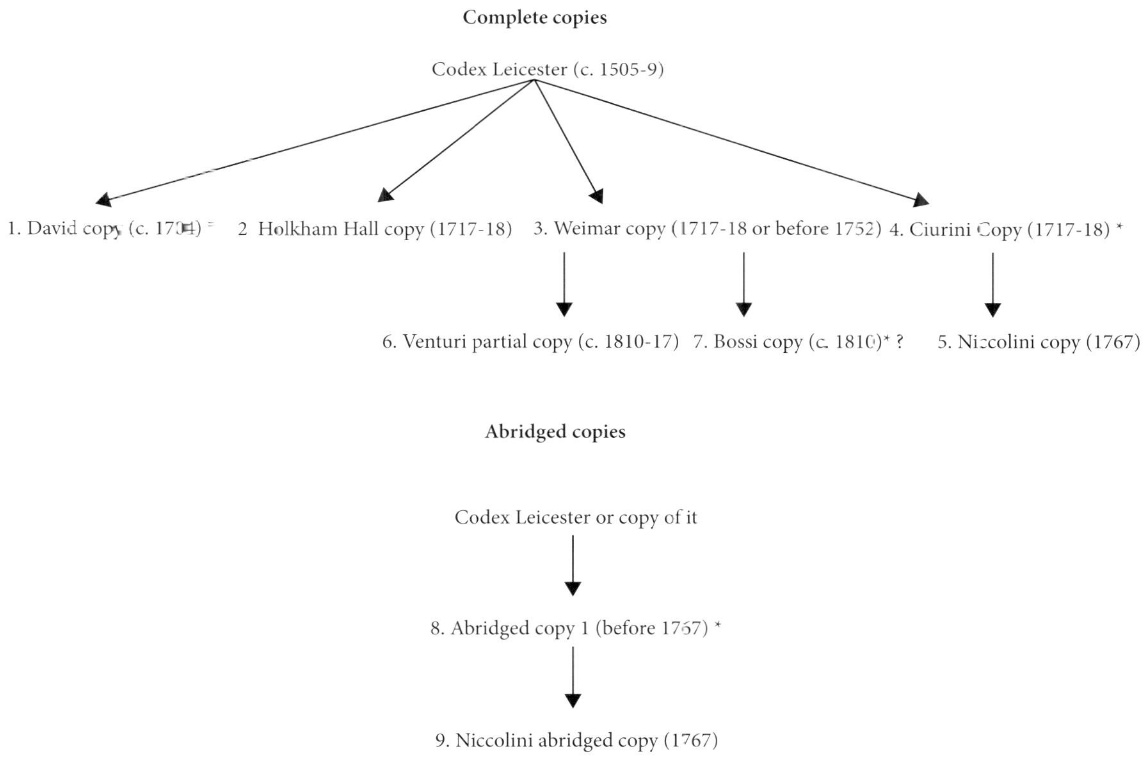

1. The David Copy

Realized in Rome, *c.*1704, by Lodovico Antonio David (1648–1720/28). Whereabouts unknown.[1]

2. The Holkham Hall copy (Holkham Hall, Ms. 700; cm. 33 x 24; Figures 54–8, 60, and 65)

Produced in Florence, *c.*1717–18, on behalf of Thomas Coke. It arrived in London in 1719 and was transferred to Holkham Hall after the middle of the eighteenth century. The text was copied by Francesco Maria Ducci (?–1718), the drawings probably by Tommaso Redi (1665–1726). Almost all the drawings lack letters and some lack captions.[2]

3. The Weimar copy (Weimar, Herzogin Anna Amalia Bibliothek, f. 326; Figures 70–2).

Produced in Florence, *c.*1717–18, if made by Ducci (d.1718), or no later than 1752, because a note added in this copy mentions Bernardino Ciurini (d.1752) as still living.[3] Between the end of the eighteenth and the beginning of the nineteenth century, it passed into the Library of Duke Serra di Cassano in Naples and, shortly after, in 1810, into the library of the painter Giuseppe Bossi in Milan, where it remained until 1817, the year in which it was purchased by Grand Duke Karl August of Saxony-Weimar, arriving in Weimar in 1818.[4]

There are many differences between the Weimar and the Holkham Hall copies. First of all, the Weimar copy (27.1 x 19 cm), apart from its generally less refined character, has a different arrangement of the drawings. In the Holkham copy, they are almost all placed in the left margin of the sheet, next to the text associated with them, each drawing either reversed or, after the first sheets, copied with the same orientation as the original. In the opening folios of the Weimar copy, the drawings on the recto and verso of each folio of the Codex Leicester, are copied together on one page which is followed by a second page with the transcription of the text associated with the drawings (Figures 70–1). Each drawing, some of which are reversed, is marked by a corresponding number inserted in the margins of the text. Then, beginning with folio 5 in the Weimer copy, the drawings are copied in the left margin of the text, some reversed and others not, while the drawings inserted by Leonardo within the main text, if copied, are always reversed (Figure 72).[5] However, after Codex Leicester folio 4, only a fraction of the drawings were copied. This is therefore an incomplete copy with respect to the drawings. The drawings, unlike the Holkham copy, were almost certainly inserted by the same person who copied the text, given the close similarity of ink and handling between text and drawings. Given the poor quality of the drawings,

1 See section on 'Rome 1689', in particular pp. 160–2 of this volume.

2 On the spine: *L. Da Vinci Della Natura & c. delle acque. Ms. Trascritto dall'originale.* See Sections on 'Thomas Coke's Acquisition of the Codex Leicester' and 'The Codex Leicester in the English Scientific Context', pp. 166 and 181 of this volume.

3 See the Ciurini copy below.

4 See Sections on 'A Copy of the Codex Leicester in Naples' and on 'A Copy of the Codex Leicester in Weimar', pp. 207 and ff. and 228 and ff. of this volume; Calvi 1909, p. xxvii n. 1, with bibliography; Pedretti 1987a, p. liv; Pedretti and Vezzosi 1983, p. 138; Laurenza 2016b.

5 e.g. the drawings after Leicester, ff. 23r, 22r, 17r. In other cases a blank space was left in the text.

they were made by someone who was not an artist. In the transcription of the text there are several differences between the two copies. For example, the drawings in the Weimar copy include letters and captions, which the Holkham copy lacks. Two words in a passage copied from Codex Leicester, f. 30v ('Sco Nicholo'), missing in the Holkham copy, are included in the Weimar copy. The evidence excludes the possibility that the Weimar copy was based on the Holkham copy, as suggested by Calvi, who had no opportunity to examine directly the Weimar copy.[6] It seems instead to be an independent copy that obeys different parameters with regard to the arrangement of text and figures. Given the presence of some text that is absent in the Holkham copy, it is almost certainly a direct derivation from the Codex Leicester. It is therefore possible that the Weimar copy was also made by Francesco Ducci, as suggested by the style of writing and by the watermark of the paper (three mountains surmounted by a cross and with a *T* and a *D* on the sides), which corresponds to one of the two more frequent watermarks in Ducci's Holkham copy. The Weimar copy could therefore be a private copy made by Ducci for his own use. There are also some corrections, as in the case of the figure of the moon copied from folio 2r, which is crossed out and redrawn. The uncertainty in the arrangement of the drawings could suggest that this is a draft copy, made before the Holkham version, which reveals no indecision about the placement of the drawings in the left margin.

4. The Ciurini copy

Produced in Florence, *c.*1717–18 by Francesco Maria Ducci. Whereabouts unknown.[7] A note inserted in the Weimar copy before 1752, on a small piece of paper pasted on the front cover, reads: 'The original and a copy of this book went to London; a copy is in the hands of Mr. Ciurini, Master of Geometry at the Accademia dei Nobili in Florence, and this copy together with the other one was made at the same time that Ducci, priest at San Lorenzo, transcribed it for London.'[8] This note mentions three copies of the Codex Leicester, rather than one as has been assumed.[9] The three are: a copy now in Weimar; a copy which, at the time of the note in the Weimar copy, was the property of the architect Bernardino Ciurini; and the Holkham Hall copy, made by Ducci and sent to London. At the death of Bernardino Ciurini (1752), the copy in his possession remained, at least until 1770, in the hands of his heirs. It is in fact mentioned in 1770 by Francesco Durazzini as belonging to the heirs of Ciurini:

> Mr. Francesco Ducci, former librarian of the Laurentian library, copied a manuscript by Vinci, which was in the hands of an Englishman, containing various thoughts, for the most part about hydrostatic issues. This copy by Ducci is presently in the legacy of the well-known architect Bernardino Ciurini.[10]

6 Calvi 1909, p. xxvii n. 1.

7 See Section on 'Scientists Versus Artists', in particular pp. 194–7 of this volume.

8 Weimar copy, Herzogin Anna Amalia Bibliothek, Weimar, manuscript f. 326:

> Di questo libro l'originale ed una copia andò in Londra; una copia l'ha il Sig.re Ciurini, M.ro di Geometria nell'Accademia de' Nobili in Firenze, e questa presente copia con l'altra furon fatte nell'istesso tempo che il Ducci, Prete di San Lorenzo, la trascrisse per Londra.

9 Calvi 1909, p. xxvii; Favaro 1917, p. 159; Pedretti 1987b, pp. liii–liv.

10 Durazzini 1770, p. 130:

> Il Sig. Francesco Ducci, stato Bibliotecario della Laurenziana, copiò un MS. del Vinci, che aveva nelle mani un Sig. Inglese, contenente varj pensieri, la più parte sopra punti d'Idrostatica. Questa copia del Ducci si trova presentemente nell'Eredità del noto Architetto Bernardino Ciurini.

A detail in the Niccolini copy, almost certainly copied after the Ciurini version (see below), suggests that for the Ciurini copy Francesco Ducci followed an organization similar to that applied in the Holkham copy. In the Niccolini copy, on the left margin of the drawing of the new moon drawing from Codex Leicester, f. 2r, is a '6', the number with which Ducci had marked Leonardo's drawing in the original Codex Leicester and in the Holkham copy (Figure 54) in order to show to the artist where to place the drawings. The position of this number '6' in the Niccolini copy does not correspond to the Holkham copy, where it is higher and further from the figure of the moon. Evidently in the Ciurini copy Ducci had inserted the '6' in the position where, in 1767, Bacherini copied it in the Niccolini copy. In the Ciurini copy, Ducci limited himself to copying the text and inserting indicative numbers for the artist who would have copied the drawings. It is therefore probable that this missing copy was of higher quality than the Weimar copy and was closer to the Holkham version.

5. and 9. Niccolini more complete and abridged copies (National Library, Rome, Ms. Vitt. Em. 1767; 27.5 × 19.5–20 cm; Figure 67–8)

These two copies are presently bound in the same volume. Briefly mentioned by some Leonardo scholars at the beginning of the twentieth century,[11] it subsequently disappeared and was considered missing.[12] It has been possible to recognize it in a manuscript recently acquired by the National Library in Rome.[13] The volume, at some point in its history, found its way into the Boncompagni library in Rome, perhaps via the prince Baldassarre Boncompagni (1821–94), mathematician, historian of science, and editor of the first journal entirely devoted to the history of mathematics. The library was put up for sale in 1798 after Boncompagni's death. The volume then found its way into the library of Mario Cermenati, geologist, science historian, and Leonardo scholar (1868–1924), and finally, in recent times, the National Library in Rome.[14]

The volume contains two different copies of the Codex Leicester, each copy with independent page numbers, a complete copy[15] and an abridged copy without figures.[16] The latter, considered by Gerolamo Calvi to be a fragmentary and confused copy, is on the contrary, as we have seen, a very significant historical document for both the scientific criteria that inspired its compilation and because it demonstrates the circulation, so far unknown, of abridged copies of the Codex Leicester, just like the known cases of the *Treatise on Painting* and of Leonardo's scientific manuscripts now in Paris. The abridged copy of the Codex Leicester is then followed by a few pages with passages copied from the *Tiber Shackled* (*Il Tevere incatenato*, Roma, 1669), a hydrological treatise by Filippo Maria Bonini (1612–post 1677).[17]

11 Baratta 1903, p. 313; Calvi 1909, pp. xxvii–xviii, 204; Favaro 1918, pp. 376–8.

12 Pedretti 1980, p. 19, 1987a, p. liv.

13 Laurenza 2005.

14 Acquisition at Mario Cermenati's library sale (Christie's Rome, *Libri, Autografi e Stampe*, Auction catalog 15 June 1999, p. 78, Lot no. 494).

15 National Library, Rome, Ms. Vitt. Em. 1767, ff. 1–140, old numbering (from f. 6 onward, modern numbering).

16 National Library, Rome, Ms. Vitt. Em. 1767, ff. 1–80r, old numbering (ff. 148–228r, modern numbering).

17 National Library, Rome, Ms. Vitt. Em. 1767, ff. 80v–82, old numbering (ff. 228v–230, modern numbering).

A note on its authorship precedes each version of the Codex. Both notes are by Giuseppe Bacherini, librarian of the Marquis Niccolini in Florence. In the note dealing with the full copy, dated 4 May 1767, Bacherini claims to be the author of the copy, while in the short note regarding the abridged copy, dated 9 May 1767, he says that he only collated it, that is he limited himself to check the copy against its source.[18] The style of the writing confirms the different authorship of the two copies. Bacherini also states that each version was copied 'after another similar copy' ('*da altra copia simile*'). Therefore, another abridged copy of the Codex did exist in eighteenth century Florence, a copy that has now disappeared.

The complete copy used by Bacherini does not seem to correspond to the Weimar copy. It is true that various aspects are common to the Niccolini copy and the Weimar copy, such as the presence of letters and captions in the drawings and the fact that the astronomical figures, after Codex Leicester, ff. 1–2, are grouped together, before the relevant texts to which they are linked through reference numbers. But, at the same time, there are differences that exclude a derivation. For example, in the copy of one of the astronomical figures after Codex Leicester, f. 1r, the interval between the two horizontal lines intercepting the waves of the earth, is faithful to the original in the Niccolini copy (Figure 67), while it is incorrect in the Weimar copy (Figure 70). It is unlikely that Bacherini corrected this mistake in the Weimar copy. It is more likely that he used a copy different from the Weimar manuscript, more faithful to the original in the case of this figure. Another difference is the incomplete transcription of a passage from Codex Leicester, f. 30v, which lacks the words 'di Sancto Nicolò', which are instead transcribed in the Weimar copy. Perhaps it was based on the now lost Ciurini copy.

6. Venturi partial copy (Panizzi Library, Reggio Emilia, MS Regg., A 35/3, ff. 276r and v; Figure 73).

This is the beginning of a copy of the Codex Leicester, realized by Giambattista Venturi (1746–1822) in Milan after the Weimar copy that was still in Milan. Venturi tried, without success, to copy more before the copy left Milan.[19]

7. A copy by Bossi (?)

An unpublished list apparently concerning a sale of manuscripts formerly in the library of the painter and Leonardo scholar Giuseppe Bossi (1777–1815) includes a copy of the Codex Leicester in Milan in 1830, which cannot correspond to the copy passed from Bossi's library to Weimar in 1817.[20] Perhaps,

18 National Library, Rome, Ms. Vitt. Em. 1767, f. 1, old numbering; f. 147r modern numbering:

> Attesto io infratto [sic] Bibliotecario degl'Illmi Sg.ri Msi Nicolini che il presente Cod Cart Ms intitolato 'Discorso di Lionardo da Vinci della natura peso e moto dell'acque e osservazioni sopra il corso de' fiumi' è stato fedelmente copiato nel presente anno da altra copia simile, e che quindi è stato da me diligentemente collazionato, talmente che in tutto esso e in ciascuna sua parte è conservata l'istessa identità, et in fede Io [?] Giuseppe Bacherini M° pa.

19 See Section on 'At the Centre of the Scientific World', in particular pp. 225–7 of this volume.

20 Ambrosiana Library, Milan, Fondo Bossi, S. P. 13/6 A, [file] 8, f. 556r:

> 5. Copia tratta dal manoscritto Originale di Leonardo da Vinci. Della Natura, Peso e misura delle acque composto in tempo di Lodovico il Moro, nel condur che fece le acque del Naviglio sulla Martesana dall'Adda a Milano = L'originale di proprietà del Duca di Cassano in Napoli.

At the end of the list (f. 557r): '1830. Nel mese di Luglio' (July 1830).

given the titles Bossi inserted in the latter (geology, hydrology, etc.), he made a copy with the texts grouped according to their content. Its whereabouts are unknown.[21]

8. Abridged copy 1

Florence, before 1767. The Niccolini abridged copy was based on it (see above). Whereabouts unknown.

21 See section on 'A Copy of the Codex Leicester in Naples', in particular pp. 207–14 of this volume.

Picture Acknowledgements

THE AUTHORS AND publishers wish to thank the following for their kind permission to reproduce the illustrations:

Biblioteca Nacional de España, 7, Plate A; Bibliothèque Nationale, Paris, 9; © The British Library Board, 64 (117.m.61), 19 and 39 (Codex Arundel 263); Gallerie degli Uffizi, Florence, 20; Library of Congress, 66; © The National Gallery, London, Plate D, Leonardo da Vinci 'The Virgin with the Infant Saint John the Baptist adoring the Christ Child accompanied by an Angel ('The Virgin of the Rocks'); photo © RMN-Grand Palais (Institut de France) / René-Gabriel Ojéda, 5, 27, 35; Royal Collection Trust / © Her Majesty Queen Elizabeth II 2018, 8, 10, 12, 14, 18, 21, 33, 38, plates B and C; © 2018. Veneranda Biblioteca Ambrosiana/DeAgostini Picture Library/Scala, Florence, 1, 2, 3, 4, 6, 11, 13, 15, 16, 17, 22; Museum and Library Casa di Goethe, Rome, 74

OXFORD
UNIVERSITY PRESS

Great Clarendon Street, Oxford, OX2 6DP,
United Kingdom

Oxford University Press is a department of the University of Oxford.
It furthers the University's objective of excellence in research, scholarship, and education by publishing worldwide. Oxford is a registered trade mark of Oxford University Press in the UK and in certain other countries

First Edition published in 2019

Impression: 1

Published in the United States of America by Oxford University Press
198 Madison Avenue, New York, NY 10016, United States of America

British Library Cataloguing in Publication Data
Data available

Library of Congress Control Number: 2019935955

ISBN 978–0–19–883288–1

Printed in Great Britain by
Bell & Bain Ltd., Glasgow